Learning from Gal Oya

Learning from Gal Oya

Possibilities for
Participatory Development and
Post-Newtonian Social Science

Norman Uphoff

With a new Introduction

IT Publications 1996

Intermediate Technology Publications Ltd
103/105 Southampton Row, London WC1B 4HH, UK

Copyright © 1992, 1996 by Cornell University

A CIP record of this book is available from the British Library

ISBN 1 85339 351 7

First published 1992 by Cornell University Press

This paperback edition with a new introduction
published by IT Publications 1996

Printed in the UK by SRP Exeter

Contents

Introduction to 1996 Edition   vii
Acknowledgments   xv

Part One   Observations

1   Of Rocks and Rivers: Discovering Possibilities   3
2   The Gal Oya Irrigation and Resettlement Project   26
3   The Program Starts Out (1981)   54
4   The Program after One Year (1982)   75
5   Moving Farther Down in the System (1983)   112
6   The Start of a Tamil Program and Sinhalese
    Farmer Initiatives (1984)   143
7   Growing Tensions and Evident Progress (1985)   173
8   The Project Formally Ends (1986)   208
9   The Program Continues and Expands (1987 to 1990)   239

Part Two   Explanations

10   Of Conjunctions and Sums: Conceiving Possibilities   277
11   Relativity, Complementarity, and Uncertainty   304
12   The Rehabilitation of Altruism and Cooperation   326
13   Social Energy as an Offset to Equilibrium and Entropy   357
14   Living and Learning with Chaos   388

Appendix 1   Diagram of Institutions and Roles
               Involved in Gal Oya Project   413
Appendix 2   Abbreviations and Acronyms   414
Appendix 3   Participants in the Gal Oya Experience   418
References   423
Author Index   439
Subject Index   443

To all the "rice plants" in Gal Oya, Narangoda, both Punchibandas, both Ratnayakes, Neeles, Gunaratne, and hundreds of other farmer-representatives; to the IOs who helped them emerge and gain recognition, Chuti, Sita, D. S., Jina, Sarath, Ratna, Keerthi, P. B., Siri, and 150 other institutional organizers; to Wijay, Sena, Lucky, Ed, Jeff, Hammond, Munasinghe, Nalini, Nanda, Bhawani, Gil, Randy, Walt, Dave, Doug, Mark, and others with the ARTI-Cornell group; to Godfrey, Joe, John, Ken, Herb, Senthi, both Manos, Godaliyadde, Kum, K. D. P., Dixon, and others in the government of Sri Lanka and in USAID—from whom and with whom it was such a pleasure to learn.

Introduction to 1996 Edition

Having just revisited the Gal Oya irrigation scheme in Sri Lanka after being away since 1989, I am pleased to report that the farmer organizations there continue to be active and effective more than a decade after outside assistance was withdrawn. These organizations, though still imperfect like all human creations, dramatically changed the socio-economic landscape of that area and gave impetus to a national program for participatory irrigation management. They show the possibilities for improving people's lives through self-help and collective action, managed and directed from below but catalyzed as necessary by supportive outside actors.

This is much more than a case study. By being involved in a complex process of developmental change that started in 1980 and is still continuing, it has been possible to gain deeper insights into how participatory development can be fostered and institutionalized. Various techniques and concepts were developed to make such outcomes more attainable, and these are discussed in this book. But the most important learning from immersion in the Gal Oya situation was that *the ways we think about social reality* affect our opportunities for making progress both with and for people. Many constraints are created through our mental constructs. So in addition to dealing with material conditions we need to work effectively in the realm of ideas. This is why this book is about "post-Newtonian" social science and not just participatory development.

Because farmers in Gal Oya responded so rapidly and constructively to the opportunities for collective action which our program offered, I had to revise much of what I previously understood the social sciences to tell us about individual motivation and institutional dynamics. A better approach is discussed in Part Two of this book. But first it tells a true story about the transformation of economic, social and political relationships in a large, disadvantaged region of Sri Lanka.

My narration takes readers through a ten-year evolution of this process of change. It lets them share in our aspirations and apprehensions, in the hopes fulfilled and those disappointed, observing false starts and breakthroughs, seeing wrong things done for the right reasons and the right things done for the wrong reasons. This "immersion in the particular," to use Albert O. Hirschman's phrase, gave "glimpses of the general" which converged to suggest a more serviceable understanding of how to promote social change. Even if one cannot control such change, one can and should try to steer.

In 1980 the Agrarian Research and Training Institute (ARTI) in Colombo and the Rural Development Commmittee at Cornell University were requested by the Government of Sri Lanka and the U.S. Agency for International Development (USAID) to introduce water user associations in Gal Oya. This irrigation scheme was reputed to be the most difficult and poorly managed system in the country (see pages 4–6). When young organizers were posted in Gal Oya communities in 1981 to start the process of farmer organization, the main reservoir was only one-quarter full; crop failures and worsened water conflicts were anticipated.

Yet within six weeks, 90 percent of farmers in a pilot area over 2,000 hectares were voluntarily undertaking a program which they devised themselves, with organizers' encouragement and facilitation. They cleaned channels, some of which had not been cleared for 15 or 20 years; rotated water deliveries, so that tail-enders would get a fair share of the available water; and saved water wherever possible to donate to farmers downstream who would otherwise receive little or no supply. Such demonstrations of altruism and cooperation are generally considered unlikely. They were quite remarkable because Gal Oya farmers, resettled into the area two to three decades earlier, had previously been known mostly for their conflictual and individualistic behaviour, "even murders over water," as one farmer told us.

When we started, we had no idea whether the project would be successful. Indeed, we avoided the word "success" and thought that the best we might do would be to document and understand where and why we failed, because everyone told us Gal Oya was such an unpromising place to work. However, our basic strategy and assumptions were well-chosen. We followed an explicit learning-process approach, and we proceeded with the conviction that development is mostly a matter of fulfilling human potential. In both these respects, our approach was validated.[1]

Revisiting Gal Oya in 1996

Can institutional and behavioral innovations last? This is always a fair question. Indeed, it is unlikely that any program can maintain high levels of effectiveness and efficiency for any long period of time. The dynamics of social life are such that performance levels are likely to go up or down rather than remain exactly the same. Contrary to Newtonian concepts of reality, nature abhors an equilibrium almost as much as a vacuum.

When I revisited Gal Oya in March 1996, I was concerned that the organizations there, after ten years on their own, might have lost much of their idealism and energy. The visit was made with C. M. Wijay-aratna, who provided so much leadership during the first years of our work in Gal Oya when he headed ARTI's water management unit. Now Wijay was head of the Sri Lanka program of the International Irrigation Management Institute. He had kept in touch with the situation in Gal Oya and had assured me that the organizations there were still working well.

We learned that farmers' efforts to use water carefully, plus improved coordination with the Irrigation Department, continued to keep water-use efficiency almost twice as high as before the project started. Coordination of planting schedules had reduced the staggering of cultivation by 50 percent, making it possible both to save water and reduce expenditures on chemical pesticides. Local merchants had reported a 25 percent decline in their sales of pesticides, and this season farmers had agreed to advance their planting date by two weeks to reduce the need for pesticides still further.

At a meeting with 25 farmer-representatives, I was told that the farmer organizations are now able to sell their unmilled rice directly to the Paddy Marketing Board. By cutting out middlemen, who generally gave short weight as well as lower prices, farmer prices were about 40 percent higher, though they complained that their incomes had not improved commensurately because their cost of inputs had risen so much (with the removal of subsidies as part of the government's liberalization policy). The organizations now buy seeds and fertilizer in bulk and sell these to members at a lower price (and with better quality) than from private merchants. The previous season, 3,800 tons of fertilizer were provided to members, worth over $100,000.

Though government budget allocations for operation and maintenance (O&M) of the main canal system are woefully inadequate, now all plans and expenditures for O&M are a matter of public record.

The formerly common misuse of O&M funds by officials is largely deterred by this transparency. A production credit scheme which charges only 10.5 percent interest and is expected to raise yields by 10–15 percent is being implemented through farmer groups.

The farmer-chairman of the Project Management Committee was particularly proud of the rice harvest festival that the organization had held at the end of the previous season. Between 7,000 and 8,000 farmers participated, each giving thanks according to his own religion, Buddhist, Muslim, Hindu or Christian. The chairman reaffirmed their continuing policy of avoiding partisan politics, citing instances where the organizations had protested against both the previous government and more recently against the present one.

Was everything satisfactory? No. Numerous complaints were voiced, as should have been expected. (I invited the farmer-representatives at the meeting to tell me about their disappointments after their chairman had reported on their accomplishments.) Farmers judged their organizations to be about 80 percent successful, quite a good achievement.

We drove down into the heart of the Left Bank system, to the Mandur area where there is now an implicit boundary between the Sinhalese and Tamil communities. (Until recently, the ethnic tensions which had tormented the rest of the country had been kept out of Gal Oya by the kind of initiatives and solidarity reported in this book.) Here we spent several hours with farmer-representatives hastily called together to meet us by one of the few remaining organizers.

The farmers said that water deliveries are now quite satisfactory, a judgment hard to imagine this far down in the system 15 years before when we started. They confirmed receiving higher prices for their paddy and getting the fertilizer discounts we had been told about, and that the production credit scheme was now operating. They also affirmed their commitment to keeping "politics" out of their organizations.

At the same time, they lamented the fact that the government has provided them very little training since 1985. There has been no continuing investment in human resource development, as was done when ARTI and Cornell were involved here. There is bound to be turnover among officers, they said, and an on-going training program is essential to maintain their organizations' effectiveness. Some government officials, the farmers complained, were ignoring the organizations, not coming to meetings as expected, but they gave the Irrigation and Agriculture Departments high marks for cooperation. This represented a big improvement even if not all departments were fully supporting the new system.

The farmers sitting with us on low benches in a tiny pre-school classroom stressed what the representatives in the larger meeting had also described as the main benefit from our, now their program: "Before there was no group action, whereas now people meet and discuss to solve their problems." Organizations have given farmers means to identify their needs and develop remedies through joint action. This was the basic aim of our program, and it is the essence of participatory development. It represents a concrete process whereby people produce their own empowerment rather than receive it from someone else.

Broader Implications

Wijay and I and our ARTI and Cornell colleagues embarked in 1980 on the task of introducing durable organizations, and of empowering farmers in Gal Oya as a practical task. But we all learned much more than we expected from this process. The experience validated most of what I had read and concluded previously about the advantages and methods of participatory development. But more important, it challenged my understanding of individual and collective motivations and capabilities.

When I compared what we were observing and accomplishing with what I had been taught to expect by the predominant theorizing in social science, I found increasing disparity and dissonance. Thousands of farmers in Gal Oya, and indeed many officials, were not behaving as they were "supposed to." They were not deterred by the free-riding of a few from mobilizing the many for collective action. They did not calculate their advantage simply in self-serving terms, not did they necessarily place greatest weight on material benefits. Notions of public good and collective interest were alive and well in an area that had been described as the most conflictual and selfish one that the Irrigation Department had to deal with. Somehow the processes of group interaction which were initiated through the organizers were eliciting a cascade of cooperative and generous behavior unanticipated by anyone, and certainly by us.

At one level, this could be explained by modifying certain assumptions that prevail in most social science today, invoking individual self-interest maximization in a reductionist way of thinking. Suggested revisions are presented in Chapter 12 on "The Rehabilitation of Altruism and Cooperation." But rethinking required reconsideration of the ontological and epistemological assumptions on which most social science is grounded: a worldview that privileges the individual over the collective, the material over the ideational, and the mechanistic over the organic, discussed in Chapter 10. This is not to say that one should embrace either focus instead of the other. What we arrived at inductively in our Gal Oya work was a "binocular" perspective. This keeps apparently opposite objects or values in view and in focus, drawing insight and benefit from their different qualities and apparent contradictions. We came to share the ancient Greeks' appreciation of paradox as a path to wisdom rather than as a failure of logic.

An introduction is not the place to retrace the experiences and logic which brought me to see how current social science is pervasively influenced by the concepts of classical physics, formulated most brilliantly by and therefore associated with Sir Isaac Newton. Such concepts have been immensely productive for several centuries across a wide range of phenomena. But in this century, we have discovered that these are not the only concepts and principles for understanding the material realm. In the Twentieth Century, as discussed in Chapter 1 and in more depth in Chapters 10, 11 and 14, physical and biological scientists have discovered other rules and relationships that are not so predictable and invariant as those derived from the laws of celestial mechanics.

There are now other ways to understand the universe which do not posit mechanistic, deterministic causation. There is a valid *post-Newtonian* view of the world that is shaped more by concern with *energization* than with equilibrium, and oriented more toward *evolution* than toward entropy. It frames relationships in terms of *open systems* rather than just closed systems or collapsing ones. It is interested more in promoting *positive-sum* dynamics than evaluating zero-sum alternatives or, worse still, being fixated on negative-sum outcomes. These observations and their implications are elaborated in Part Two.

The most important argument made in this book is that promoting participatory development, while it can be furthered with conventional thinking, will be more successful and effective with a more contemporary understanding of the nature of social as well as material realities. Insights from those scientists who have developed relativity theory, quantum theory, chaos theory, and now complexity theory, can help us to move beyond the limiting assumptions about social structures and human motivation which have been influenced, usually unwittingly, by classical physics.[2]

"Post-Newtonian" social science as presented in Part Two understands reality as embracing both objective and subjective factors, related in less linear and less deterministic (though still causal) ways. Whereas defining *essences* is central to a reductionist Newtonian worldview, appreciating *contingencies* is more important within a post-Newtonian framework. Values and personal factors find a legitimacy in post-Newtonian considerations which is denied them in any scheme modeled after classical physics, where objective and subjective factors are considered entirely separate (a view which could also be considered as post-Cartesian.)

An introduction should launch a book, not attempt to give its arguments or its answers. These are laid out in the chapters that follow. This preface confirms that the remarkable social experiment started in Gal Oya, Sri Lanka over 15 years ago is still intact and moving forward. It is not a unique occurrence. There are still faults and shortcomings. Such is the nature of all human ventures. But neither farmers nor engineers who have discovered the advantages of more collaborative modes of operation than they knew before would like to return to the situation before participatory management was introduced.

NORMAN UPHOFF

Ithaca, New York   April 14, 1996

Notes

1. K. D. P. Perera, a senior Sri Lankan engineer who served as first director of the Irrigation Management Division (1984–86) and then as Director of the Irrigation Department (1986–89) and State Secretary for Irrigation (1989-90), wrote to me upon receiving this book: "I feel proud to have been able to be associated even in a minute way in this learning process. I think we in Sri Lanka should be grateful to USAID and persons like you for the success of this venture. Frankly, I must confess that I was skeptical about any successful results in this programme when it was proposed at the beginning. But now this has made a tremendous impact in improving the performance of our irrigation schemes. . . . Now the ID [Irrigation Department] accepts this concept [of farmer participation] and actively supports it." (Personal communication, Oct. 5, 1992)

2. When this book was being written, complexity theory was only beginning to gain attention beyond the circle of scientists, many of them associated with the Santa Fe Institute in New Mexico, who were moving "beyond chaos theory" into more encompassing modes of analysis. If rewriting this book, I would include complexity theory as one of the strands of "post-Newtonian" thinking that can strengthen contemporary social science. See, for example, M. M. Waldrop, *Complexity: The Emerging Science at the Edge of Order and Chaos* (New York: Simon and Schuster, 1992); Murray

Gell-Mann, *The Quark and the Jaguar: Adventures in the Simple and the Complex* (New York: W. H. Freeman, 1994); John L. Casti, *Complexification: Explaining a Paradoxical World through the Science of Surprise* (New York: Harper Collins, 1994); and Peter Coveney and Roger Highfield, *Frontiers of Complexity: The Search for Order in a Chaotic World* (New York: Fawcett Columbine, 1995). For the most imaginative application of quantum theory to social issues, see Danah Zohar and Ian Marshall, *The Quantum Society: Mind, Physics and A New Social Vision* (New York: William Morrow, 1994).

Acknowledgments

This is a "both-and" book, trying to contribute to both practical and theoretical concerns, drawing inspiration from both the field and academia. Its origins are manifold, and many persons have contributed, knowingly or not. It deals with possibilities that may be brought together to make the desirable more probable. My involvement in the work in Sri Lanka that led to this book, and in the various intellectual enterprises that enriched it, was itself a string of possibilities for which I am grateful, the more so because they came together in ways that could be useful to others.

My colleagues in the Gal Oya enterprise are acknowledged in the dedication and throughout the book, so I will not mention them here. Four institutions have contributed in different ways. The Agrarian Research and Training Institute in Sri Lanka invited me as a visiting researcher during 1978–79 on a Social Science Research Council postdoctoral fellowship that got this all started. ARTI is a rare "both-and" institution, having academic research and training functions but operating as a government agency under the Ministry of Agriculture. It had some fine staff who made possible the introduction of farmer organizations in Gal Oya. The United States Agency for International Development was also essential for its financial support throughout most of this period. Its Office of Rural Development, especially under its first director, Charley Blankstein, sought to get academics involved in real-world problems, and its cooperative agreement with Cornell University supporting the Rural Development Participation Project, 1977–82, was a unique venture in government-university cooperation, sadly now less engaged in by USAID. Several USAID staff members in Sri Lanka—John Eriksson, Ken Lyvers, and Herb Blank—were essential partners in this enterprise, appreciating how their tasks could be assisted by academics. The Social Science Research Council's South Asia Committee not only got me engaged in Sri Lanka but also gave direct intellectual stimulation, as reported in Chapters 10 and 11.

Most important, throughout this whole enterprise the faculty and student members of the Rural Development Committee at Cornell University were a continuing source of ideas and friendly support. Cornell itself, with its Center for International Studies, served as an ideal base for both-and work. The Government Department, which is my academic home, accepted my comings and goings, and the Center's directors, Milton Esman, Davydd Greenwood, and Gil Levine, gave unstinting support—ideas as well as encouragement—for this work.

Two contemporary social scientists, Kenneth Boulding and Albert Hirschman, have given me much inspiration over the years, through their writings and in person. Some of my debts to them are noted in the book. Several readers of the manuscript made helpful suggestions: Forrest Colburn, Arthur Goldsmith, David Leaman, and Douglas Merrey, as well as readers for Cornell University Press. Doug deserves special mention because he has spent more years than I in Sri Lanka promoting the development of participatory irrigation management there. Finally, when all is said and done, this project could not have come to completion without the invaluable assistance of Virginia Montopoli, who kept the Rural Development Commmittee functioning, and of my wife, Marguerite, who could not spend much time in Gal Oya because of her medical practice but whose support made my extended work there possible. Like the world itself, this book is a highly contingent enterprise, with many connections and reciprocal influences. The challenge is always to make wholes more than the sum of their parts.

NORMAN UPHOFF

Ithaca, New York

PART ONE

OBSERVATIONS

Immersion in the particular proved, as usual, essential for the catching of anything general.

Albert O. Hirschman, *Development Projects Observed*

1

Of Rocks and Rivers:
Discovering Possibilities

The rock which the builders had rejected has become the cornerstone.
                                                                Ps. 118:22

It is not possible to step into the same river twice.
                                                                Heracleitus

In the Sinhala language of Sri Lanka, Gal Oya means Rock River. This odd name juxtaposes solid, stable matter with something that is fluid and ever changing. There was a Rock River not far from the farm in Wisconsin where I grew up. By a curious coincidence, I came to spend several months a year between 1980 and 1987 in a place similarly named, halfway around the world, working with an irrigation project to improve water management and agricultural production through farmer organization and participation. This experience in Gal Oya, and what was achieved there under adverse conditions, changed my understanding of social science and of the possibilities for promoting development in ways that should be helpful to others as well.

The linear logic and the mechanistic models usually employed to comprehend and act on social realities abhor both ambiguities and paradoxes. Unfortunately, many opportunities for change are obscured by the kinds of social science we have fabricated by our reliance on simple notions of causation and parsimonious explanations. I myself underestimated such possibilities when I began working in Sri Lanka with colleagues from that country's Agrarian Research and Training Institute (ARTI) and my university.

As I became immersed in efforts to establish farmer organizations so that some of the poorest families in an underdeveloped country could improve their standard of living, I found myself rethinking much of what I had learned as a student and as a professional social scientist.

3

Fortunately, as the conventional explanations of individual motivation and collective action began to appear less satisfactory, I was able to find many helpful ideas in writings at the cutting edge of various disciplines ranging from physics and business management to cognitive science and philosophy.

The spirit of the times seems to be encouraging strikingly similar changes in the dominant paradigms of numerous disciplines, pointing toward what Thomas Kuhn (1962) called scientific revolution. Some of these new conceptions and orientations will be introduced in this first chapter and then explored in part 2, which tries to explain the unexpectedly constructive processes and possibilities observed in Gal Oya. Because this effort proceeded more inductively than deductively, part 1 shares with readers the unusual experience that unfolded in Gal Oya, which is instructive and exemplary in its own right.

The account of our efforts to introduce farmer organizations in Sri Lanka will show why I had to unlearn much of what social science now offers as explanations and prescriptions. The insights that emerge from the Gal Oya experience suggest how we might recast the ways we think about ourselves and about what we call the real world. These alternative ways of thinking open up avenues for understanding and action that are now inhibited by "either-or" formulations and disparaged by norms of value-free analysis. The new ways establish positive-sum possibilities where present zero-sum assumptions constrain both options and outcomes. My conclusions are based on a melding of inductive and deductive reasoning, combining experimentation and conceptualization. They connect insights from the microcosm of Gal Oya with innovative macrocosmic analysis from several disciplines. Let us first consider the empirical situation.

The Rejected Rock

In January 1980 our ARTI-Cornell team made an initial field visit to Ampare district in southeastern Sri Lanka where the Gal Oya project is situated. As we traveled back to Colombo, the senior deputy director for water management in the Sri Lankan government's Irrigation Department tried to encourage us with these words: "If we can make progress in Gal Oya, we can make progress anywhere in Sri Lanka." We were concerned that Gal Oya was probably the most difficult irrigation system in the country to improve, being the most in need of rehabilitation both physically and socially. Normally one does not begin with the most adverse settings for finding new techniques and approaches. The problems were many:

1. Gal Oya was the largest and most complex scheme in the country, covering 125,000 acres. Even its Left Bank (LB) subsystem, which was to be rehabilitated with aid from the United States Agency for International Development (USAID), was bigger than any other irrigation system in Sri Lanka. Owing to uneven terrain, the channel network was long and dendritic. Because of its size, it overlapped the boundaries of two administrative districts, which caused further complications.

2. It was probably the most deteriorated physically, with channels silted up and their banks eroded. About 80 percent of the LB gates and other structures were broken or inoperable (FAO 1975). Water flow was controlled and measured at only seven points in the whole Left Bank (Murray-Rust 1983). Moreover, most of its soils were not particularly suited to growing irrigated rice, the main crop, contributing to low yields and low farmer income (Government of Ceylon 1971).

3. To accommodate the growing population in Gal Oya, the cultivated area had been expanded by 30 to 50 percent, far beyond the original design specifications. It had been government policy in the 1950s to bring large families to Gal Oya for settlement. This policy gave immediate benefits to more people from land-scarce areas and furnished a larger labor supply for producing rice, but it also created population pressures later that made managing and maintaining the system more difficult, when demand for water so greatly exceeded its supply.

4. Supply itself was a problem because the water yield from the catchment area was less than expected. In the previous twenty-five years, the main reservoir with its 770,000 acre-feet capacity had filled only twice. At the start of the dry season it was seldom more than half full, so typically there was not enough water during the dry season to irrigate the full area farmers needed to cultivate to support their families. The lower third of the system seldom received irrigation water and relied entirely on rainfall.

5. The difficulties presented by insufficient supply and a deteriorated physical system were compounded by unreliable management of the main system by Irrigation Department (ID) personnel. One could sympathize with their situation, not having enough water or adequate staff and structures to distribute it. But their haphazard performance of duties and their aloof, often antagonistic attitudes toward farmers added to the unpredictable day-to-day operation of the system.

6. Conflicts among farmers over the scarce supply of water contributed to breakage of structures, lack of channel maintenance, and irregular distribution, which in turn made for more conflict among farmers. Settlers here were thought to be particularly unruly and uncooperative.

Some had been selected for transfer to this remote and unattractive area by village headmen who wanted to get rid of troublemakers or undesirables in their communities. Indeed, some were former convicts.

7. Ampare district was regarded by government personnel as an undesirable location to be posted to, sometimes assigned as a punishment. As a frontier area, it lacked most of the amenities found elsewhere in Sri Lanka. (Even today, wild elephants can be a menace there.) In travel time by road, Ampare was the district headquarters farthest from the capital city, Colombo. So getting good staff to come to Gal Oya and stay there was difficult.

8. On top of all this, water distribution had an ethnic dimension. The upstream (head-end) areas reclaimed from the jungle were settled by Sinhala-speaking families from villages in the central highlands and from the western and southern coasts. Downstream (tail-end) allotments were given to Tamil-speaking farmers from villages along the eastern coast (see map 1). When water did not reach the tail, Tamils could attribute this to maliciousness of Sinhalese settlers upstream rather than to geographic factors. The minority's sense of grievance was not allayed by the fact that a majority of Irrigation Department engineers were Tamil.

No wonder the deputy director regarded Gal Oya as the toughest challenge of any system in Sri Lanka. But the government of Sri Lanka and USAID had agreed that it would be their pilot area for improving irrigation management. Almost as an afterthought, they decided that farmers should be organized for this. Our task was to help in that effort.

In early 1981, thirty-two young college graduates, otherwise unemployed, were recruited and trained to live and work in Gal Oya communities, to act as catalysts for farmer organization. When these institutional organizers (IOs) had completed their training, the Government Agent, as the top civil servant heading the district administration, spoke to them at a small ceremony. He tried to encourage the IOs with these words: "If you can bring even ten or fifteen farmers in Gal Oya to work together, that will be a big achievement." This reflected the view most officials had of the population as uncooperative and recalcitrant, as difficult as the area and the irrigation system. Our program was asked to organize ten to fifteen *thousand* of them.

Moving from Solid to Fluid State

The settlers, as the government agent's comment indicates, were regarded as rejects from their natal villages. The IOs were hardly an elite

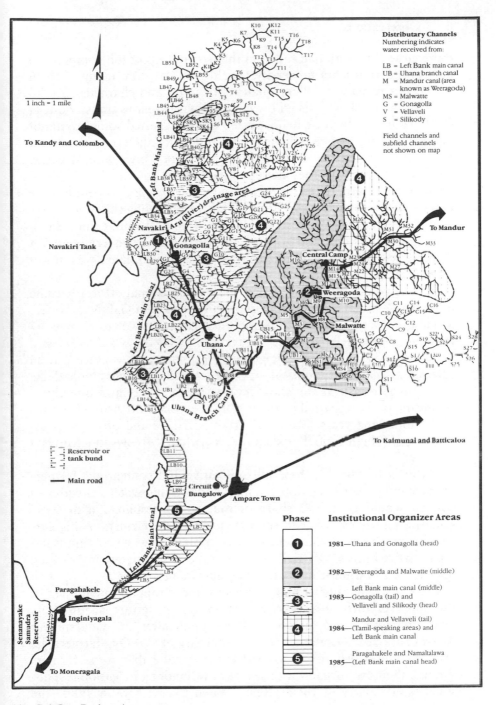

The Gal Oya Project Area

cadre either, since all had failed in the first rounds of job competition after graduation from university. Yet like the farmers, many of them turned out to have excellent qualities of mind and character. So did many of the local officials who were otherwise eager to transfer out of Gal Oya, if only because being posted there suggested some stigma of inferiority. They had been criticized for indolence and even corruption, but even if they had been "part of the problem," once a constructive effort was under way, they became "part of the solution."

There were hard, objective reasons why Gal Oya and its people were in such an unproductive condition. Over the preceding thirty years, relations and expectations had settled into a kind of equilibrium, satisfying few of the needs and hopes of the settlers or the government. Yet within a few months after the new program started, there was movement. Things started to flow.

No one could say exactly when the organizational efforts became impressive enough to change people's thinking about Gal Oya farmers and about the possibilities for improving irrigation—or to change my thinking about social science. We tried to avoid the word "success," knowing from experience elsewhere that almost any advance can be reversed by economic, political, or other setbacks. Yet what looked like an impossible task at the start became (both slowly and quickly) a promising process praised by practically everyone who visited Gal Oya in the following years. More important, farmers and officials themselves spoke highly of the program's benefits (Merrey and Murray-Rust 1987).

Within a few weeks, these uncooperative and contentious farmers were cleaning their irrigation channels and rotating water deliveries so that all would get a fair share of that scarce resource, with some groups even saving water to send to downstream farmers. A big surprise came when Sinhalese head-enders "donated" water to Tamils living in the tail areas. Officials began to take some interest in working together with farmers, in what was seen as a new start.

There were, of course, many setbacks and disappointments, yet in the midst of difficulties and shortcomings the effort acquired its own momentum and direction. Three years after the Government Agent cautioned IOs about the incorrigibility of Gal Oya farmers, he told an interviewer for a government magazine that these farmers through their own organizations were now handling irrigation responsibilities effectively by themselves: "When I came here in 1980," he said, "about 100 people would come to my office on Mondays and

Wednesdays [his days for meeting the public] to speak to me about water problems. Now not a single farmer comes to complain to me about water problems."[1] This may sound like an exaggeration, but on several occasions the District Minister said essentially the same thing: "Before there were farmer organizations, out of every ten farmers I talked with, eight had problems getting water. Now I hear practically no complaints about irrigation distribution." The deputy director of irrigation for Ampare from 1981 to 1985, who became a crucial supporter of the farmer organization program, told me before he left to do graduate studies abroad: "I used to get hundreds of complaints about water each season, in the form of registered letters with copies sent to the minister, the prime minister, the president, everybody. These took a great deal of my time to answer officially. Now with farmers cooperating among themselves and with the Irrigation Department staff, such letters are reduced to a handful."

These statements, even if they come from authoritative sources, may not satisfy readers who want hard data. Within five years, water use efficiency had almost doubled in the Left Bank system. Before the project, between 8 and 9 acre-feet of water per acre was issued in the dry (*yala*) season when there are no rains and farmers depend entirely on the irrigation system; 5 to 6 acre-feet of water was given per acre in the wet (*maha*) season when irrigation is supplementing rainfall (FAO 1975).[2] In the 1985 and 1986 dry seasons, water issues were down to 5 to 5.5 acre-feet, meeting the national target (commonly exceeded), while in the corresponding wet seasons issues were about 2 acre-feet, well below the national norm of 3 acre-feet.

Such numbers are gross indicators and reflect many factors. The qualitative assessments by officials quoted above and the widespread expressions by farmers of satisfaction with the water distribution are probably more valid measures of improvement. The Irrigation Department certainly deserves much credit for this change through its physical rehabilitation of the system and its management improvements.

1. *Desatiya* (Colombo), no. 15 (October 1984), 19. Since not more than half of the physical rehabilitation down to the secondary level had been completed by 1984, only part of the change could be credited to physical improvements in the system.

2. Frequent reference will be made to the maha and yala seasons. Sri Lanka has two periods of monsoon rain a year. The first covers the whole island and produces the rainy maha (major) season when crops can be grown with just supplemental irrigation or with rainfall only. The second monsoon falls only on the southwest quadrant of the country, so any crops grown elsewhere—including Gal Oya—during the dry yala season must have irrigation.

But farmers could have sabotaged the reduction in deliveries, as they had done in the past. Instead, they actively cooperated in increasing water use efficiency.[3]

For farmers, more adequate and reliable water supply not only translated into greater production and income, it also contributed to better quality of life. On M5, an unusually complex command area with 2,500 acres served through a maze of channels, the lower 40 percent had not received water during the dry season for many years. Farmers there told me that their system of farmer organization, using improvised rotations, had gotten water to practically the whole area during the 1983 dry season. Their chairman told me proudly: "There used to be lots of fights among farmers here over water, even murders. You can check the records of the police if you don't believe me. Now there are no more."

This claim sounded too good to be true, but it reflected the kinds of changes that emerged in Gal Oya. I have followed this evolving situation for over ten years, working with a dozen Sri Lankan colleagues who spent much more time in the field than I could and who could delve more deeply into what was happening because they knew the culture and Sinhala language better than I did. We worked with over 150 young organizers who lived in the villages and with more farmers than I can count. My understanding of the program, its impact, and the reasons underlying this change came mostly through them.

The program, it should be said, did not have a stable cadre. IOs averaged only nine months with the program before moving on to more permanent jobs. For various reasons, the Sri Lankan government took longer establishing the institutional arrangements than had been agreed to in the project design, long after IOs had proved their value.[4] Most organizers were reluctant to leave the program, and a few

3. Within two years, the Ampare Irrigation Department deputy director stated: "The non-damaging of irrigation structures is the acid test of IOs and the success of the farmer organizations. This time we were able to close the channel and not have it broken open" (*Sunday Observer* [Colombo], March 20, 1983, 11). An engineering study found very little water flowing in the Left Bank field drains, showing that water in the canals was being used efficiently (Zolezzi 1986).
4. The USAID project design called for an Institutional Development Division, but this got sidetracked in bureaucratic disputes, budget crises, and union objections. USAID/ Colombo reported in an unclassified cable to the State Department on January 19, 1983, that the government of Sri Lanka had agreed in principle to create a permanent group of organizers within the Ministry of Lands and Land Development. A small core group of veteran IOs was finally appointed as permanent staff in 1987 to oversee a larger cadre of short-term IOs working on contract.

returned after finding their new jobs—even if easier—less satisfying. But of the 169 IOs trained and fielded by ARTI, only 8 were left at the end of 1985, an attrition rate of 95 percent.

Moreover, ARTI's supervision of the program was unsteady after some solid direction given by its research and training officers during the first three years. During the next three years, six different staff members were responsible for the IO's work, including one who had only the status of a research assistant. Such turnover would normally doom a program, but this one kept moving ahead, despite the heightening ethnic conflict in Sri Lanka, with terrorist activities intruding more and more on the project area.[5] From inside the program, we saw many defects and shortcomings. We always felt shorthanded and beset by budget and time constraints. Yet from outside, people described the program as a great success, with IOs seen as near-mythical wonder workers. How could these conflicting views be reconciled? One could conclude that we were simply lucky. I thought this myself for a while. On each visit to Gal Oya I expected to find the program regressing. But after some time, it appeared that the approach we had devised was helping make things move in desirable ways.

Our work could be characterized by several orientations consistent with new thinking emerging in the social sciences. First, we adopted a learning process approach, which contrasts with the more conventional "blueprint" approach underpinning most planning and implementation. David Korten, whose influential article (1980) had introduced the learning process approach to a broad constituency in public administration, had worked with a similar program using young organizers to introduce water user associations in the Philippines. His analysis drew on an emerging literature that reassessed learning processes and behavior at both collective and individual levels.[6] Korten was a member of the ARTI-Cornell team that visited Gal Oya in January 1980 to map out an initial strategy.

5. Some continuity and direction, it should be said, was provided by Cornell consultants. I was visiting every six months, and three other faculty members came less often. More important, several Americans and Sri Lankans worked with ARTI during this time as resident advisers, and several graduate students, American and Sri Lankan, did field research in the area. Cornell's involvement is discussed in chapter 2. From 1984 on, "Tiger" terrorists operated fairly openly and freely in Batticaloa district adjacent to Ampare, including the lower portions of the Gal Oya scheme. There were ambushes and massacres in nearby areas, but no mass killings in the actual project area until 1990.

6. Some of these sources include Dunn (1971), Morss et al. (1976), Argyris and Schön (1978), and Sweet and Weisel (1979).

A learning process approach is appropriate for most areas of human activity. It presumes that neither the ends nor the means of social interventions can be fully known in advance, and that understanding and consensus on them must be built up through practical experience. Mistakes are unavoidable and some failures are bound to occur, but with ongoing evaluation, results can be improved. Programs that seek to promote learning must "embrace error," modifying actions so that ultimately they meet socially defined objectives. This can be a rocky road, but it is a preferable one.

Second, we proceeded on the assumption that drawing on human potential is more important for developmental progress than either technology or capital investment, the factors most often stressed by governments and donor agencies. As seen from the unfolding story of Gal Oya, we were not disappointed in our expectation. Organizers, farmers, and eventually most officials performed well above the level most people, including themselves, thought possible. The ARTI and Cornell participants were part of this catalytic process in which people brought out the best in each other and in themselves.

While I found myself continually pointing to human factors when seeking explanations, I came to conclude that ideas—the way we think about our goals and constraints, about our strategies and about ourselves—are ultimately crucial determinants. When I started working in Gal Oya, I had been trying for twenty years to make standard social science concepts and premises work for me and others. I was prepared to exclude values from analysis and to emphasize materialistic and individualistic considerations when explaining behavior, as most social scientists do. Though familiar with the long-standing philosophical debate between "materialist" and "idealist" conceptions of the world, I equated the latter with a purely normative (and therefore naive) social science, incompatible with the empiricism and pragmatism that almost all teachers and researchers endorsed. But ideas and normative influences kept arising as explanations for tangible effects evident in Gal Oya. These factors did not displace or replace materialistic phenomena, yet they demanded consideration as valid sources of explanation. To continue to ignore them would disqualify me as an empirical social scientist.

Understandably, I was reluctant to move away from the prevailing paradigm in contemporary social science, which emphasizes measurable things (Lincoln and Guba 1985). Like others, I had invested much in learning it and using it. But just as important, I expected my conclusions would be more comprehensible to others and also more acceptable to the extent that I followed it. Still, as I read about changes in the thinking and concepts of other disciplines, pointing in the same direction my thinking was moving, I came to appreciate their relevance for what we were seeing and accomplishing in Gal Oya.

Changing Ideas of Reality

Sharing with readers the remarkable data of the Gal Oya experience is a large undertaking in itself, so I will reserve an exploration of explanations for part 2 of this book. Some of the streams of theory that can enrich our understanding of the empirical flow of events in Gal Oya will be introduced here, however. Inductive and deductive modes of thinking are not best employed in the strictly sequential or mutually exclusive manner that the distinction implies.

The New Physics

Social scientists often wish that their objects of study were as tangible and exact as the matter and energy physicists deal with. Those who work in the "soft" sciences like sociology or political science often envy colleagues in the "hard" sciences, who we imagine are measuring and experimenting with distinct, unambiguous things like horsepower and falling objects.[7]

Such a view of physics, shaped by Newton's laws of celestial mechanics, applies above the atomic level. Over the past seventy-five years, however, a "new" physics based on quantum mechanics has been devised to deal with the most elementary levels of physical reality. Once physicists seek knowledge at the atomic level or below, they encounter elusive and ambiguous phenomena that put them more on a par with social scientists. At least the latter can converse with their subjects, as one cannot do with electrons, muons, gluons, quarks, and other inhabitants of the subatomic world.

Social science has proclaimed the norm of objectivity—detachment between the observer and the observed—as a guiding rule. Yet as physicists press closer and closer to what may be seen as ultimate reality, the components of the atom, the dichotomy between observer and observed becomes questionable. Physicists' firm faith in

7. Johnston and Clark (1982, 19–20) have referred to this as "physics envy."

objectivity fades as they find the measurement and prediction of material microphenomena uncertain.

The classification of phenomena into mutually exclusive categories is challenged by the puzzling observation that electrons, ubiquitous constituents of atoms, present themselves alternatively *and simultaneously* as particles and as waves, as matter and as energy. Einstein's famous formula, $e = mc^2$, presented matter and energy as different things that can be converted into one other. Yet at the same time it linked them forever. Physicists now assert that in principle, not just in practice, they must remain uncertain not only about the measurement of electrons' mass and velocity but also about their very nature. Perhaps social scientists should feel less ashamed about their lack of precise measurements and iron-clad classifications.[8]

The implications of the new physics for contemporary social science will be explored in part 2. Einstein's concept of relativity has its corollary in social science. Moreover, the work of biological scientists on evolution and self-organization offers significant ways of understanding social processes that show some escape from the rule of entropy that casts so broad a shadow over the social sciences (Jantsch 1980; Prigogine and Stengers 1984). Perhaps the most important development in the natural sciences has been the emergence in recent years of chaos theory (Gleick 1987). This new body of theory explores different kinds of order and causation that are nonlinear and only loosely determinant, finding surprising patterns in the dynamics of open systems that match human realities better than the closed-system reasoning of classical physics.

If natural scientists find their theories still evolving, this should be true also for social scientists. The materialistic ideas that have shaped our perceptions of the social universe may not be as valid as we think, being based on too-limited notions of what the physical universe is like. The Gal Oya experience continually challenged my understanding of what I thought was reality. I found that the writings and musings of physicists could help me construct a more coherent way of thinking about and dealing with people.

---

8. This reference is to Werner Heisenberg's "uncertainty principle." Some very accessible presentations that I found helpful on quantum physics and relativity were Zukav (1979), Wolf (1981), Briggs and Peat (1984), Rae (1986), and Bohm and Peat (1987). Penrose (1989, 149–301) gives an excellent summary of the contrast between classical Newtonian physics and the new physics; on uncertainty, see 248–50. See also Coveney and Highfield (1990), Davies (1989), and Gribben (1984).

The New Business Administration

When it comes to dealing with people, one could argue that the most solid elements of current social science theory have been incorporated into the analysis of business enterprise—how to get the greatest productivity out of a work force, how to predict and profit from consumer preferences, and generally how to manipulate economic behavior. Schools of business management have concentrated for decades on quantifiable and materialistic subjects like accounting, portfolio analysis, and programming techniques. Many achievements have been credited to this approach, which regards capital as the all-important independent variable and profitability as the criterion of success.

These preoccupations have been challenged by recent writers, however, particularly those comparing United States business management unfavorably with that in Japan (Ouchi 1981; Pascale and Athos 1981). The most widely read alternative approach for businesses has been the best-selling book by Thomas Peters and Robert Waterman, *In Search of Excellence* (1982). This tries to explain how some of America's most dynamic and profitable companies, such as IBM, 3M, Xerox, General Electric, McDonald's, and Boeing, have been managed. These authors got ideas from the new physics just as I have done (1982, 90). Their generalizations, derived from studying American corporate giants, matched our emerging ideas about working with farmers' organizations: maintain "a bias for action," learn from clients and customers and also employees, increase productivity through relying on people more than on technology or capital, and operate a management system with "simultaneous loose-tight properties."[9]

This last prescription, like combining competition and cooperation, is contradictory on its face. Yet case studies of successful companies showed that capitalizing on both kinds of relationships produced the best results. Speaking analytically, looseness and tightness are not necessarily zero-sum. Having more of one does not mean having less of the other if they can be combined in a positive-sum manner to increase overall benefits.

Within these large materialistic enterprises, Peters and Waterman pointed to the productive role of innovative ideas and humanistic values, not just the physical assets a firm owned. These authors showed

9. While I was completing this book, a friend put me in contact with Michael Best, whose new book (1990) analyzes successful business management in Europe and Japan. The key to success was a matter of combining the principles of competition and cooperation in a positive-sum way. The kind of thinking that helped small enterprises in Italy was consistent with what was working in Sri Lanka.

that treating people with respect and expecting them to have worthwhile ideas could be profitable, and that caring, something I observed to be important in Gal Oya, should not be ignored as outside the realm of serious investigation. If major American corporations could by similar means increase their profits in the marketplace—the persuasive "bottom line" for methodological and ideological hardliners—perhaps our experience in Sri Lankan rural communities was not simply fortuitous and might be explained in nonconventional terms that others could accept.

New Concepts in Economics and Political Science

Albert Hirschman has been one of the most wide-ranging thinkers in economics. Having written one of the major works on economic development in 1958, he moved from the macro level to the micro level (e.g., Hirschman 1967, 1970), and he has become increasingly a philosopher not only of economics but of human values, ideals, and behavior (e.g., 1977, 1982). His 1977 book, for example, showed how the idea that men are driven primarily by material interests emerged among European intellectuals in the seventeenth and eighteenth centuries in response to the mechanistic images that their contemporaries who were delving into physics were making popular.

In 1983 Hirschman made a brief trip to six countries in Latin America, visiting forty-five grass-roots development projects supported by the Inter-American Foundation. His short report, Getting Ahead Collectively (1984), described many improvements achieved through people's self-directed and creative efforts. Reading it evoked repeated echoes of the Gal Oya experience, as the imagination and perseverance of impoverished slum dwellers, small farmers, micro-entrepreneurs, tenants, and dedicated professionals managed to break through economic, political, and social barriers.

The successes Hirschman documented were admittedly on a small scale, but the main constraints to expansion were not so much a lack of capital as a lack of confidence that such changes could be achieved on a wider scale. To make sense of what he saw, Hirschman introduced the concept of social energy, which spoke powerfully to me, since I had just returned from Gal Oya when I read the book. To find an economist invoking a "soft" variable like social energy was encouraging. It was the best explanation he could come up with for the hard results he was observing. His noting the emergence of networks of social relations that were "more caring and less private" (1984, 97) reflected just what I was seeing in Sri Lanka. It corresponded exactly to the incipient theorizing I was embarking on to explain changes in behavior and productivity there.

About the same time, a book titled *The Evolution of Cooperation* appeared. Robert Axelrod (1984) used the power of computers to test dozens of strategies for maximizing individual and collective gains against each other. He found that a strategy of cooperation—formally called reciprocity or simply "being nice"—was more advantageous over the long run than systematic or opportunistic exploitation of others. This strategy could not easily be dismissed, not just because of its robustness in repeated computer tournaments, but because it illuminated real-life relationships such as the conduct of trench warfare in World War I.

Axelrod's development of altruism as a rigorous concept for social science analysis showed how it could emerge and survive within a population of rational, self-interested individuals. This orientation does not require the unqualified surrender of one's own interests to benefit others; rather, it attaches some positive value to others' well-being concurrently with one's own. This work extends the conceptual analysis of other social scientists (e.g., Sen 1977, Boulding 1978, and Margolis 1982) who have called into question the validity of narrow self-interested, materialistic models of behavior and explored alternative systems of interaction based on valuing mutual welfare.[10]

To combine the language of welfare economics with that of game theory, altruism can represent a Pareto-optimal solution which if reciprocated leads to positive-sum outcomes. These in turn can yield a stable and productive dynamic equilibrium. Such formalistic terminology may be uncongenial to some readers, but it represents as rigorous a set of concepts as economists and political scientists have devised. More important, it captures the logic of individual choices and collective action observed in Gal Oya. In the chapters that follow, readers can themselves experience vicariously "the evolution of cooperation" and

---

10. It was helpful to receive from Albert Hirschman a small book on "the political economy of love" by one of his colleagues at the Institute for Advanced Studies at Princeton (Drescher, Esser, and Fach 1986) while Kenneth Boulding was elaborating on his analysis of "integrative" exchange (1989). In November 1986 I was asked to be a discussant of a paper that drew on both Boulding and Axelrod to assess international economic problems (Streeten 1989). A Cornell economist colleague, Robert Frank, was similarly concluding that "on the strength of the evidence, we must say that the self-interest model provides a woefully inadequate description of the way people actually behave" (1988, 256). See the collection of essays in Mansbridge (1990) including ones by Margolis and Frank.

"getting ahead collectively" in an impoverished, unpromising situation in Sri Lanka.

Cognitive Science

Having started with a model that tried to explain human behavior in terms of material interests and incentives, I was increasingly struck in Gal Oya by the power of ideas. Although they could not eliminate material interests or all conflicts that arise from them, they could redefine interests in people's minds and alter objectives from competing to compatible. Ideas could evoke principles of choice and action that transcended narrow individualism and could encourage initiative by clarifying options and giving new rationale. They could enlist cooperation where none had existed before and even modify behavior by evoking ideals and norms that were otherwise dormant or discounted.

Increasing concern with the energizing effects of ideas did not exclude the influence of material factors. Rather, there appeared to be a dialectical relation between the realms of matter and ideas, with people brokering between the two. Ideas might even be a more potent force in social relations than material things, because things by themselves, unrepresented by ideas, lack value and thus motive attraction. Einstein's formulation suggested not just that matter and energy are mutually convertible but that energy is the more powerful. Perhaps ideas bear a similar relation to matter.

As such thoughts took shape, Howard Gardner's book *The Mind's New Science: A History of the Cognitive Revolution* (1985) became available, reviewing parallel developments in six disciplines—philosophy, psychology, artificial intelligence, linguistics, anthropology, and neuroscience—contributing to what can be termed cognitive science. Gardner found each of these fields moving from relatively closed, mechanistic, and reductionist models to more open, contextual, and integrative ones. In these disciplines there is increasing evidence of the significance of ideas as phenomena worthy of study in their own right. Conceptions of people as discrete units simply responding to others' stimuli and making mechanistic calculations are repeatedly found inferior to explanations that invoke both context and purpose. Bringing thoughts into the realm of social science opens up promising new lines of explanation.

Thoughts not only shape our understanding of reality but constitute an undetermined share of it. As Pearce (1971, 85) put it, "What a thing *is* to an unknowable extent is determined by or influenced by what we *think* it is." This growing realization was explored by Jerome Bruner, a leader in the field of cognitive science, in his book *Actual Minds, Possible Worlds* (1986). The dualistic conception of mind and matter, taken for granted in the realist conception of the world, is challenged by current psychological research results and new philosophical formulations such as appear in Goodman (1984). I was in no position to resolve the controversies, but emerging propositions within cognitive science and constructionist philosophy made much sense in light of the Gal Oya experience. It was also encouraging to find that fellow political scientists were starting to study ideas as explanatory variables in the arenas of national security and economic policy.[11]

Social Science Methods

There have been increasing challenges to the mainstream methodologies in social science, which deliberately imitated those used in the physical sciences, where repeated identical experiments can be done with exact measurements—at least in principle.[12] For three-quarters of a century, however, physicists have been modifying their view of what constitutes reality and how we can know about it, owing to the surprises encountered when we try to understand the atom. Still, most social scientists cling to methods of research and theory building extrapolated from notions about scientific laws and proof more suitable to celestial mechanics than to quantum mechanics.

Mainstream methods may be appropriate to many kinds of problems, but they are not the only valid means of deriving and organizing knowledge. Systematic efforts to make sense out of observations need not be depreciated as an art rather than a science because no measurements or a priori hypotheses are involved. One of the leading figures in

11. Such analyses I came across included Jervis, Lebow, and Stein (1985), Derthick and Quirk (1985), Reich (1988), Goldstein (1989), and Hall (1989).

12. The possibilities of exact measurement in the physical sciences have themselves proved illusory, as is discussed in part 2. The axioms of mainstream social science have been summarized by Guba (1985, 82–83) as deriving from the assumptions of classical Newtonian physics. They include: (1) there is a single, tangible reality "out there," fragmentable into separate variables and processes that can be studied independently and then predicted and controlled; (2) the inquirer can maintain a discreet distance from the object of inquiry, neither disturbing it nor being disturbed by it; (3) the aim of inquiry is generalized statements independent of both time and context; (4) every action can be explained as the effect of a cause that precedes the effect (or is simultaneous with it); and (5) inquiry is value-free, and "facts speak for themselves." As Kuttner points out, Adam Smith's concept of equilibrium in market economics was itself "a variation on eighteenth-century Newtonian mechanics. Physics has served ever since as a model to which economics should aspire" (1985, 76).

political science has challenged the applicability of positivism as follows: "We believe that this 'normal science methodology' is inappropriate in rapidly changing situations, where those very parameters of political action are in flux" (Almond 1990, 4). Maybe it is not so suited for conditions of slow or little change, either.

Going with the Flow

The intellectual trends sketched here are ones I think the Greek philosopher Heracleitus would have been comfortable with, contrasting with the more Newtonian notions of cause and effect. One can never step into the same river twice—it is always changing. But that does not mean there are no causes or effects. Relationships, to follow Heracleitus's metaphor, are more hydraulic than mechanical. The challenge is to be able to discern and utilize causal associations or, better said, influences in a world that is more like a river or an ocean than like a machine.

The old French saying, "The more things change, the more they stay the same," can also apply in reverse. Things may seem to stay the same but nevertheless be changed because their context has changed and they cannot exist or be understood in isolation from their setting. In the words of a more recent Frenchman, Jean Monnet, "If you change the context, you change the problem" (cited in Oye 1992, 31). Or if people's purposes change, this modifies the thing itself, in subtle ways if not in obvious ones. NATO and Warsaw Pact weapons stationed in European theaters are not the same "things" they were five years ago.

One of the most powerful and at times overwhelming impressions we got from being involved in this project in Gal Oya was the *changefulness* of reality, along with its multifaceted and frequently paradoxical appearance. Although we saw intrinsic merit in the learning process approach as an organizing concept for our actions, we also found that "blueprints" simply were not tenable means for dealing with an ever-evolving situation. Even well-made plans and clear-cut agreements became outmoded or were superseded. Yet some planning was both possible and necessary.

Faced with flux, it was easy to sympathize with Plato's solution, to grasp at the ontological straw of assuming that only ideal forms, being changeless, are true, being more "real" than what can be observed in the surrounding world. But such ideas are not operational, since they are either too abstract or simply tautological—true by definition.

Heracleitus perceived in the world around him an ultimate unity of opposites that were evidently but productively in tension with one another. The school of thought to which he contributed found modern expression in the idealist formulations of Hegel and the materialistic dialectical principles of Marx. In Heracleitus's worldview, changes in one direction are balanced by changes in opposite or at least different directions, with underlying connections among all things giving ultimate coherence to the world. Division and conflict are real, but they coexist with forces of unity and cooperation, which are also real.[13]

Such considerations, obviously metaphysical, may seem a world apart from the practical concerns of getting an adequate share of water to farmers at the end of their field channels so that crops will survive the hot, dry summer season. But there were many connections that needed to be made—between ideas and material phenomena, between macrocosms and microcosms, between principles and practice. This need was not well understood when we first undertook the concrete tasks of improving irrigation performance, though the effort ultimately proved fruitful for the development of broad theory.

There is good reason to share the field experience with readers in some detail. It is instructive in its own right, giving an inside look at a complex case of developmental change. But as the conclusions to be drawn are linked as much with the process as with the results, the story becomes important and is not replaceable by a summarizing chapter or a final report. Thus part 1 shares what was attempted and what happened, what was possible and what was not. Part 2 offers an analysis and construction of explanations that have clarified for me what was occurring and why. I hope that both will contribute to an improved social science of action.

The Gal Oya Experience as Text

Some readers may have reservations about embarking on so large an intellectual enterprise based on a single case. Yet case studies can produce explanations if they proceed from an adequate theoretical framework (Yin 1989), and this one meets practically all the criteria

---

13. These issues are very readably presented by Stone (1988, 68–78). Heracleitus was no purely qualitative philosopher. According to the *Encyclopaedia Britannica*, he calculated the angles of elongation for the orbits of Mercury and Venus and concluded correctly that they revolved around the sun two thousand years before Copernicus proposed the idea.

specified by Eisenhardt (1989). Still, I had the reservations that any contemporary social scientist must feel about having "an N of one." To be sure, good results had been achieved in the Philippines from a similar approach to participatory irrigation management (Bagadion and Korten 1985; Korten and Siy 1988), and in 1986, with colleagues in Nepal, we had started a similar experimental project there and found farmers, organizers, and officials quickly making the same kind of progress as in Sri Lanka. A program following this approach in Thailand produced familiar outcomes (Laitos, Paranakian, and Early 1987). So the process and results observed in Gal Oya were not unique.

But how valid could I or others consider my working from one case study in detail and in depth? In 1987, during a visit to Ghana, an unexplained change in bus schedules stranded me for a day at the university guesthouse in Kumasi. Serendipitously, I had with me a book whose title had intrigued me in a bookstore, *Beyond Objectivism and Relativism* (Bernstein 1983). This work reviewed contemporary debates in philosophy and introduced me to the concept and methods of hermeneutics, increasingly advocated and used among philosophers. Exasperation over missing a bus and having to get up at 4:00 the next morning to get back to Accra was quickly replaced by the excitement of discovering ideas that elaborated on the centrality of human agency and initiative and that challenged the hegemony of the scientific method as the sole means of establishing valid knowledge.

Hermeneutics is usually ambiguously defined, but it represents the theory and methodology of establishing meaning from what is written, said, or observed, by dealing with words and ideas in a serious and systematic way. There is to be no exaltation of "subjective" awareness over a presumed detached scientific objectivity (Rabinow and Sullivan 1987, 5), but neither are objects and purposes granted a self-evident, autonomous existence. Through agreed conventions of inquiry and discourse, meaning and coherence are sought from given data, whether written or observed. This approach proposes that we listen to the data being presented, participating and sharing in them, opening ourselves to the claims of truth that they make upon us (Bernstein 1983, 137–38, citing Gadamer). This approach to knowledge, coming from philosophers in continental Europe, has found slow but increasing acceptance in Anglo-American circles.[14]

14. "Interpretive" methodologies are becoming more accepted among social scientists, as seen in Rabinow and Sullivan (1987, 6). They emphasize the all-pervasive effects of cul-

It occurred to me that I was attempting a kind of "barefoot herme-neutics," to borrow a metaphor from Chinese development. The trip reports on my semiannual visits to Gal Oya, totaling over eight hun-dred pages by June 1987, set forth a detailed record of change in a complex environment. To make the reports more readable, I had writ-ten in the present tense in a first-person, narrative style. They thus con-stituted a text to be explicated. It had seemed to me valid to derive some learning from our field experience without subjecting it to the kind of statistical tests social scientists utilize with large and randomly drawn samples of data. Some might regard interpretive efforts as un-scientific, but I was pleased to find that philosophers had constructed an extended justification for such an approach to knowledge.

Those who believe that hypothesis formulation and testing provide the only ways to gain valid knowledge should expose their presump-tions to the challenge of philosophers who have reflected and debated on the nature of reality. Most social scientists presume that what they are observing, measuring, and explaining exists "out there" as some-thing absolute. Cognitive and learning theory as well as experimenta-tion, however, show that purely objective observations independent of preconceptions are not possible. Observation and explanation, though commonly perceived as activities of separate individuals and as our own doing, are cumulative, collective social enterprises.[15] Interpreta-tion deserves as much "scientific" standing as experimentation, espe-cially in a world of change such as Heracleitus described.

---

ture—those shared meanings, practices, and symbols that constitute the human world. This "does not present itself neutrally or with one voice . . . both the observer and the observed are always enmeshed in it. . . . There is no outside detached standpoint from which to gather and present brute data." This approach rejects the stark either-or views of Descartes and Kant, preferring "more historically situated, nonalgorithmic, flexible understanding of hu-man rationality . . . sensitive to unsuspected contingencies and genuine novelties encountered in particular situations" (Bernstein 1983, xi). It actually has much in common with Max Weber's writing (1947, 9–10) on the search for Verstehen ("understanding") in social sci-ence. Popper (1972, 162) defines hermeneutics as the theory of understanding. Geertz (1973) has written instructively on hermeneutical analysis in anthropology. Kuhn (1977, xv) says that for himself, "the term 'hermeneutic' . . . was not part of my vocabulary as recently as five years ago. Increasingly, I suspect that anyone who believes that history may have deep philosophical import will have to learn to bridge the longstanding divide between the Con-tinental and English-language philosophical traditions."

15. McCloskey (1985) shows that even in economics, quantitative methods do not neces-sarily produce more scientific and therefore more valid results than would more qualitative lines of inquiry. Shared assumptions and conventions are more important to the scientific endeavor than data and statistical tests. The conclusions arrived at through positivist meth-ods do not meet their own requirements, McCloskey demonstrates, and therefore they do not properly substantiate their conclusions. Conventional epistemological approaches in eco-nomics thus have much in common with the emperor's new clothes.

Many themes emerge from the empirical world of Gal Oya, enriched and organized by concepts from the literature of various social sciences and other disciplines, including the new physics and what is known as chaos theory. The latter has special significance for the question whether individual action can make a difference. If our social universe operates in a clockwork, Newtonian manner, we can disavow all personal responsibility and can justify inaction and fatalism. But this attitude is challenged by the contention of chaos theory that small causes can have large effects. Although we cannot know for certain what the consequences of individual and group efforts will be, because of the probabilistic and uncertain nature of the world around us, we can reasonably presume that it is possible for individuals to alter the course of events and thereby to affect outcomes. Insights from chaos theory thus reclaim a role for personal action and responsibility in the social universe.

Both stability and change are characteristics of the world we live in. Which facet we regard as more prominent, as more real, will shape our choice of social science assumptions. If the social world as well as the physical world is seen as continuously evolving, our vocabulary and theories need to reflect this. Such concepts from twentieth-century physics as relativity, uncertainty, and complementarity should amend the dominant thinking of earlier eras—for example, regularity, invariance, and linearity. The largely nonlinear world we live in lends itself more to the insights and patterns of chaos theory—better characterized as the theory of nonlinear dynamics—than to the constructions of Newtonian thought.[16]

Too much of our social science is reductionist in its conceptualizations and explanations, treating social reality in purely individualistic, materialistic, and mechanistic terms. Such simplified images of social phenomena and relationships are not so much wrong as inadequate, needing to be enlarged by the complementary conceptualizations considered in part 2. The aim of social science is seen here as opening up possibilities for improving the human condition, not simply explaining it—though that is necessary—through positive-sum thinking and

16. Some economists are now using novel, nonlinear dynamic concepts inspired by chaos theory, finding them more illuminating than the mechanistic, deterministic ones underlying conventional economic analysis (see Anderson, Arrow, and Pines 1988). The elegant workings of neoclassical economics, with its unique, stable equilibrium points, require the assumption of diminishing returns. Wherever there are increasing returns, which are inherently positive-sum, many alternative equilibriums are possible. The irregular dynamic paths that ensue are very sensitive to initial conditions and to subsequent interactions with other nonlinear phenomena (Arthur 1990).

analysis. Although zero-sum alternatives may be more rigorously compared, they too often unnecessarily constrain us in carrying out developmental tasks.

Analytically, choices can be reduced to simple "either-or" alternatives, but these are usually creations of the mind. Where more expansive possibilities framed in "both-and" terms are conceivable and
tenable, we will derive greater benefit from existing resources, human
and material. This applies particularly to the emergence and reinforcement of altruism and cooperation, two propensities that were seen unexpectedly but abundantly in Gal Oya and that produced social energy
with remarkable effects. These qualities derive more from prevailing
structures and norms than from innate human nature, as discussed in
chapter 12. Greater satisfactions can be created from limited means if
value orientations can be realigned in positive-sum directions.

These highly condensed conclusions summarize what I learned from
Gal Oya. They open up opportunities for promoting development in
more participatory ways, in the Third World or anywhere. At the same
time, they reorient social science in directions that support this goal
and can be characterized as post-Newtonian.[17] It may be said that
these are only ideas, but by the end of the twentieth century, when so
much has been achieved through the unfolding of ideas, this is not a
very substantial criticism. Ideas should be examined and explored for
what utility they can produce in practice, and in the area of Third
World development, as in other human endeavors, we need better
ideas more urgently than ever before.

17. It could also be characterized as "post-Cartesian," for rejecting the highly deductive
and dualistic reasoning that marked Descartes's contributions to Western science. The
emerging alternative relies more on observation and induction, learning to deal empirically
with the multifaceted, changing reality around us. It is also more appreciative of ambiguities,
complementarities, and paradoxes in the real world than was Descartes.

2

The Gal Oya Irrigation
and Resettlement Project ˙

The ancient ruins and inscriptions discovered in the Gal Oya Valley
appear to indicate that the Valley was first populated with people pos-
sessing a knowledge of irrigation in the Third Century B.C. They
probably irrigated their paddy fields from small reservoirs and anicuts
across the tributaries of the Gal Oya [River]. In the course of time, the
forests were felled and a vast network of tanks was set up with the Gal
Oya as the principal source of supply of water.

Unfortunately this progress was not maintained. The Dutch found
the irrigation works in ruins in the 18th Century and restored part of
them. Later, the works once more fell into disuse but this was remedied [by the British colonial administration] towards the end of the
19th Century. At the time the [1948] Project was being investigated,
only the areas near the coast were being regularly cultivated. The remainder of the region was in forest, inhabited by a few small groups
of jungle tribes.

                                    I. M. de Silva, "The Gal Oya Scheme"

When one visits the Gal Oya project area today, it is difficult to
imagine that much of it was uninhabited jungle forty years ago. Cultivation is now seen almost everywhere, and the settlements look like
those throughout most of Sri Lanka's dry zone. Yet a retired surveyor
general who led the first reconnaissance team into the area after World
War II to begin mapping its vastness for irrigated development recalls
encounters with wild elephants, buffaloes, leopards, and bears, to
whom he lost several of his crew. He tells of brushes with violent out-
laws seeking to elude the police. A 20 percent casualty rate for malaria
had been allowed for, but the incidence was almost double this. The
team managed to complete the survey and get out just before monsoon
rains flooded the area and made in impassable (Chanmugam 1976).

Such perils are no longer common. Floods in 1984 and again in 1986 caused crop losses and some property damage, but life went on with relatively little disturbance. One night in 1983, a wild elephant killed a student several hundred yards from the Irrigation Department's circuit bungalow on the edge of Ampare town, where we always stayed when working in Gal Oya. Otherwise the occasional sighting of wild elephants foraging around the reservoir was a matter of entertainment, not danger.

Irrigation has long been a means of taming the wilderness in Sri Lanka. Twenty-five hundred years ago, Sri Lankans were already constructing irrigation systems, large and small, to provide the material basis for an impressive civilization. Fifteen hundred years ago their evolved technology enabled them to build a main canal seventeen miles long with a drop of only six inches per mile, a precision difficult to match today. For almost two millennia, a series of kingdoms rose and fell in Sri Lanka, with occasional dynastic ruptures and periodic invasions from South India. Then, in the fourteenth century, a political, economic, and cultural decline set in that is still not satisfactorily explained. From 1500 on, Portugal, Holland, and Britain, one after the other, each controlled as much of the island as it could for about 150 years in turn. Nearly half a millennium of European mercantile hegemony leading into colonial rule gave way to internal self-government in 1931 and full independence in 1948.[1]

The first prime minister, D. S. Senanayake, previously Minister of Agriculture, is credited with the decision to start the original Gal Oya project.[2] Design work began in the year of independence, and a huge dam creating the main reservoir, an inland sea named Senanayake Samudra, was completed in 1952. The project was financed by the newly independent government without foreign aid, though

1. This compresses 2,500 years of history into one paragraph. Readers will find it laid out lucidly in de Silva (1981). Sri Lanka's irrigation history is documented in Brohier (1933) and Gunawardana (1971). Nobody knows whether the eclipse of Sinhalese civilization should be attributed to the repeated invasions from India, to disintegration of the ruling elite, to the spread of malaria, or to decline in agricultural productivity owing to poor maintenance of the irrigation network or loss of soil fertility. See Indrapala (1971).

2. The idea of constructing such a large scheme is attributed to a former director of irrigation who visited the area in 1936. He wrote, "I have come up for three days through the jungle. . . . I stand now on the rock of Inginiyagala and visualize in my mind's eye the site where a future dam should come." He added that, if this were built, "I would urge a government of the future to examine the problems from the human angle; for here in the lower reaches under the [existing irrigation scheme], hard-working peasants live in a state of semi-serfdom, earning the barest pittance as recompense for the severity of their servitude." Cited in Abeyratne (1982, 1).

construction was done by an American engineering firm. (The reservoir and canals are shown in map 1.)

The project area covered six hundred square miles, and 5 percent of the country's area came under the authority of the Gal Oya Development Board, modeled after the Tennessee Valley Authority in the United States. The board was to plan and promote integrated economic and social development, but this was never adequately achieved (Abeyratne 1982). In the mostly unpopulated Left Bank area, six thousand families were settled in forty "colony units" that were numbered, not named. The first settlers were each given four acres of irrigated land to grow paddy (the word used for rice before it is processed for eating) and three acres of unirrigated land for a house and garden. Households that came later received three and two acres of each type of land.

Clearing land, providing water to fields, and settling the planned communities was completed by the mid-1960s. Each colony unit had built for it a school, a village hall, a recreation center, a cooperative store, market stalls, and other facilities. The first job our institutional organizer supervisor had when he was assigned to Gal Oya in the early 1960s as a young colonization officer was to help settlers adjust to their new environment. In those days this was a real frontier area, he told me: "Wild boar meat was freely available in the market for only two rupees a pound; you practically had to give it away."

The Gal Oya scheme was planned to irrigate 125,000 acres, but the numbers never quite added up. Colony units in the Left Bank area where most of the settlers lived accounted officially for 24,000 acres, and the total area served by the Left Bank main canal was supposed to be 42,000 acres, including areas already settled. The rehabilitation project planned in 1978 used these figures. But four years later the estimate was revised, based on aerial photos, to about 65,000 acres in the Left Bank, not all of which got irrigation water. Large additional areas had been brought under cultivation by settlers or their offspring or by new immigrants. Usually this involved encroaching on land that had been officially reserved, as right-of-way along channels to facilitate maintenance work, for drainage channels, for cattle grazing, for buffalo wallows, for forests, and so forth. A 1975 study by the United Nations Food and Agriculture Organization (FAO) reported that, all together, 95,000 acres were being cultivated, but in fact nobody knew the actual area covered by the scheme.

Because sugarcane was grown in part of the Right Bank area to supply a sugar mill built there by the government, a cabinet decision had given the Right Bank preference in water allocations. The River Division, which lay between the Left Bank and Right Bank systems, could claim priority in water allocations because it had received water diverted from the river before the Gal Oya project was initiated. Fortunately it needed less water per acre because its soils retained moisture very well, and it received some drainage water flowing to it from the Left Bank and Right Bank systems.

The growing population pressure and the problems listed in chapter 1 contributed to a drastic deterioration in system capacity and performance by the mid-1970s.[3] An assessment of the typical cultivator's water management practices was nearly as dismal: "He probably did not have any previous experience in farming or at least in irrigated farming before he was resettled. . . . He plants where, when and if he wants. . . . He is dependent on the Irrigation Department for water but also on the cultivators above him; even if he wanted to practice intermittent irrigation he could not because his neighbors above him would continue to pour water on him. His irrigation efficiency is generally very low" (FAO 1975, 5–6). The actions of cultivators affected not only their own fields and crops but others' also. It was common for upstream farmers (head-enders), by taking more than their share of water, to deprive those downstream (tail-enders). If planting dates were staggered, not only did this reduce the efficiency of water use (because water had to be delivered over a longer period), but crop losses increased, since insects and diseases could move from field to field when crops were at different stages of maturity.

There was a tradition, sanctioned by modern law, that all cultivators should meet before each season to agree on dates by which all land would be plowed for sowing, planting would be completed, the first irrigation water would be issued, fences would be erected to keep stray animals out, and so forth. But these seasonal meetings in Gal Oya were planned to cover as many as 2,000 farmers, and their conduct could be confusing, even farcical (Murray-Rust and Moore 1984). Few farmers attended, and compliance with decisions reached or announced at the meetings was minimal.

---

3. "Many canals have scoured out and many others have silted in; others, some in the primary system, are in almost immediate danger of washing out completely. . . . Drainage facilities earlier installed have become so clogged and closed, some purposefully by cultivators and encroachers, that they have largely ceased to function . . . nearly all gates leak 24 h[ours] a day. Padlocks have disappeared, gates have been damaged and destroyed, extra pipes and siphons have been installed, and canal banks have been cut by the cultivators; there is very little discipline or enforcement" (FAO 1975, 4–5).

Water distribution problems between head and tail areas can be found in most irrigation systems throughout the world, and staggered planting was not due simply to farmers' ignorance of its consequences. They faced cash, labor, and draft power constraints at planting time that made it difficult or costly to meet the target dates. But apart from such explanations, there was a palpable lack of cooperation among Gal Oya farmers, whether due to psychological factors or to more rational barriers to collective action (see chap. 12). The FAO experts' report on Gal Oya did not suggest working with or through farmer organizations to remedy the situation. Perhaps they thought this approach impractical. Instead, among their recommendations, the experts said the government should "strengthen discipline and enforcement measures" (FAO 1975, 14).

The USAID Water Management Project

When I first visited Sri Lanka in January 1978, making plans to spend a sabbatical year as a visiting researcher at the Agrarian Research and Training Institute, I stopped by the office of the U.S. Agency for International Development in Colombo. Its representative told me that USAID and the government of Sri Lanka had recently reached an agreement to make improvements in water management one of the priorities for American economic and technical assistance. Knowing this could not be accomplished quickly, USAID was thinking in terms of a twenty-year commitment to help build up knowledge, competence, and institutional capacity for the irrigation sector. Sri Lanka's Irrigation Department needed to replace its preoccupation with design and construction, common to the engineering profession worldwide, with a greater concern for operation and maintenance (O&M) to increase the efficiency of water use and boost agricultural production.

A long journey begins with a single step, and USAID recognized that the knowledge base and professional skills, not to mention institutional capacity, for such a redirection were slim. The effort would begin with a project to rehabilitate the Gal Oya scheme, developing skills, techniques, roles, and organization for improving water management that could be applied to other run-down systems. The means for this were to be worked out by consultants in the project design process. Obviously no one could know how many unanticipated consequences, how many accidents for better or for worse, would befall the project. The naive presumption was that a project process could be designed.

It was not stated why the Gal Oya scheme was chosen for this pilot effort. The largest and most deteriorated system would require a substantial amount of donor resources coming into the Irrigation Department. A political consideration could have been that the system's condition was an embarrassment to the recently elected United National Party government, since Gal Oya had been built by the first UNP government thirty years before. Also, the major criterion for American aid in 1978 was helping "the poorest of the poor," and there were many thousands of potential beneficiaries in the Left Bank area.

Two predesign consultant reports were commissioned. The first, by an agronomist with experience in Pakistan, focused on the observable waste of water at field channel level. He regarded farmers' behavior— water theft, lack of field channel maintenance, staggered planting, breaking gates, cutting bunds to get water directly from larger channels—as the main problem. His report echoed the FAO experts in calling for more "discipline" among farmers, enforced by legal measures. The proposed physical rehabilitation was to restore both channels and structures to their original designed capacity.[4]

The second study was by an agricultural engineer who had trained at Cornell and subsequently was appointed the first director general of the International Irrigation Management Institute, headquartered in Sri Lanka. He saw farmers' behavior as largely a response to the unreliability and inadequacy of water deliveries. The Irrigation Department had practically given up trying to follow the schedule of water issues announced at the start of each season. Political pressures were mobilized to get ad hoc deliveries made, either legitimately to meet emergencies when crops' survival was threatened or not so legitimately to service influential farmers' fields. This consultant identified "main system management" as the chief problem to be addressed, viewing farmers more as victims than as culprits.

As it turned out, the first diagnosis and prescription prevailed, since the agronomist was chosen to head the project design team. His interpretation of the problem matched that of the Irrigation Department.

4. This approach to rehabilitation did not allow for the fact that irrigation systems themselves evolve and change. The Gal Oya system now commanded 30 to 50 percent more area than when designed, and the water supply had been greatly overestimated. There was now much more knowledge about soils and hydrology, much of it in farmers' minds, that could be brought to bear on improving system performance. But the USAID project idea was that of "restoration." For lack of a better idea, efforts and resources were thus misdirected at the start. The ID and its foreign consultants were unable to restore the structures and channels to original specifications, however, because no copies of the original blueprints could be found.

The project design team included a British sociologist (doubly marginal on a team of American technicians) who had been a researcher at ARTI and who was a respected colleague and friend. He did not, however, push for organized farmer involvement in irrigation management as might have been expected. When he visited Gal Oya in December 1981, nine months after the IOs had started work, he reversed his position, writing that when the project was being designed,

> [I] argued that work on farmers' organisations would tend to divert attention from the primary goal of reforming the Irrigation Department. Hindsight reveals that this was wrong. The farmers' organisations under the IO programme have at least the potential to become an effective catalyst and pressure group to reform Irrigation Department practices.
>
> Even if, as is not impossible, the physical rehabilitation is a complete failure, a strong IO programme may increase the chances of doing much better next time, above all by showing that Irrigation Department staff and farmers can interact for their mutual benefit. If this can be done in Gal Oya, it can probably be done anywhere. (Moore 1981a, 1–2)

Even though the consultants did not recommend investing in the establishment of farmer organizations as "social infrastructure" for irrigation, the final project document provided for this. A staff member who was brought in by USAID to manage the Gal Oya project had seen good results from farmers' participation in irrigation improvement in Pakistan, so he wrote an organization component into the project. I endorsed including such an effort, though I can see now how little I understood about what this would entail. This was a good example of "the hiding hand" in development projects (Hirschman 1967).

USAID asked the Irrigation Department (ID) whether it would like, with donor support, to build up a socioeconomic unit to deal with farmer organization, farm management studies, and so on. The ID's response was that it was a technical department and that the Agrarian Research and Training Institute might more appropriately handle such activities.[5] ARTI, however, had no experience in water management. Cornell's Rural Development Committee (RDC), which

5. Though ARTI was administratively under the Ministry of Agriculture and the ID worked under the Ministry of Lands, the director of irrigation was a member of ARTI's board of directors, and ARTI staff had a reputation for competent research. The ID, which was to implement the Gal Oya project, agreed to give ARTI a subcontract to do the farmer organization and other socioeconomic work.

had collaborated with ARTI on some research several years before, did have such experience, and my spending a sabbatical year at ARTI had strengthened collegial relations. Serendipitously, USAID in Washington had entered into a cooperative agreement with Cornell in 1977 to assist AID missions and Third World governments in introducing participatory approaches to rural development, so funding for a collaborative effort was already available. Several Cornell faculty members had been involved for some time in interdisciplinary research on water management.[6] Agreement to collaborate with ARTI on Gal Oya work could be reached easily and quickly.

One problem was our initial lack of appreciation for how difficult the task would be. We had accepted at face value the task assigned by USAID, to develop and test various models of farmer organization. The best of these was to be used to organize, by the end of the project (within four years), all 19,000 farmers thought to be cultivating throughout the Left Bank system. The actual total turned out to be maybe 50 percent greater, but this discrepancy was the least of the difficulties.

Upon some group reflection, it became apparent that testing alternative models in any formal scientific sense was impractical and probably unjustifiable. A number of variables such as size of organization and mode of decision making could be hypothesized that would probably affect the performance of water user groups. But the number of combinations to be assessed quickly became too many, since even four variables with only two alternatives each produced sixteen different options to be tested in the field.[7] To test the statistical significance of

6. Gil Levine, an agricultural engineer who had pioneered work on water management, was joined in the mid-1970s by Randy Barker, an agricultural economist who returned to Cornell after eleven years with the International Rice Research Institute in the Philippines; Walt Coward, a rural sociologist who had done some of the first research on indigenous organization for irrigation management in Southeast Asia; and Milton Barnett, an anthropologist with similar experience. The Rural Development Committee was started as an interdisciplinary group of faculty and graduate students in 1971, with me as chairman and Gil Levine as vice chairman. He and I, together with an anthropologist, taught an experimental course on rural development in 1972. The next year, Randy Barker and Walt Coward, with a colleague in nutrition, taught another version of the course. These half dozen persons served together on the RDC executive committee for most of the years after 1975. Our close personal and professional relations certainly contributed to the effectiveness of Cornell's work with ARTI in Gal Oya.

7. In our first analysis, we identified five probably significant variables: packaged versus modular approach; small versus large organizations; decision-making structure; delegation of authority; and federation. The last one appeared desirable based on previous studies (Uphoff and Esman 1974), but experimentation was appropriate to determine what would be the best structure, pace of federation, and issues to emphasize (memo to Gil Levine from Norman Uphoff, September 12, 1979, 8–9). Gil Levine was going to visit Sri Lanka at the

observed variation, we would need to establish at least twenty farmer organizations of each type and evaluate them within two years. How could we decide on a model and replicate it throughout the rest of the scheme by year 4? There was no way we could properly field test alternative organizational models in the conventional social science mode. Yet just as daunting was the responsibility we would be taking for farmers' lives if we were to try to implement difficult models for the sake of testing them rather than for the farmers' sakes. It also seemed unlikely we would get cooperation if farmers understood this was only an experiment.

As we got further into our task, the ideal of a detached social scientific endeavor receded. Additionally, we had to ask ourselves how we would introduce farmer organizations even if we were ready to do so. This in itself would be a major undertaking. ARTI had done many fine field studies in rural Sri Lanka, but it had neither the staff nor the experience to carry out the operational tasks required to test alternatives. Fortunately, ARTI's director assigned three very capable staff members to constitute a new Water Management Research Group: an economist to head the group, C. M. Wijayaratna, known as Wijay; a sociologist, Lakshman Wickramasinghe, nicknamed Lucky; and a statistician, R. B. Senakaarachchi, called R. B., who had grown up in a colony unit in the Gal Oya Left Bank.

This group was small for such a large task, and it was not yet experienced in water management. Cornell could provide short-term consultants who would come out for several weeks at a time between semesters. The director asked Cornell to provide with USAID funding a consultant who would be in residence, working full time with his irrigation group.[8]

Further discussions recast our approach to the task of organizing farmers. We could:

---

end of September 1979 and could discuss these variables with our ARTI colleagues. With their feedback and further discussion, we added three more variables: single-purpose versus multifunctional organization; statutory body created by law versus voluntary association set up by a nongovernmental organization; and residential versus hydrological basis for membership. Several of these probably did not need to be tested because there was considerable evidence on them in the literature, but this still left a complex problem of multivariate analysis.

8. The Institute already had such relations with the universities of Reading (United Kingdom) and Wageningen (Netherlands), whose researchers worked together with ARTI staff "as one team . . . (with) a sense of equal partnership." ARTI would welcome "collegial interaction" with closer collaboration than usually found in an "advisory capacity" (letter from T. B. Subasinghe to Norman Uphoff, November 11, 1979).

1. devise several models with different structural features that could be introduced in six to ten locations and then monitored, to "see what works" (not exactly testing in a scientific sense);

2. adopt some overarching concept, such as creating organizations as vehicles for giving farmers some voice and power in planning and implementation, encouraging farmers to organize themselves in whatever manner they liked;

3. present communities with a set of options they could choose from (like a do-it-yourself organizational kit), and then assist and monitor the organizations in their activities; or

4. introduce one or more models or let communities design their own organizations, and then, rather than evaluating the characteristics of the models themselves, "focus on what things can make *any* model work better, e.g., what kind of training, or what level of government support . . . contributes to organizational success."[9]

This last approach was judged the most justifiable ethically and the most practicable. It was redefined operationally as trying to develop a process rather than a model for farmer organization to help improve irrigation management. We concluded that *how* organizations are conceived and introduced may be more crucial than particular structural features. Our thinking had evolved this far when we had our first encounter with the situation in the field.

First Field Visit: January 1980

Gil Levine and Randy Barker had already visited Sri Lanka and ARTI briefly during the fall of 1979. Walt Coward and I came out in January 1980 for several weeks, together with David Korten, who was a Ford Foundation adviser in Manila. David and his wife Frances had helped develop water user associations in the Philippines that could cooperate with the National Irrigation Administration (NIA) in planning, constructing, and paying for capital improvements in user-managed community irrigation systems. David was putting the finishing touches on an article that presented a learning process approach to development planners and administrators. This conceptualization made a lot of sense as we surveyed the task before us.

9. Memo from Gil Levine and Norman Uphoff to Randy Barker, November 6, 1979. Randy was going to Sri Lanka to help ARTI set up a record-keeping program to monitor hydrological, agronomic, and economic variables in Gal Oya and to discuss these ideas.

Wijay, Lucky, and R. B. had already made several trips to Gal Oya to start ARTI's program of gathering data on farm-level and irrigation channel operations. For Walt, Dave, and me, the ten-hour drive to Ampare over worn-out roads was our introduction to the posterior price one had to pay to work in Gal Oya.

The deputy director of irrigation (DDI), whom we met in Ampare the next morning, impressed us both with his charm and with the ambiguous way the system was being operated. When asked about the allocation of water among the Left Bank, Right Bank, and River divisions, the DDI furtively pulled a small notebook from his briefcase and in hushed tones read some numbers in thousands of acre-feet. He asked us not to divulge them to anybody, because they were "still being experimented with and adjusted," saying that his predecessor had left him no data to work with. Since water means livelihood in a place like Gal Oya, it seemed incredible that a public resource could be distributed so secretly. Did farmers know this allocation? Walt asked. No.

When we spoke with farmers, we were struck by how much hostility there was toward the Irrigation Department. We saw examples of utterly run-down channels and inoperable gates. Some of this could be blamed on farmers, but their grievances against ID staff as unreliable, capricious, and even dishonest were deep and apparently often well founded. The harshest words were directed toward the technical assistants (TAs), the lowest level of professional staff, who with some postsecondary technical training were each responsible for operating and maintaining an area of about 5,000 acres.

We visited one TA's office and found him asleep in his nearby house at 2:00 in the afternoon. Though he had been posted here for two years, he knew little about the acreage within his area. We asked what he did if farmers in the area had a water shortage. He said he would get approval for a special issue sent from the main reservoir, which might take up to a week. Since all the gates in the channel we had just driven along were missing or broken, we wondered out loud whether any water would reach this point. He said he could put sandbags in the gates to block them up, but he conceded, when asked, that the bags would probably be removed by upstream farmers. So water was unlikely ever to arrive this far down in the system. We could sympathize with the difficulty of his task, but we could also see why farmers regarded him (at best) as useless.

We came away from our field visit with several conclusions. First, we could not and should not try to establish farmer organizations through the existing Irrigation Department staff. Even conscientious ID personnel who might have training and aptitude were overburdened and lacked time for such a responsibility. The lazy or dishonest officials who were disliked could not win acceptance or cooperation. So we decided to work through specially recruited and trained organizers, following the precedents of NIA in the Philippines, which deployed community organizers, and the Small Farmer Development Program in Nepal, which employed group organizers to bring together poor rural households in productive activities. This strategic decision to use social catalysts (Lassen 1980) was crucial for our Gal Oya work, and we never regretted it.

The task would be extremely difficult. Our trip report stated, based on observations and conversations: "The system seems socially deteriorated as well [as physically run-down]. The visual evidence of lack of field channel maintenance and of uncultivated land even in *maha* season suggested social disorganization, which farmers verified in discussions. . . . Most places there is virtually no social organization for local-level water management, as head-enders make no common cause with tail-enders and all avoid responsibilities for maintenance, which they regard as the government's job."[10] Our second main conclusion, based on observation and discussion, was that no significant changes in farmers' behavior and attitudes should be expected unless and until there were changes in the behavior and attitudes of government staff. This turned out to be overstated, fortunately.

We had agreed to help set up farmer organizations in Gal Oya thinking this was something appropriate for social scientists to do, complementing and supporting the separate technical work done by engineers to rehabilitate the physical system. The more we learned about farmer-official relations during our visit, however, the clearer it became that farmers' behavior at field channel level was a response to the way the main system was managed. Moreover, farmers were alienated by the

10. We noted that while our view might be biased by unrepresentative visits and interviews, two field studies done the previous summer for USAID based on longer visits had come up with similar findings. "Trip Report: Cornell Rural Development Committee Team to Work with Agrarian Research and Training Institute, Irrigation Department, and USAID/ Colombo, on Socio-economic Aspects of Implementing Water Management Project, January 5–19, 1980," by E. Walter Coward, Jr., and Norman Uphoff, February 6, 1980, 5. The studies referred to were by a Cornell agricultural engineering student and a Sri Lankan–born sociologist who spent time in Gal Oya during the previous summer assessing irrigation problems (Murray-Rust and Cramer 1979).

indifferent, condescending, sometimes even insulting way officials treated them. (It was dismaying to hear engineers refer to farmers, even jocularly, as "those donkeys" or "those farmer buggers.")

If our second conclusion was correct, "organizing farmers" was itself a misconceived and probably infeasible task. We should be trying to influence the Irrigation Department instead, which had been the design team sociologist's earlier conclusion (Moore 1981b). We had no mandate to work directly with engineers, however. It was good that our conclusion was only partly correct. In the course of our work, we found that improvements made by farmer organizations could make engineers' attitudes and performance more positive, and such changes from the official side in turn encouraged water users to take more responsibility, which helped challenge and further change officials' negative stereotypes about farmers. Such two-sided change could be reinforcing and accelerating in a dialectical manner. But appreciating this required us to get out of a "set piece" frame of mind and recognize the possible fluidity of social relations.

Difficulties in Starting Work

We had to get started in the field to see how much potential there was for movement. Sadly, it took a year before institutional organizers were recruited and trained, instead of the six months initially expected. We were getting support from the director for water resource development in the Ministry of Lands, Joe Alwis, who understood and accepted our way of thinking. The government had available a large cadre of Development Assistants, recent university graduates hired with little experience but also few responsibilities. Joe suggested we select thirty of them to be posted in Gal Oya. I planned to return to Sri Lanka in the summer and could participate in their training.

This plan fell through because after talking with some of these Development Assistants Joe concluded that they would not be able or willing to withstand the rigors of living in villages in Ampare district. As university graduates with enough political pull to get government jobs, they seemed unsuited for the task: those who came from villages themselves were disinclined to return to the countryside, which would suggest that their education had gained them little, while those who were city-born seemed unlikely to adapt and be accepted. Being politically well connected, they would probably spend much of their time trying to get transferred to more desirable locations.

I came out at the end of May as planned with Ed Vander Velde, a geographer who had experience with irrigation systems in northern India and had interacted with our Cornell group when he was teaching in Binghamton, New York. He had gotten leave from the college in Michigan where he now taught so that he could spend a year at ARTI as the resident consultant for Cornell. We made another field trip to Gal Oya in June with Wijay, Lucky, and R. B., during which time we met with the Government Agent and with the District Minister, a member of Parliament appointed by the President with cabinet rank to oversee district administration. The political and administrative heads of the district both expressed support for what we were setting out to do.

Our discussions with farmers and Irrigation Department staff confirmed our January observations. I remember spending several hours in a mud-and-stick lean-to with several tail-end farmers in unit 39. Much of their meager water supply was being stolen by persons cultivating in drainage areas. When they reported this to the police, they were beaten up by thugs, since the police apparently were in cahoots with the powerful encroachers.

There was little we could do to help these farmers, but we told them the planned rehabilitation project should at least improve their water supply. As we left the house, one farmer tugged at Wijay's sleeve and asked whether there was any way some agency other than the Irrigation Department could be given responsibility for implementing the project we had told them about. A bold question to ask a visitor from Colombo. It suggested the depth of distrust our program would have to contend with.

Back in Colombo, we tackled the problem of how to initiate the program. We leaned toward hiring secondary-school rather than university graduates as organizers but learned we would have to draw them from what the government called the Job Bank. In principle this "bank" was a good idea. Unemployed persons would enter their qualifications and job preferences on a form, and employers could get a long list of appropriate candidates for job vacancies. All government departments were required to fill positions through the Job Bank. The hitch was that application forms were available only through members of Parliament, who usually gave them out only to party supporters (Perera 1985, 170–71). Such persons might make good organizers, but as with the Development Assistants, we felt we could not count on them to work hard and persevere under difficult conditions. And if they ever used their positions in a partisan way, this could kill the cooperation needed among water users.

Since the Job Bank had thousands of unemployed secondary-school graduates on file but only hundreds of university graduates, there was a chance it would not have enough candidates who met our criteria, which included coming from a dry-zone district and knowing irrigated agriculture firsthand. We could justify requiring university education at the outset because graduates could absorb social science concepts more quickly, but we were apprehensive that social distance between university graduates and farmers might impede progress in the program. As it turned out, we made the right decision for the wrong reason. We were wrong about the matter of social distance, which proved to be no problem. (We were right to try to keep partisan politics at arm's length.)

We also knew we needed a supervisor in the field who could keep on top of the program as neither ARTI nor Cornell personnel could do from afar. Since the program was supposed to be absorbed into the government within two years' time if it was successful, we wanted it to operate in consonance with official personnel procedures. The Ministry of Lands arranged for us to interview a number of senior colonization officers who had already been screened for possible appointment as irrigation project managers. One of our most fortunate decisions was to select (and then not lose, when the ministry delayed assigning him to ARTI for five months) someone with fine personal qualities and previous experience in Gal Oya. S. Munasinghe, known as Mune, was almost not chosen because he was then president of the Colonization Officers' Union, and the ARTI director thought this could cause some controversy. But Mune was finally preferred because he expressed strong interest in working with young people and showed genuine concern for the problems of farmers in irrigation schemes.

The team was further expanded when Hammond Murray-Rust arrived in June from Cornell to begin thesis research as an agricultural engineer on the operation of the Gal Oya main system. He had spent the previous summer getting acquainted with the system and could contribute a needed technical perspective to the ARTI-Cornell team. To draw on Philippine experience, we arranged for short-term consultancies later in the summer from Ben Bagadion, NIA's deputy administrator for operations, who had given outstanding leadership in introducing participatory irrigation management there, and Carlos Isles, NIA's organizational specialist who helped ARTI plan its training for IOs.

Another good piece of luck was to have Piyasena Ganewatte (Sena), an experienced Sri Lankan rural sociologist, come our way. He had been a UNICEF staff member in Colombo for four years but had resigned to go to Zambia as an FAO consultant. Without explanation that assignment was canceled, and he was suddenly without a job. I suggested he talk with Ed Vander Velde and the ARTI director, T. B. Subasinghe, about joining the Water Management Group. It turned out Sena and Subasinghe were old friends from time together in the university and Health Ministry. Also, Sena had been a temporary consultant for ARTI before joining UNICEF. Unfortunately, the ARTI board refused to appoint him at our proposed salary level, less than he had been receiving from UNICEF and *much* less than he would have been paid by FAO. Not giving up, Ed and Subasinghe worked out an agreement for Sena to become a Cornell consultant paid for by USAID and assigned to ARTI's Water Management Group. (R. B. had left to do graduate study in England, so ARTI was definitely short-handed.) I mention these personal associations so that we don't lose sight of such factors, which are usually not reported.

Between the time I left in July and returned in January 1981, Ed, Wijay, Lucky, and Sena had to wrestle with many difficulties and often unseen resistances. Eventually they got approval from the Job Bank (which did not have enough qualified candidates in its files) to advertise nationally. About 800 applications were received, reflecting the volume of unemployment more than enthusiasm for the life of an IO. The ARTI group screened the number down to 150 to be called for interviews.

We were still not beyond the reach of politics. About fifty applicants who did not receive invitations to come for an interview were able to bring letters from MPs, ministers, or other notables so that they had to be put on the interview list. The ARTI director, to his credit, did not identify these persons as such, however. Instead, he kept their letters of introduction in his pocket. The group interviewed two hundred applicants and selected for training the thirty-two with the highest marks. Only after selections had been made did he pull the letters out of his pocket "in case the group was interested." Only two of the fifty sponsored applicants had made it through the interview, and they turned out to be some of the least effective of the first batch. Having initial selection based on our best estimates of merit was thus very important.

Training and Deploying Organizers

Before returning to Colombo in January 1981, I heard of tensions within the group and feared some changes in personnel might have to be made. Fortunately, relations had begun to improve by the time I arrived, and a two-week training program for IO recruits began January 15 in Colombo, followed by four weeks of field training in Ampare district.

The IO trainees were eager and enthusiastic, though only somewhat less clear than we were about what they would actually do in the field. To show that we expected self-reliance and self-management from them, on the first day of the training program they were divided into four committees to handle training sessions, food and refreshments, social activities, and any discipline problems.

The sessions were in retrospect too formal and too theoretical, but we were feeling our way. One of the highlights was a presentation by Godfrey de Silva, at the time deputy director of irrigation for the Kandy range, about his introduction of a three-tiered structure of farmer organization for the 15,000 acre Minipe irrigation scheme (de Silva 1985). This experiment was not approved (indeed, it was tacitly disapproved) by the leadership of the Irrigation Department, but it gave us encouragement that such organizations could be introduced with farmers in Gal Oya. Godfrey quietly played the role of guardian angel for our program through the years, seeing in it a chance to move beyond his experiment with participatory irrigation management.

Soon I myself experienced the Irrigation Department's active lack of interest in farmer participation. The day our training program started, there was a meeting with A. J. P. Ponrajah, senior deputy director of irrigation and soon to become director. He was accompanied by two deputy directors. The meeting was called by USAID's Gal Oya project manager, Ken Lyvers, and I attended along with Ken's Sri Lankan assistant, Dr. Kariyawasam (Kari).

The meeting began with Ponrajah grumbling about the Ministry of Lands' initiatives to improve water management. "We do water management already," he fumed, referring to his department. Based on what I had seen and heard in Gal Oya, I tried to persuade him that there were still some shortcomings in water management, at least in Gal Oya. His response was that if only the farmers would obey schedules and rules, there would be no problem. This line of attack could be deflected by pointing, delicately, to certain deficiencies in the ID's operation of the system but then suggesting how farmers' cooperation with the ID could be improved if they were organized. It took about an hour to get agreement that water management could and should be improved and that it was at least worth experimenting with farmer participation through organization.

But then Ponrajah made the argument that the Irrigation Department could and should handle this task itself. My heart sank. He seemed not to know that his department had previously agreed to pay for the IOs and their training as part of its local-cost contribution to the Gal Oya project. ARTI had recruited IOs and was training them at that very moment based on an oral agreement with the ID; nothing had been received in writing. Would we have to cancel the IO program before it got started? USAID could pressure the ID to implement its previous agreement, but this would likely make the department less willing to accept our program later as a permanent part of its structure even if IOs produced good results.

We argued that ID personnel were already overburdened and were unprepared for such responsibilities, stressing the need for "technical" training of IOs in social science methods for organizing farmers. This appeal to professionalism seemed to gain some ground. Finally Ponrajah said that he accepted the logic of our arguments and was prepared to try the program and to use IOs, but that they should come immediately under the ID rather than under ARTI. This called for a different tack. We pointed out the need to remain flexible and experimental with the program. Nobody knew yet how large the areas of responsibility should be, how long IOs should work in different modes, and such. Agreement was finally reached, after almost three hours, that ARTI would be responsible for supervising and evaluating the program, though it would come under the Irrigation Department administratively. Within two years the ID would review the results to determine whether it wanted to incorporate the program fully into its structure. Such an agreement was as much as we could hope for.

When the IOs arrived in Ampare for the field phase of their training, they found that the Police Reserve Academy where they were to be based was an absolute mess, without even brooms to sweep up the accumulated bird droppings and dirt. The IOs and their trainers spent two days cleaning the place enough to make it tolerable. Although this was unplanned and undesired, it was a good initiation for the IOs if any was needed. It established that Wijay, Lucky, Sena, Ed, and Mune did not intend to play the usual role of superiors. Adversity built a stronger camaraderie, one of the many unanticipated outcomes we were to encounter.

At the end of four weeks, the IOs were ready to take up their work in the field. There was a memorable ceremony marking the end of their training, addressed by the Government Agent. The organizers were deployed in groups of four to six IOs, with collective responsibility for a large command area (one or more distributary channels) coexisting with individual responsibility (each IO was assigned a certain number of field channels, decided and adjusted by the group itself). This arrangement could have led to confusion or to shirking of duties, but instead it mobilized a more active sense of responsibility from IOs. Self-management turned out to be one of the keys to good IO performance in the field. Only in retrospect did we realize that we had again done the right thing for the wrong reason. Teams were valuable not just because they permitted us to employ women as IOs.

The previous June we had considered whether to take women as well as men into the IO program. In the Philippines, women were found on average to be better than men in these roles (eventually 80 percent of NIA's community organizers were women), but Wijay and Lucky were doubtful that young, unmarried women could be accepted and effective in rural communities. Organizing work required a great deal of traveling around, often at night. The farmers we asked about this on our June visit approved of having women IOs. ("They're less likely to get drunk," said one farmer, himself somewhat tipsy.) We though that by deploying IOs in teams, each with at least two women organizers who could live and travel together, we could avoid breaking social taboos. The team concept of IO deployment proved to be a major source of strength for the program, but we were chastened when we realized later that we happened on it fortuitously rather than by design. In the planning process we made a number of wise decisions, but this could be known only in retrospect, and some were made with little or wrong justification.

We expected the IOs to spend the first three months just getting acquainted with and accepted in their respective communities. Specifically, they were to prepare descriptive profiles of the farmers and farms served by each field channel. This would give them reason to visit each household and to establish rapport, as was done in the Philippine program.

Soon after fielding the IOs, however, we feared that the climate might destroy our program before it got started. The 1981 yala season was the most water short in many years. The amount of water in the main reservoir at the start of the season was only 180,000 acre-feet, less than one-quarter of Senanayake Samudra's capacity and only about half the usual level of recent years. We were having to begin our efforts to organize farmers when water was extremely scarce.

Most social scientists, myself included, would have predicted that these were the worst conditions for eliciting cooperative efforts, because resource scarcity is thought to engender more conflict than cooperation. As seen in the next chapter, however, this was not our experience, and this surprise kindled my first suspicions about conventional social science formulations.

Collision Course?

While we were considering the effects of climate, it became clear that the atmospherics within the Irrigation Department could be even more adverse. The department did not follow through on its January agreement to take administrative responsibility for the IO program, so if we were to continue, ARTI had to take on substantial transportation, accounting, and other duties for which its staff was not prepared. This work absorbed much of Wijay's and Lucky's time, cutting into the research, training, monitoring, and evaluation they were expected to do.

We had agreed to link the organizers' work to the ID's schedule for physical rehabilitation of the system, to encourage farmers to contribute ideas to the redesign effort and so they would be willing to work on restoring field channel capacities. This entailed the danger that organizers would become (or would be seen as) work supervisors mobilizing farmers to do unpaid labor. But we had no choice. We desperately needed support from USAID to keep the ID from dispensing with the program altogether, and USAID seemed obsessed with farmers' contribution of labor to the rehabilitation as a tangible but questionable form of "beneficiary participation."

No budget had been provided for manual labor to repair the field channels. If farmers did not contribute labor, rehabilitation at this level would not be completed, and all the work done to improve higher levels of the system would have little impact. We were afraid our program would be discredited if farmers, understandably, balked at this levy of free labor. We objected to this requirement, pointing out that Gal Oya farmers had never been consulted and had not agreed to it.

It served an unexpected beneficial purpose, however, by enabling us to persuade the engineers to be more solicitous toward farmers, to discuss system deficiencies frankly, and to seek suggestions for improvement, thereby gaining farmers' support for the work. In retrospect our objections were misconceived, because the requirement of labor contributions, though unfair and nonparticipatory, encouraged engineers to go out and meet with farmers as had seldom been done before. This was another instance where the right thing happened for the wrong reasons.

The Gal Oya Water Management Project was officially inaugurated in March 1981, with the Minister of Lands cutting a ribbon and operating a line dredge. The only rehabilitation work to be shown off was some field channel cleaning that a farmer group had done along the main road for visitors to see. The Irrigation Department put up a sign at the project office calling it the "Gal Oya Rehabilitation [sic] Project,"and its invitations to the inauguration were similarly titled. When Ministry officials noted this, the ID staff in Ampare passed it off as an unintended slip in wording. But this revealed the engineers' view that this project was to be concerned mostly with design and construction, not taking seriously USAID's and the ministry's operation and maintenance objectives.

After the inauguration, both Ed and Sena wrote confidential memos detailing the many weaknesses they saw in the program's position, though they did see some bright spots.[11] It was not clear how firm was the commitment of the Ministry or USAID to keeping this a water management project, with a participatory role for farmers. ARTI itself lacked the resources to sustain the IO program; all funds for this work came from the ID. Already the training budget had been slashed so that no second batch of IOs could be fielded until 1982. After our IO supervisor, Munasinghe, had been assigned by the ministry to work full time on the program, the Government Agent (GA) decided he needed Mune's services at least half time as a district land officer in the district headquarters. Since we did not want to offend the GA, we

11. The bright spots Sena added after listing nine pages of problems plus suggestions and recommendations included: we have a superior batch of IOs; the Water Management Group has grown as a strong, close-knit unit; Munasinghe has developed as IO supervisor and is an asset to the project; there is considerable interest and concern in intellectual and professional circles in water management and farmer participation; the first training course for IOs was a great success; and Lucky and Wijay have grown in stature and are becoming quite confident (letter to Norman Uphoff, April 10, 1981).

reluctantly went along, hoping that Mune would still be able to carry out what should be a full-time job.[12]

We certainly needed someone in the field with Mune's experience (and with his access to the Government Agent and District Minister). Wijay and Ed both wrote during the second week of April to say that relations with the ID in Ampare were getting worse. Ed said there was "substantial noncooperation" from the ID and a "growing willingness to actively oppose" farmer participation. Wijay explained in his letter:

The key engineers with whom we are dealing are not "happy" with our IO program. I am pretty sure that this is mainly due to two reasons:
1. They suspect that they cannot continue with malpractices (corruption) once the farmers get organized.
2. It is not their aspiration to share the "water rights" with farmers and they do not wish the farmers to be empowered with "control" of water even at the farm level.

Wijay reported that the ID had made a number of unilateral decisions about how the system would be rehabilitated, without any consultation, and the first meetings with farmers about redesign and reconstruction had been unsatisfactory: "Therefore we have decided to 'educate' and 'motivate' the farmers (through the IOs) to question the engineers to clarify all these issues. . . . Please do not get the impression that we have already started a battle with the ID. We are well aware that we cannot implement the IO program without ID's cooperation and therefore we think we should be very diplomatic at this juncture."

In early May, Hammond drafted a note based on his contacts and discussions with ID personnel, "Some Observations on Farmer Participation in Gal Oya." It concluded: "There is no indication that the Irrigation Department, from top to bottom, has any interest in getting genuine farmer participation beyond the draft guidelines [requiring meetings to inform farmers of plans] for labor input at field channel

12. We were apprehensive that Mune's becoming a district land officer would tarnish the IO program, because DLOs had a widespread reputation for corruption. But, he was accessible and helpful enough in his concurrent role as DLO that this boosted farmer acceptance of our program. It also gave the program direct access to the GA and district minister, which proved valuable. Once again, we misjudged consequences, but who would not have been more concerned with the likely negatives than with the possible positives in those circumstances?

level. Without a change in attitude here, it is very hard to see how there is going to be any significant involvement by farmers in deciding, for example, on redesign, on water scheduling, on operation, and so on. The signs all point to yet another scheme that within a few years will be substantially the same as it was before rehabilitation."

Hammond's pessimism was confirmed by a meeting on May 5 that the deputy director of irrigation held in his office in Ampare with his senior engineer and foreign technical consultants to decide when farmers' work on field channel rehabilitation would begin. Since nobody from ARTI was in Ampare at the time, Hammond was asked to attend, though he had no authority to speak for ARTI, the IOs, or the farmers. The DDI drafted minutes of this May 5 meeting on May 13 and posted them to Colombo on May 20. The letter arrived at ARTI finally on May 28, informing the Water Management Group that IOs were expected to have farmers ready to begin work on June 8, having received the design plans just one day in advance, on June 7.

When I arrived in Colombo on May 31, we discussed how to respond to this gauntlet thrown down. ARTI colleagues had not taken the request seriously because they knew how far behind schedule the ID was in preparing any field channel redesigns for farmers to follow. The guidelines Ed and Hammond had been trying to get the ID to agree to called for at least two meetings between engineers and farmers on each channel before designs were made final. So farmers should have an opportunity to make suggestions or voice objections before they were expected to implement the plans. These guidelines had not been officially approved, however. In ARTI's view, there was no way the ID would be ready to start field channel work on June 8.

But this looked like a trick where the deputy director could win either way if he managed to get at least some designs ready by June 7. If farmers were not on the job June 8, he could condemn the IO program for failing to deliver. If somehow some work did start on that date, he could take credit for getting the rehabilitation under way—at the farmers' expense. (There was tremendous pressure on him from USAID to get some work, any work, started; no improvement was likely to begin on the main channels or secondaries before the end of 1981.) We did not like the idea of farmers' and IOs' being used, at best, as pawns in a game between the ID and USAID, and at worst as scapegoats. The program was being led down a dangerous path.

The Water Management Group drafted a letter that queried "the technical advisability of beginning field channel excavation as the *first* step of rehabilitation." When work was done later on the main, branch, and distributary canals, silt would be stirred up and deposited in the field channels. Since farmers were not being paid for their labor, we suggested they might be unwilling to do a second time the field channel cleaning necessitated by the ID's desilting upstream. We raised some other questions about the scheduling and logistics of the request, but mostly we wanted to make clear to others, who got copies of the letter, how unreasonable was the plan being proposed.

Ed initially opposed what might be seen as a confrontational response to the deputy director of irrigation's request. But during a field trip to Gal Oya the first week of June, Ed was told flatly (though off the record) that the ID did not want farmer participation, and this was confirmed by the ID's expatriate engineering adviser. The only exception was Manoharadas, the chief irrigation engineer in Ampare who was responsible for redesigning and refurbishing the channels. He was sympathetic to a larger farmer role but had to be careful about breaking ranks with the DDI, his boss. In a phone call from Ampare, Ed said the atmosphere was so unfavorable we might as well send the letter and see what happened.

The response from the Irrigation Department in Colombo was not hostile. The senior deputy director for design and construction (who became director of irrigation in 1985) agreed it made no technical sense to start at the field channel level. He was not willing to countermand the request, however. Instead he arranged for a meeting to be held in Ampare to discuss the matter. The ID's deputy director for water management would attend, and Joe Alwis from the ministry also made the long trip down. The deputy director in Ampare, surprisingly, did not show up for the meeting, pleading "ill health." (We were told he was in good spirits the night before at a rip-roaring party at the home of one of the foreign engineers.)

The meeting was chaired by Manoharadas, known as Mano, who was personally agreeable to a farmer role. His willingness to move away from the obstructive position of his superior helped get the process of change started among farmers and engineers. How much of his openness to farmer participation was a matter of his own convictions and how much derived from his being responsible for getting quick results, maybe not even he could say. (If farmers refused to cooperate, the rehabilitation effort would bog down at field channel level.) In any case, the meeting was an amicable one, producing consensus among the ID, the ministry, ARTI, and Cornell consultants on involving farmers in a more active role in rehabilitation.

First Meeting of Engineers and Farmers

There had already been some apparent movement in the ID's thinking by the time our letter to the DDI would have arrived in Ampare.
On June 10, Hammond attended a meeting between engineers and
farmers to discuss redesign of the first distributary channel off the
Uhana branch canal, UB1A. This was a short channel that supplied
only seven farmers, but about fifty farmers attended the meeting because this event was such a novelty. Because the engineers were Tamil
and not very fluent in the Sinhala language, a Sinhalese technical assistant handled introductions and did some translation.

This channel, situated at the head of the system, had more difficulty
in draining off excess water than in getting enough for irrigation. At
one point the chief irrigation engineer, Mano, said in simple Sinhala,
only half in jest, "You have few water problems here. You have enough
water for crocodiles to live in while others downstream don't have
enough for their crops or drinking water." Various problems were
brought up and at least acknowledged if not solved directly. The highlight of the meeting was one farmer's suggestion to spare a row of coconut trees and save construction costs by leaving and using the
existing channel for drainage and putting in a new irrigation channel
on the other side of the road, instead of building two new parallel
channels, one for irrigation and one for drainage. The suggestion
would require reclaiming some of the land within the right-of-way that
was being illegally cultivated by a few farmers, but consensus supported the proposal.

At the end of the meeting, the TA thanked the farmers for coming
and said the meeting had shown that farmers have more experience
and know more about problems and their solutions than officers do.
He went over the proposal again so there would be no misunderstanding. He said the ID wanted to bring farmers and officers closer together and credited ARTI for "the project," apparently referring to the
IO program rather than the rehabilitation project.[13]

Hammond had been trying to broker appropriate procedures and
design criteria that would bridge the gap between farmers and ID officials, in the middle of which stood the IOs. He thought this meeting
represented a "dramatic change" in the thinking of the technical staff.
Instead of trying to redesign the entire channel system, there was a

---

13. This account is from two undated memos by Hammond Murray-Rust, "Design Meeting—UB1A Channel" and "Comments on the UB1A Design Meeting" prepared for the
ARTI-Cornell Water Management Group shortly after the June 10, 1981, meeting.

move toward a more adaptive and pragmatic approach. He was not sure why this "almost complete reversal of policy" was occurring, with attentiveness now to farmer needs and suggestions, but it might have come from a realization that technically elegant solutions were costly and difficult and might not be accepted by users anyway. He saw Mano and the TA as now prepared to use such meetings as a genuine learning experience.

Continuing Visits to the Field

This redesign meeting occurred just one week before I made my first field trip to observe the IO program in action. As it turned out, I was able to make such visits roughly every six months for six years. I got the idea of writing my trip reports in narrative style, without attempting to analyze or evaluate what was going on. That could be done later. I would try to capture and communicate in a cinema verité style the unfolding flow of events in Gal Oya. At the time we started, I had no idea whether this venture would turn out to be successful or disappointing. Either way, it was appropriate to document in a direct, detailed, even intimate way the evolution of the program, its advances and retreats, its blockages and breakthroughs, even trying to convey some of the occasional tedium. The response to these trip reports was very positive. Government officials said it made the program and the process come alive, and academic colleagues found this kind of process data very stimulating. Though the reports became longer as the program (and my powers of observation) expanded, people still read them and asked to receive the next one.

All this while, I was trying to make sense of what was happening before my eyes. It was chastening to find out how often I and my colleagues could be wrong about many things, despite our formal learning and not insignificant experience. At the same time, it was gratifying to learn how many of our ideas were appropriate, or how the program could progress in spite of our shortcomings. As I began to think about how to share this learning with others—especially as IOs, farmers, and officials were making evident progress in Gal Oya—I decided that this sharing should be of the process and not just of the conclusions. Summary analysis could not do justice to the changefulness of the situation, the contingency of outcomes, the significance of individual initiatives and contributions, and the balance between material interests and the power of ideas.

As a social scientist I wanted to develop appropriate theory—that is, explanations—to account for what was occurring. I did not believe the outcomes were purely a matter of chance, though clearly many things did unfold by chance. Nor did I want to accept the reason advanced by some that the outcomes depended mostly on my own involvement and leadership. Although my role in all this was central and visible, I quickly came to appreciate that it could just as well be overestimated as understated. So many persons were crucial to the progress of the program that nobody could be identified as sufficient for its success, and by the same token, if one is trying to unravel and attribute causation, nobody was truly necessary. Indeed, the concept of necessary or sufficient causation began to disintegrate under operational scrutiny in Gal Oya.

It seemed there must be some principles we were operating on that could be generalized and made useful to others. At times my main contribution seemed to be taking the initiative, not being inhibited by certain precedents or social expectations. Changing others' ideas and anticipations could enable them to act more boldly and effectively. At other times my role seemed to be keeping alive a spirit of mutual trust, confidence, and cooperation. Anybody should be able to do that if it is understood as important for energizing people's best efforts. Finally, it appeared that some structural relationships between incentives and performance could be illuminated.

A book like this is simultaneously very specific and also oriented toward broader explanations and effects, seeking "a glimpse of the general through immersion in the particular," to use Hirschman's phrase. The trip reports in the next seven chapters have been edited and considerably shortened, to eliminate redundancy and information extraneous to the learning that ultimately resulted. To limit the length, I have after the first year included only one report a year. I have not excised misconceptions or mistakes, since they need to be seen as part of the learning process, though I have sometimes written delicately around situations so as not to reflect unnecessarily harshly on others. There is no need to embarrass anyone in a story so full of acts of personal generosity and creativity.

I have tried to preserve the essence of this experience without going into excessive detail. To have left in everything I observed and noted would have more than doubled the length of this book. Matters dealing with irrigation or land tenure have been reduced or eliminated, for example. Some may still think there are too many details, but that is part of the reality of the situation.

The original trip reports are available from the Cornell International Institute for Food, Agriculture and Development if anyone wants to see what was left out of these chapters from the longer reports (which are as long as eighty-seven pages) or to see the reports not included here. The exposition has been polished and tightened to make the book more readable, but the narrative style has been kept to give readers a sense of the concreteness and immediacy of the situation. The first part of each of the seven chapters that follow takes the reader into a point in time, seeing what I was seeing and hearing what I was hearing. The trip reports were written hurriedly each time in Colombo after I returned from the field, and statements translated from Sinhala were rough. In editing the reports, I have tried not just to show readers the nuggets but also to give some sense of the process of panning for gold, leaving in at least some of the rocks, sand, and fool's gold.

Appendixes provide the full names of participants and also spell out abbreviations and acronyms, with identifying or clarifying information. Organizational relationships among the various actors are shown in appendix 1. A map of Gal Oya has been supplied in chapter 1. Following each narrative trip report, I include some reflections on events and some bridging information that leads into the next reported visit to the field.

3

The Program Starts Out
(1981)

After driving to Ampare from Colombo yesterday, June 17, we are headed toward Uhana to meet the first group of institutional organizers (IOs). With me are Wijay, coordinator of ARTI's Water Management Group, and Sena, the training consultant for our program. We have just picked up Munasinghe (Mune), the district land officer who serves as our IO supervisor, when we see two young women pedaling bicycles energetically down the road toward us. They are coming to meet Mune before starting their daily rounds, to discuss with him a problem they came across yesterday. They stop and tell us about an area with eight farmers and fifteen acres not served by any field channel. Because it is not shown on the Irrigation Department's map, it is being bypassed in the planning for rehabilitation.

I last saw these two women IOs three months ago, at the start of their training. They were bright, attractive, and eager, but I could not then envision them as field workers. A real transformation has taken place. They have tackled their job with enthusiasm and show considerable competence as they rattle off acreages, field channel locations, and drainage flows. One of the IOs, because she is only four feet six inches tall, is nicknamed Chuti. (The best English translation might be Bitsy, an affectionate diminutive.) They have floppy white broad-brimmed hats and wear blouses and skirts, long enough to be respectable, short enough to be practical for work. (Initially the women were asked to wear saris, to establish respectability in the community.)

As we drive on, I comment that Chuti, the smallest IO, is obviously big enough for the job. Wijay says the women are generally doing very well (something farmers will tell me also). He says both Chuti and Sita, the other IO, have thrown themselves into the work, spending lots of time in the field with farmers. Chuti has gotten rather sunburned, losing her fair complexion. This is quite a sacrifice to make for the program, since skin color affects people's judgment of beauty in Sri Lanka, and thus marriageability. Wijay tells me that farmers have commented on the women's willingness to compromise their appearance for the sake of their work.

First Meeting IOs: Encountering the Rural Elite

Five IOs are gathered at the house near UB9 distributary channel, where the two women in this group are renting a room.[1] We sit under a huge mango tree, and though the sun is withering, a fairly steady breeze makes the shade quite comfortable. Our discussion begins with the team coordinator, Gunasena, telling us about a conflict he is having with an influential farmer in his area. This man is the *yaya palaka* (YP) for his area, and because YPs are officially recognized farmer-representatives under the Agrarian Services Act, this may present a serious problem for our program.[2]

As the situation is described, it appears that the conflict has been almost unavoidable. The YP in question was formerly the traditional irrigation headman in this area. He is not particularly rich, but he is politically influential. When YPs were elected in 1980 for this area, he and his brother were both unopposed to become the YPs representing this colony unit. This does not mean he is popular. His allotment of land happens to be at the end of the field channel, UB9.4. Instead of using his influence to get more water for the whole channel (at some other channel's expense), he got the ID to raise the levels of the pipe inlets installed to serve fields above his on UB9.4 so more of the water delivered to the channel runs down to his allotment at the end. Because he is influential, the other farmers cannot get the situation rectified, especially now that he is yaya palaka.

1. The numbering system for distributary channels is based on the canal they draw their water from: LB = Left Bank main canal, UB = Uhana branch canal, G = Gonagolla branch canal, M = Mandur branch canal, and so on. So UB9 is the ninth distributary channel (D-channel) to take water from the Uhana branch canal. Field channels are similarly numbered using decimals, so UB9.4 mentioned below is the fourth field channel off UB9. Main and branch canals are shown and labeled on map 1.

2. *Yaya palakas* are to help farmers get seeds, fertilizer, and such from Agrarian Service Centres situated throughout rural areas. They are supposed to be elected by fellow farmers, but often the choice is not very democratic. Farmers are expected to pay them a quarter bushel of paddy per acre per season as compensation for services, whether rendered or not. Especially when responsible for large areas, YPs cannot give very intensive service, but they gain even greater income from their position.

Five weeks ago, Wijay agreed to the IOs' request that they begin working with farmers to undertake desilting and cleaning of field channels on a *shramadana* basis to improve the flow of water within their channels.[3] This dry season in Gal Oya, farmers are faced with a serious water shortage owing to a failure of the rains. The water level in the reservoir is very low, and efficient water use is urgently needed to reduce the probability and extent of crop failure. With silt and weeds removed from field channels, it is possible to rotate water deliveries more efficiently among all the fields along a channel and to give all fields a more equal share. If there is poor maintenance of channels and no rotation, most of the water goes into fields at the head of the channel and little reaches the tail—unless the pipes have been tampered with.

When the IO, Gunasena, suggested doing shramadana work on this channel, the yaya palaka objected, making various arguments and finally leaving the meeting in a huff. The other nineteen farmers on the channel wanted to clean it (and adjust the pipes to their proper level), so they went ahead and did it without the YP's participation. He has begun criticizing Gunasena, however, saying that the IO is interfering with the YP's job.

This puts the IO in a bind—whether to accede to the YP's pressure, since he is politically influential and is currently the recognized farmer-leader, or to assist the other 95 percent of farmers on the channel. I express concern that this may embroil the IO program in partisan controversy, but Gunasena says that the other farmers are also supporters of the ruling party, as is the yaya palaka, and are in favor of our program. When the YP threatened to get Gunasena transferred out of the area, the other farmers said they would all sign a petition to the District Minister supporting the IO. This was apparently the first time they have been willing to "go public" against the petty tyranny of this member of the rural elite.

So far the YP has not taken the dispute to higher levels but has resorted to rumor mongering. First he suggested that the IO was a *ganja* (marijuana) dealer. Farmers came to Gunasena about this, and he assured them the charge was false. Then the YP said that our program

3. Shramadana, which I will refer to often, is a traditional institution of voluntary group labor to produce some collective good. In the Buddhist and Hindu traditions, one earns spiritual merit by such contributions of labor, and these are usually accompanied by religious observances and social festivities. Shramadana as an expression of solidarity has been common in old Sri Lankan villages, but it is less frequently mobilized in resettlement schemes like Gal Oya, where traditional social bonds and values are weaker.

was a plot to get farmer support for the IOs by raising production so that they can start an insurrection in two years' time.[4] How boosting production in Gal Oya would create revolutionary fervor is unclear. Gunasena says he has allayed farmers' fears on this score as well.

I understand the yaya palaka's hostility toward our program better when another IO reports that farmers have come to him and offered to give *him* the quarter bushel of paddy per acre they are supposed to pay the YP under provisions of the Agrarian Services Act. Farmers say the IOs have helped them more within a few months than the YP ever did in a year.

Wijay, Sena, and I reiterate what we said in training, that IOs are to maintain good working relations with all local-level officials and that they are not to be replacing anybody. (If farmers decide to elect a different YP next time, that is their affair.) IOs will not be permanent fixtures in the community but will be moving on once the foundation for farmer organizations are laid. They should involve all officials as much as possible in activities, helping them to work more effectively in conjunction with farmers.

The IOs report that farmers have been more willing to work together to improve water delivery where their water problems are more serious, as would be expected. Shramadana work has been done on field channels 9.3, 9.4, 9.8 and 9.9 as well as in the tail section of UB9. The other field channels off UB9 either have not had siltation problems or have been kept relatively free of silt and weeds by individual farmers' each clearing an assigned length of channel, the officially prescribed system.

Farmers on most of the field channels have recently undertaken some kind of rotation to ensure that a fair share of water reaches those at the tail of UB9. Farmers at the head of a channel close their pipes for the first two days of each five-day water issue to the channel, sending water to the tail first, then unblocking their pipes so that the head and middle get water during the remaining three days. Water is supposed to be delivered on a schedule of five days on, five days off, but the Irrigation Department does not always adhere to this. So giving up water on the first days of an issue requires considerable faith in the ID or in other farmers to help out if head-enders get shorted. The rotations

4. In 1971, about 10,00 youths under the banner of the JVP, a self-proclaimed leftist insurrectionary movement, tried to overthrow the government but were put down by security forces after several weeks of bloody conflict. Gunasena has a beard, so he might be cast as a revolutionary like Fidel Castro or Che Guevara.

have been much appreciated. One farmer told an IO that shramadana work had not been done to clear his field channel for fifteen years.

What problems are they having? The IOs say they gave gotten close to the farmers and now get invited to many social events such as weddings and temple services. They are expected to make contributions on each occasion, and they have no allowance for such obligations. They feel their salary is not so much that they should be expected to cover these expenses. We see their problem but can only say we will consider it (though this is something very difficult to sell back in Colombo).

I ask the women IOs, Sanda and Asoka, if they are having any particular problems, and they say no. Gunasena says the women get better attendance at the meetings they organize, and there is general amusement in the group. The girls are wearing saris, and both are strikingly handsome, hardly looking like field workers at the moment. (I whisper this to Wijay, and he says they wear skirts when in the field.) I can't resist asking the women whether farmers accept them as able to help with problems of water management. They answer self-assuredly, "If farmers ask what we can tell them about irrigation, we answer that we are farmers' daughters," which is true.

We discuss various other matters, for which my notes get sketchy. I am still new to this and cannot comprehend all that Wijay and Sena extract from the conversation. The meeting lasts three and a half hours. Two other IOs have arranged with the household where they are living to give us rice and curry for lunch, so we drive there. Anyone who has had such a meal in a rural home knows that this can be the best cooking in Sri Lanka, and this meal is really good. We must eat in a less leisurely manner than we would like because there is a meeting we want to observe on UB2.2.3 between ID engineers who are planning the rehabilitation and farmers cultivating on that channel.[5]

Redesign Meeting: A New Start

As part of the program to rehabilitate the Left Bank, ID engineers are now meeting with farmers to identify location-specific problems that need to be addressed in the redesign and improvement of distributary and field channels, thereby supplementing information

---

5. UB2.2.3 designates a subfield channel, the third channel branching off field channel UB2.2, which is the second one off UB2, the second distributary channel off Uhana branch canal.

obtained from surveys. To help in gathering information, the IOs in the area where first-year construction is planned are arranging a series of meetings between design engineers and cultivators. This particular meeting is being held in a school classroom near the road between Uhana and Gonagolla.

Because the meeting has been postponed an hour, we are early rather than late. We talk with some farmers gathering outside the school, asking their ideas about the best structure for water user associations. They offer insightful views of the inadequacy of any organization limited to the field channel or even D-channel (distributary channel) level because many issues such as water allocation need to be addressed at a higher level for the whole project.

The IOs call the farmers together to brief them before the meeting and to encourage them to speak up. Two jeepfuls of ID officials arrive, fifteen minutes early, and we talk with them for a while. Practically all who cultivate along UB2.2.3 are present by 3:00. Wijay, Sena, and I sit at the back of the schoolroom with Kumarasamy, newly appointed ID deputy director for water management. Kum is the government's project manager at the Colombo level.

Four IOs are moving around, getting farmers seated at the front of the room and passing around an attendance list for them to sign, indicating their allotment number and acreage. The sight of an aged farmer in sarong and faded shirt grasping the pen with crooked fingers and signing his name slowly is gratifying. He squints at the paper and adds his appropriate numbers. The farmers are attentive as the meeting begins.

A technical assistant who is Sinhalese begins with a polite introduction. The engineers speak Tamil as their mother tongue, and these farmers are all Sinhala speakers, so it is best that the TA begins. As the discussion proceeds, the chief irrigation engineer, Manoharadas, takes a more active role. Speaking in Sinhala is not easy for Mano, but I admire the fluency he has attained, having worked hard to learn some of the language myself. He communicates a sense of respect for farmers, and his sense of humor, expressed in Sinhala, builds rapport. Having heard less polite words used by some engineers in English, I am pleased to hear him address farmers as *govi-mahatturu*—literally, "farmer-gentlemen." (Sena need not translate this for me.)

Mano acknowledges that relations between engineers and farmers have often been unsatisfactory, and he says he hopes the project will make a new start. "We are coming to listen to farmers, not to tell you about the project. This is not a meeting but a discussion. You must tell us your problems. We engineers will take heed of them." He says the IOs are there to help farmers get organized so they can take more responsibility for irrigation.

He adds that there is no need for introductions because he knows most of the farmers already. The informality of the event is jarring to us as observers. There is no chairman, no agenda, no attempt to limit speaking to one person at a time—sometimes four or five are talking at once. But Mano has said he wants this to be a discussion rather than a meeting. About thirty persons are in attendance, not all of them farmers along this channel. Several are quite old, with colorful towels around their necks. One woman comes in a splendid blue and purple sari and sits quietly but seriously.

The most active participant is the former Cultivation Committee secretary, whom we met before the meeting and who speaks often on behalf of farmers.[6] Also outspoken are a young shopkeeper who rents and operates two acres, the schoolmaster, who rents some land and carries himself stiffly with white clothing and slicked-down hair, and the *jala palaka* for the area. (This is the lowest-level ID employee, whose designation means "water manager"—JPs mostly open and close channel gates.) This JP has worked here for many years, and he tries to impress others with his knowledge. He is quite obtrusive until farmers finally ask him to keep quiet.

One farmer argues to get not only water from the nearby channel but also drainage water across a neighbor's field. The JP finally tells him to sit down: "You will get enough water for your four acres." When the man persists, the JP asks, "Why are you insisting? We know you have ten acres and a tractor elsewhere. You don't need all this water." The man sits down indignantly. (An IO tells us this request reflects a long-standing feud between neighbors; the man doesn't really need the second water source.)

The schoolmaster now complains that cattle wander onto the school grounds because farmers do not control their beasts. He says he put up a fence to stop the trespassing, but the farmers took it down. He is unceremoniously contradicted by a farmer who says, "The teachers themselves remove the barbed wire and sell it, and then use the fenceposts for fuelwood." The schoolmaster sits down with wounded dignity. Another farmer taunts him further: "You are more interested in

6. Between 1958 and 1977, elected committees of farmer owner-cultivators and tenants were established by the government to implement tenancy reforms and improve agricultural cooperation. These were abolished in 1977 and replaced by Cultivation Officers (Uphoff and Wanigaratne 1982).

your farming than in the welfare of the children anyway." All this makes the meeting lively and humorous, but there is serious business to be transacted.

One farmer has arrived late. Since the attendance list has already gone around, a woman IO goes over to him with the list and a pen. She doesn't want him sitting at the edge of the room, so she picks up a chair and sets it within the area where the rest of the farmers are seated and guides him to it when he is finished signing. She is a university graduate, and he is barely educated. Yet the ease and graciousness with which she does this is gratifying. He does not speak up, though he follows the discussion earnestly.

Mano is handling the meeting with friendliness and frankness. He explains that the ID will give farmers full responsibility for managing the field channel and eventually the D-channel once they get together in organizations and share water effectively. This must be surprising to many farmers, who are accustomed to being treated by officials as incapable of taking any responsibility.

The shopkeeper now makes some sharp criticisms of the ID and the project, expressing farmers' fears that he hears in his shop. "The money for rehabilitation will be all spent before the work actually starts. So much money has been spent already on big machines, and we see ID vehicles driving up and down the road all the time burning up fuel." Mano tries to explain the constraints the project is working under, particularly the shortage of trained staff. He explains why it takes time to plan thoroughly, so the work can be done well. He says they do not want just to move in with machines, since this might cause unnecessary damage to houses, channel bunds (banks), coconut trees, and such. (He may be referring to what was learned from the meeting on UB1A reported at the end of chap. 2). To do the job properly, he says, will take some time; he asks for farmers' cooperation and patience.

As the meeting is drawing to a close after two hours, the JP, who has been quiet for some time, gets up and says that if all the ID staff worked as hard and as well as the IOs, there would be no problem getting things improved. He adds that waste of water in the area has already been considerably reduced at the initiative of the IOs. Heads nod in agreement. (The IOs later tell me that this JP previously was not very popular with farmers because he had not gotten close to them. Once the IOs started working in the area, they took him around and introduced him to farmers. Now he is working harder and the farmers speak favorably of him, which is probably why he likes the program.)

There are many impressions of the meeting, but the most enduring one is of Mano, patiently and humorously making an effort to build rapport, to learn about farmers' problems, and to make "a new start." If his attitude and approach can become operative for the whole project, its prospects for success should be greatly improved.

Second Meeting with IOs: Confrontation with Local Officials

We arrive later than planned at the Uhana school, where eight IOs are waiting. Some local officials have been objecting to the IOs' work. A work supervisor from the Irrigation Department confronted Anula, one of our women IOs, at a public meeting. He challenged her in what she considered an insulting way, saying that she was not an official and he was, so farmers should not listen to her or cooperate with her.

These officials, it seems, have been engaged in irregularities and thus fear farmer organization. The work supervisor reportedly told farmers that for fifty rupees and a bottle of liquor he could assure any farmer of water. When farmers circulated a petition against him for this boast, he blamed the IO for his trouble. The JP who has been obstructing Anula's work happens to be the brother of the JP we just heard praising IOs at the other meeting, so something is ringing false.

We spend a long time discussing this situation, and Munasinghe says he will meet the officials himself. Where there are such irregularities and vested interests, some conflict may be unavoidable, though perhaps the IO could have worked with more tact. We have to think about how our training and supervision for the program can better cope with such situations.

Practically all the other IOs are already having considerable success in getting farmers to take responsibility for water management. Indeed, there has been generally good cooperation from local officials, and one who has been challenging Anula has been cooperating with another IO on a channel where he does not own land. A good deal of field channel cleaning and water rotation has started. Farmers on two field channels toward the end of UB7 came together to do shramadana not only on their own channels, which had not been cleaned for six or seven years, but also on the tail of UB7, which had never before been desilted. The troublemaking JP complained that they had desilted the channel too much, but this seems unlikely, since it would involve the farmers in extra work and would reduce the flow of water to their fields.

The IOs tell of the realignment of one field channel that had originally been routed around a piece of land rather than directly across it. This slowed the flow of water and created siltation problems. The farmers got together and persuaded the man whose land was causing the problem to let them rebuild the channel across it by shramadana. On another channel, twenty farmers cleaned a mile and a half in one day, three times as much as gets done by hired laborers. Various government staff members participated in the work, a good sign.

Farmers here are also adopting a system of rotation, though usually with the head-end fields getting water for the first two days, then closing their pipes and sending water down to the tail for the last three days of water issue. This reflects a lower degree of trust among farmers on UB7 than reported this morning on UB9, where the tail gets water first in the rotation. This should be studied in more detail. On one channel off UB7, when it was pointed out at a meeting that tail-end fields were drying up, farmers at the head agreed to give the tail the first day's water issue, a welcome sign of generosity.

As we conclude our discussion, the team coordinator raises a valid issue, asking whether the IOs can avoid creating a feeling of dependency among farmers, which would defeat the purpose of the program. We discuss how to work with farmers so that they always act as their own spokesmen. The IOs must make it clear that they are in the area to help get farmer organizations started, not to become permanent staff here. We are pleased the IOs recognize this as a potential problem, and we have to trust their judgment and ingenuity in coping with it.

Our discussion goes on well past sundown. As it gets quite dark, a teacher living at the school quietly brings out a lantern, lights it, and sets it down next to our circle of chairs. The IOs say he has been helpful to them many times before as they often hold meetings at this school. As we break up the meeting, without anybody suggesting it, the IOs not only carry their own chairs back into the classroom but move about straightening the desks into neat rows. Such considerateness, like bringing a chair for the farmer at the meeting on UB2.2.3 earlier today, may partly explain the success with farmers we have been hearing about. It helps to retain the teacher's goodwill. Wijay, Sena, and I get back to the circuit bungalow near Ampare about 9:00, surfeited with impressions and notes from the day.

Third Meeting with IOs: Sinhalese Helping Tamils

After spending the morning in a meeting with officials and project staff (described in chap. 2), we drive to Gonagolla taking with us Joe Alwis and two other officials from the Ministry of Lands who want to see the IO program in the field. The group of six IOs posted here north of Uhana are outside the first year's construction area. We want to see how they fare when trying to improve water management not tied directly to rehabilitation. We begin with the two women IOs, who have met with all the farmers along their assigned field channels. In most instances they have gotten ready agreement from farmers to begin rotation and water saving, sometimes with "too much" success.

The gates controlling water flow from the Left Bank main canal and the Gonagolla branch canal into distributary channels are all inoperative. If farmers want to reduce or stop the water flowing into D-channels, they have to put their own planks into the broken gates or put stones or tree trunks into the openings. Other farmers who want to sabotage the water-saving effort can easily do so because the gates are not proper ones that can be locked.

Officials have not been able to get such reductions, but with IO encouragement, farmers at the head of Gonagolla branch canal have been trying to save water for tail-end areas. Farmers have cleaned their field channels by shramadana, and various officials—even the technical assistant, work supervisor, and JPs—have been cooperating. The IOs say that when they came, they found some farmer hostility toward the TA and work supervisor. They organized a meeting between farmers and all the officials of different departments, including the TA. It was very useful because farmers explained their problems, and the TA explained his. He promised to provide cement for temporary repair work on damaged structures, since the rehabilitation project will not reach this area for some time.

Farmers on G9 distributary channel, not within the IOs' area, have asked for help in setting up an organization for water management there. The IOs have agreed to participate in an initial meeting on that channel, and we will see how far those farmers can go on their own. On LB27, a D-channel off the Left Bank main canal, the JP has taken the initiative to get farmers organized, though it too is outside the IOs' area, an encouraging step. On LB31, the yaya palaka is going to form an organization for water management following the IOs' example.

The IOs estimate that the flow of wastewater into the drainage canals has now been reduced by about 50 percent, but this is presenting a problem because some farmers depend on drainage water for their cultivation. Some of these encroachers are from the second generation of settlers and have only small plots, but others are well-to-do persons who have gotten control of ten, twenty, thirty, or more acres and are renting the land out to tenants or cultivating it with hired labor. These are not typical weak tail-enders. There have been instances where such tail-enders or their subordinates come and steal the planks that farmers put in the gates to stop the water flow.

We ask whether these encroachers should be included in the groups being formed for each field channel. Most large encroachers have tried to keep their tenants from even coming to the meetings organized by IOs. (One Survey Department employee has as much as eighty acres under his control, a small empire with twenty-seven tenants.) We have no ready answer for problems arising from land tenure disparities; the maximum holding here is supposed to be four acres. The legal and administrative measures needed to resolve these issues are beyond our responsibility, but we will have to find ways to cope with them.

Now an IO tells us how head-end Sinhalese farmers voluntarily reduced their water offtake in order to send more down the Gonagolla branch canal to Tamil farmers at the tail. When talking with farmers on G1, the IO had encouraged them to close off their channel after four days, sending one day's water downstream. This is not a large amount, but anything is welcome to reduce the severe water shortage in tail-end communities.

One farmer at the meeting resisted this decision, so a young farmer offered to take the holdout downstream by bicycle to see for himself the situation of the tail-enders. I imagine to myself a grown man sitting sideways on a bicycle crossbar while another pedals him down the bumpy bund road. The IO reports that the hesitant farmer was deeply moved by seeing how miserable was the plight of the tail-enders. (Sinhalese seldom travel down into the Tamil area.) Upon returning, he suggested they could manage with only two days of water and should send down three days' worth.

"At the tail they don't even have water for drinking or bathing, let alone growing a crop," the converted farmer is reported to have told his neighbors to explain his change of mind (or heart). In a country that has seen ethnic tensions, this is surprising and encouraging. (Within two months communal violence will break out in Ampare district, even in Gonagolla town, but we do not know this). Unfortunately, farmers' water-saving efforts to aid Tamil tail-enders have been impeded by certain persons cultivating the drainage area. Feeling the pinch of reduced water supply, they came at night and broke open the makeshift closure of the G1 gate. But at least this shows water was being saved.

With this IO group, as with the others, we stress that in this season of water scarcity and uncertainty there could be widespread crop failures in spite of everyone's best efforts at water management, owing to unreliable water delivery by the Irrigation Department or simply an absolute lack of water. Acts of generosity between D-channels and even within field channels may not be rewarded by all farmers' getting successful crops. There is a danger that crop failures could be blamed on IOs' efforts, especially if the decisions are not left entirely to the farmers. Even if the water-saving measures were taken voluntarily, some farmers might blame our program for a poor harvest. The IOs seem confident that they have not been imposing water rotations and conservation and that they will not be blamed. I hope they are right.

There is no good or easy course to take if one cares about the welfare of the farmers, not just our program's reputation. If there is no crop failure this season, there are grounds for optimism about the long run. One IO reports that some farmers are already saying there should be stricter laws and punishments for violation of water management rules.

We now get into a discussion with Joe Alwis about a question very much on the IOs' minds: What will be done to make an IO cadre permanent within the Irrigation Department or the ministry? Since the program is already beginning to show some results, it is appropriate to address this question. If the IOs can initiate similar improvements in water management on a wider scale, which seems possible, and if they elicit better performance from local officials, as has started to happen, the cost of the program could be repaid several times over. We cannot resolve this matter now, however, sitting in a farmer's yard.

Wijay, Joe, and other ministry officials have to leave. Sena, Mune, and I stay behind to discuss IO program "housekeeping" matters such as process documentation, supervision, and accounting for another hour and a half. It is again completely dark by the time we finish. The roosters who have interrupted us with their crowing off and on for several hours are finally asleep.

The IOs' request to be issued flashlights and bicycle lights so they can travel more safely at night is dramatically supported by the darkness around us. They tease us with comments about snakes as we walk back to our vehicle in the dark. They repeat a request for bicycle locks to protect their government-issued bikes. Sena explains the difficulties we are having with red tape on this matter, though we hope to have some supplies for them within a week.

The IOs are working hard and effectively with little material support. For example, they have no mosquito nets in a malarial area; one IO has already had a bout of malarial fever. The bicycles took a long time to procure, but they are proving adequate. (We earlier considered motorbikes, but they would have been much more expensive and socially distancing from farmers.) Bicycle lights are a legitimate request, since IOs receive no extra pay for their work at night.

Fourth Meeting with IOs: An Emphasis on Learning

The next morning we meet with the rest of the IOs, who work in three small groups because of the configuration of the channels they have been assigned to. The four IOs on UB2 are covering 442 farmers on 816 acres. Along one field channel, UB2.9, there has been resistance to doing shramadana work. There are several fairly large and rich farmers who insist it is an ineffective method. They say they send able-bodied persons from their households but that other households send only youngsters, and no serious work gets done, only socializing. They prefer that the jala palaka allocate to each household a length of channel to clean that is proportional to the area it cultivates; then each can do its own work when convenient.

This is the system currently prescribed by law, but in most channels some farmers evade or greatly delay their work, and this defeats the purpose of getting the channel cleaned so water will flow evenly throughout the entire length. Farmers on most channels prefer the shramadana method because it means the work is done more quickly, and all at the same time. In addition, the work is more pleasant because it is a social occasion, with refreshments collectively provided and shared.

It appears there are reasons for the resistance beneath the surface of discussion on UB2.9. The channel is near Uhana town, and settlers have come from many parts of the country. The farmers have competing political allegiances, mixed caste affiliations, and different-sized allotments. It is no wonder cooperation is not readily forthcoming in these circumstances or that the two richest farmers dominate decision making. An IO reports that the richest farmer, whose field is at the head of the channel, has agreed to an informal rotation, closing his pipe after two days to send more water down to the tail end. So cooperation even here may not be out of the question. Farmers are going to give the individualistic method of channel cleaning one more try and will consider doing shramadana if the present method fails. Getting such voluntary group work done is a problem in communities around the world. (When we introduce a similar program in Nepal, the most difficulty is similarly encountered in the farming area adjacent to the bazaar town.)

The IOs working on UB3, 4, and 5 have been able through discussions to get farmers there to agree to conserve water, reducing offtakes by 20 to 40 percent. But they are opposed, sadly, by one of the JPs who participated in our IO training program; he has the support of the work supervisor in sabotaging the effort. Farmers decided to close the gate off Uhana branch canal after four days of water issue, to donate their remaining one day's water to downstream users. But the WS closed it after just three days, before the tail-end farmers had gotten enough water. He threatened to jail farmers for eighteen months if they tampered with the gate to help the tail-enders. Apparently he wants to create dissension among the farmers who had forged enough solidarity to save water they could donate to others.

This is a serious matter, because the work supervisor has no authority to shut the gate before five days without approval of the technical assistant. Munasinghe will meet the TA to discuss this. We are disappointed that someone we worked with in the past is being obstructive. Wijay asks about the IOs' working relations with other lower-level officials, and they say there is no problem. It appears that previously most of the officials did not spend much time with farmers, and IOs were the first persons from the government to do so, at least in recent years. But now most of the officials are making an effort to get to know the farmers and to be helpful. Some are even taking part in the channel cleaning, a significant act of solidarity with farmers.

I ask the IOs how they were initially received by farmers. One IO says: "Some farmers at first thought this was just another program for unemployed university graduates who were being given a government job and who would be a burden on farmers." One day as the IO was riding his bicycle past some young farmers, one said this loudly enough that it was obviously intended to be heard. The IO stopped his bike and stepped into a nearby field where an old farmer was sowing paddy seed and asked if he could help. The old man was surprised but handed over the shoulder sack of seed, and the IO proceeded to broadcast it correctly enough that the skepticism among farmers in the area dissolved. Sena commends the IO for his presence of mind (and I silently congratulate us for having insisted on taking as IOs only persons with knowledge of agriculture).

I say we are pleased with their progress but stress that we expect there to be problems and shortcomings. "If you tell us you have no problems, we will not believe you." Effective rural development programs such as the Taiwan farmers' associations and the Comilla cooperatives (in their early years in Bangladesh) were purposefully self-critical. Mistakes will be made, I say, but most need not be repeated if they are discussed openly and learned from collectively. I remind them of what was said in the training program: "Nobody will be punished for mistakes made from lack of foresight or lack of skill. As trainers and supervisors, we must share responsibility for any shortcomings in performance. But we will be very unhappy if problems are neglected or concealed. Everybody makes mistakes, and that includes foreign consultants. Anybody can be wrong. I can be wrong. We do not expect perfection. But we do expect sharing of both positive and negative learning."

I add that our program rests on two assumptions. First, that young people, carefully selected and trained, can and will work effectively with farmers, showing imagination and dedication in their work.[7] Second, that farmers will respond to IOs' sincere, informed efforts to be helpful, taking initiative that goes beyond what the IOs suggest. After only three months, we have seen demonstrations of cooperation, willingness to work, even generosity, that are very impressive. This provides more basis for establishing water user organizations at the field channel level and beyond than we had evidence for a few months ago.

The coordinator for one of the teams, when we ask if there are other problems to be discussed, ventures that they may be moving too fast: "Our group has felt some pressure to press ahead with shramadana and rotation efforts because other groups are doing this. We fear we may be evaluated negatively if we do not keep pace." Sena thanks him for voicing this concern and says that this IO group, unlike the others, must help get the rehabilitation work started. We do not expect them to be doing as much water management work as the other groups. "Each group must work at its own speed, according to the circumstances the IOs encounter. Your work plans are 'not made in Colombo,' " I reemphasize. Each team must do its own planning week to

---

7. This was the experience in the early years of the community development program in India (Mayer, McMarriott, and Park 1958).

week, assessing its experience to date, with members helping each other solve the problems that arise.

After leave-taking, the IOs disperse on their bicycles, cycling off in the noonday sun. Seeing the fellows take off on their bikes is a familiar sight, but I am again a bit startled to see the three young women in this group head off on their miniwheel bikes, white hats flopping in the mild breeze as the sun beats down. They are serious yet cheerful, ready for the rest of the day's work. Sena tells me with a laugh how much difficulty and fun they had teaching the women recruits to ride bicycles during the training program.

We take Munasinghe back to the house near Uhana where he stays. He is sharing it with the area's colonization officer, an acquaintance from previous government service, who confirms what we have been gathering from the IOs' reports, that they have become very popular with farmers. The CO says they have very quickly earned a good name. On my next visit to Gal Oya, I will have to spend some time with farmers to form my own assessment of this and, if it is true, to try to figure out how and why.

On the drive back to Colombo, Sena, Wijay, and I assess our observations, thinking about how our training, supervision, process documentation, and logistical support can be improved, for this group of IOs and for the groups we hope will follow. We need to determine, with others, how a cadre of IOs can be developed and institutionalized in support of water management, and how to enhance the performance of local officials, as the IOs seem to have started doing, at the same time that farmer organizations take root and progress. We have much to learn and decide about this approach to creating local capacities for development, but the effort seems to be well launched.

This first report was written June 21, 1981, the day after I returned from Gal Oya. I reconstructed the full and exciting days from scribbled field notes and memory. It presented an upbeat picture of the program in the field, buoyed by favorable signs of cooperation from the Irrigation Department. I thus left Sri Lanka on July 1 feeling optimistic about the program.

August 1981: Ethnic Disturbances

Six weeks later, communal conflict erupted. It started in Batticaloa (over a disputed volleyball game) and spread to Ampare and elsewhere around the island. Sena wrote that nearly a thousand Tamils from the lower reaches of the Gal Oya system were sheltered in a refugee camp at Vellaveli. How some Sinhalese farmers tried to contain the violence I would not learn until my next visit.

Rehabilitation work was brought to a standstill. The saddest news was that Manoharadas, who had taken initiative in cooperating with farmers, suffered a back injury when beaten in a mob invasion of the ID headquarters. (Credible rumors had it that this was not an ethnically motivated attack on Mano but rather was instigated by disgruntled former ID employees whom Mano had gotten sacked for corruption, who got revenge under cover of communal conflict.) Mano was transferred out of Ampare, and Sena wrote in a letter, "We have lost a good man." The design engineer who had attended many farmer meetings was in Jaffna hospital receiving treatment for injuries he received. Another engineer's house had been ransacked, and two of his colleagues not surprisingly refused to return to Ampare. The small cadre of engineers who had at least some experience and inclination for working with farmers was disbanded.

The deputy director of irrigation was transferred out of Ampare and replaced as project manager because of dissatisfaction in the ministry, and finally in the ID head office, over his performance. He had been obstructive to our program, but still he was a likeable fellow who suffered the personal tragedy of having his wife commit suicide shortly after the violence. With his leaving, the Irrigation Department in Ampare was in shambles.

Difficulties with USAID and Evidence of Positive Farmer Response

The Gal Oya project had been slow getting started anyway, and it had gotten a bad reputation within the USAID mission. The project manager Ken Lyvers was "out of favor," and rumor had it that he would be leaving soon (not true). The mission considered moving the project from Gal Oya and starting over somewhere else.

ARTI and Cornell planned to host a "retreat" in September at a vacant beach resort where we could get ID, ministry, USAID, and other key actors together informally for several days. We wanted to improve communication and strengthen personal ties that could help get the project on track. Gil Levine was going out to assist this process as a resource person and as a socially oriented engineer. Invitations had already been sent by ARTI's director when USAID arbitrarily decided to cancel the event. There had been no meeting for several months of the project's steering committee, of which a number of the retreat invitees were members. So the USAID director took the position that if the Sri Lankans didn't have time to hold a steering committee meeting, how could they have time for a retreat? This was an unneeded blow to our efforts to help the mission and the project. It showed how little importance within USAID was attached to good interpersonal relations.

In September I received channel-by-channel data from the IOs on the extent and methods of water management during the 1981 yala season. The statistics, covering seventy-one field channels and 1,686 farmers cultivating 3,142 acres, showed that some combination of shramadana, water rotation, and/or water saving had been adopted in 80 percent of the channels, serving 90 percent of the area. On 30 percent of the channels, one or more days' worth of water was sent downstream without taking any other measures (often not needed), and on another 30 percent it was combined with shramadana or rotation or both. Total voluntary savings of water for downstream farmers were estimated from the figures at 16 percent, not bad in a watershort season.

These figures were circulated in a report prepared for ARTI and Cornell on the initiation of water management activities. As significant as the numbers was how quickly IOs had gotten farmer cooperation, within a few weeks of entering the communities. It was hard to believe these were the same farmers the Government Agent had spoken so warily about.

Signs of Progress

The program was getting a good reputation in the area. In late September a letter came from the Government Agent for Batticaloa district, Dixon Nilaweera, who had participated in a workshop at Cornell in 1974. His district included the lower reaches of the Gal Oya system. He wanted me to know (in Ithaca) that at a meeting he attended the previous week in his role as GA, farmers from Gonagolla had spoken very highly of the IOs and the new organizations. So despite the troubles there were signs of vitality in the program.

A meeting of the project steering committee was finally held in November, attended by the Additional Secretary of Lands, the Secretary of Agriculture, the director of irrigation, the Ampare GA, and the new deputy director of irrigation for Ampare, among others. Sena wrote that the GA "was full of praise for the IOs and indicated the role played by the IOs in the last yala season," a welcome commendation before a high-level audience.

The question came up, Who should be responsible for administering the program? The Ministry of Lands representative suggested the Land Commissioner's Department. This proposal came out of the blue, though it was plausible; Munasinghe was after all from the Land Commissioner's cadre. Wijay and Sena, however, spoke in favor of having the ID take responsibility for the program. This surprised the director of irrigation, Ponrajah: "He apparently did not expect such support from us," wrote Sena. "He agreed that the IOs had done a good job, but his major problem was that he did not have adequate staff and capacity to administer the program." There was no resolution of the matter, so ARTI retained its role, though not very happily. It was not set up to manage a large and growing field operation. No other "home" was available, however, so the institute remained responsible for the IO program to keep future options open for it.

The best news was that the new deputy director of irrigation in Ampare, Senthinathan, appeared receptive to ideas of farmer participation. Senthi had served as chief irrigation engineer in Gal Oya a few years earlier and knew the area and the farmers well. He had not been particularly close to farmers then, but he was known to be energetic and conscientious.

End of 1981

After Ed Vander Velde finished his year as our representative at ARTI in July and went back to teaching at Aquinas College, there was nobody "resident" from Cornell. Mark Svendsen, who had done research in the Philippines on irrigation systems with farmer participation, was available half time while writing his thesis for a Cornell Ph.D. in agricultural engineering. Mark could assist Wijay and others in analyzing the voluminous data being generated from their record-keeping program with over five hundred farmers, so he became our Cornell consultant in Sri Lanka in December.

Mark's first report said the American engineering consultants advising and assisting the ID with the physical rehabilitation had come up with a new design process that was "almost a verbatim recitation from Gil's September trip report." Gil Levine had proposed relying heavily on farmers' suggestions for identifying and remedying irrigation problems pragmatically. We finally seemed to be making some headway on the technical front. As encouraging was a comment by the consultant team leader that "if no other consultancies are extended [beyond the next year], that of the institutional adviser [Doug Merrey] should be." When questioned, he ranked this position even ahead of having an operations and maintenance adviser, Mark wrote—a high priority.

The ID was having tremendous problems staffing its positions in Gal Oya, and at the end of 1981 USAID lost the Sri Lankan engineer who was its deputy project manager. (Kari subsequently became a half-time consultant for Cornell at ARTI, complementing Sena.) Ken Lyvers, the USAID project manager, planned to leave sometime in 1982.

Just at the end of December, Mick Moore, from the Institute of Development Studies at Sussex in the United Kingdom, who had been on the Gal Oya project design team, sent me a report of his visit to Ampare, quoted from in chapter 2. From his observation of IOs at several meetings where farmers were choosing field channel representatives, and from his conversations with them, Mick said the IOs "appear to live up fully to the good reports I had previously heard." He described some of the meeting dynamics that impressed him and revised his earlier doubts about whether initiatives on the farmer organization front could have an impact overall on improving irrigation management. This favorable assessment of the field situation, by someone in a good position to be a critical observer, made me look forward even more to my next visit to Gal Oya.

4

The Program after
One Year (1982)

Sena and I have an all-day drive from Colombo, reaching Gonagolla about 7:00 in the evening. It is a quiet trip because today, January 15 is an official holiday, the Tamil festival Thai Pongal. There is little traffic on the roads, apparently because it is feared that ethnic conflict might break out today if secessionists based in London carry out their threat of declaring independence for a separate Tamil state in the north. Everything has been quiet, however.

First Meeting with IOs: Evidence of Cooperation

The IOs working in the Gonagolla area have been waiting for us since 5:00 P.M. After exchanging greetings in the dark by the roadside, we settle into cane chairs on the veranda of the house where two IOs are staying and start discussing their work.

My first question is whether farmers have continued the practices started last summer, rotating water delivery within field channels to ensure that farmers at the tail end get their needed share and reducing offtakes from distributary and branch canals to save water for downstream farmers. It appears that about three-fourths of the time these measures were sustained.

The reasons for suspension where this occurred were usually beyond farmers' control. Saving water reduced the flow in drainage channels, which hurt encroachers who rely on such water for their crops, and these persons sometimes sabotaged the water-saving efforts. Or the absence of operable channel gates made water control too difficult. Or unpredictable water deliveries from the Irrigation Department made it hard to adhere to schedules arranged among the farmers. We review the situation channel by channel, and the cooperation and generosity farmers are demonstrating under unrehabilitated conditions is impressive. In the current maha season rotations are not necessary, but some channels are continuing special efforts to distribute water to the tail end where needed.

I ask about relations with officials, since in June there were troubles with a few ID personnel at lower levels. Now everything is going smoothly, they say, and officials are quite willing, with one exception, to meet regularly with farmer-representatives to try to solve problems. The technical assistant for Gonagolla, a Tamil, was sharply criticized by farmers before the start of the 1981 dry season for "not doing his job properly" (Murray-Rust and Moore 1984, 55), but now he is quite popular with farmers, they tell me.

Then I learn something extraordinary. During the communal riots last August—when several shops in Gonagolla run by Tamils were burned by an unruly mob—half a dozen of the farmer-representatives (Sinhalese) took it upon themselves to guard the homes of the TA and the work supervisor for this area (Tamils) to make sure no harm came to them. This act of solidarity must have impressed ID engineers and may have contributed to the improved working relations reported between farmers and officials.

One IO reports that the previous month, when a canal below Navakiri reservoir was in danger of flooding, the TA gave a farmer-representative the key to operate the gate that regulated the flow. By frequently adjusting the gate, overflowing could be avoided. The key was not given formally, but the farmer, widely respected for his conscientiousness, could handle the task better than ID personnel, who live some distance from the area. Also, when there was a breach in the LB29 channel bund, at the farmer-representative's request the TA sent some machinery to help repair it properly, solving a problem that had existed since the 1978 cyclone. Farmers contributed labor to repair the breach.

IOs say that they have learned this area is better served with water than farther down the system. The biggest problems here come from the economic gap they encounter. All settlers were originally given the same amount of land, but there have been subdivisions and sales, leasing and mortgaging, none of them recognized by law. The result is that holdings now vary from one-quarter acre to several dozen acres. Indeed, one person in this area controls as much as eighty acres. He is a member of the G3 field channel organization but not a very co-operative one. He is unpopular, having done many unlawful things, including taking his brothers' land and appropriating drainage land. Neither the IOs nor the farmers have been able to deal with him so far because he is a power locally with good connections in high places.

A recurring problem is that farmers who want to save water lack the means of control. They have difficulty reducing flows from the branch canal because all the gates along it are missing. ID officials say the gates were removed for "safekeeping" to protect them from being damaged by farmers, but others concede that at least some gates were removed after it was learned the Left Bank area would be rehabilitated with foreign assistance. The gates were reportedly reinstalled on the Right Bank, where some ID personnel have private land. Farmers here have promised not to damage new gates, and the new deputy director of irrigation has instructed the TA to make and install new gates. If gates are put in, farmers will see this as an indication that they can get action from officials. Then it will be up to them to show they can manage water properly themselves.

I ask the IOs what they have learned since I saw them last. One says he has learned that "our concepts are not 100 percent implementable." This calls for some elaboration. He explains that the ideas we started out with have had to be modified as they proceeded. This, I respond, was our expectation: "We did not have a blueprint to implement but rather an approach that would be modified as we learned more about the situation and our task."

One IO reports that farmers in his area want to use the organization to buy fertilizer collectively for next season's crop. This season they were not able to get fertilizer because when they had money, none was available, and when it became available, they no longer had money. The idea has been put forward that farmers should collect funds from each member at the end of each season and deposit them with the Agrarian Service Centre as an advance payment to be sure they can get fertilizer when they need it for the next season. The richer farmers have no interest in such a plan because they can get fertilizer anyway, but the poorer ones like the idea.

Is this an appropriate activity for us to encourage? one IO asks. A debate ensues, because our guidelines thus far have emphasized water management as the priority. The consensus, after some time, is that especially where farmer groups do not have serious water problems, it may be important, even necessary, to build group strength by getting involved in other activities like this one.[1] To restrict group efforts only to water management will limit the development of groups in better-watered areas. The principle is that all activities must be chosen by the farmers, not imposed by the IOs. Other examples of farmer-proposed undertakings are a savings scheme where each member would contribute two rupees per bushel at harvest time to build up a group fund for interest-free loans and for collectively buying sprayers to protect their crops.

I ask about the structure of organization farmers have created thus far, since we did not specify in advance what roles—or how many—should be established. One IO explains that in all instances farmers have preferred to select a representative from each field channel or from some portion of the channel (representing ten to fifteen farmers) when it is fairly long. They have not gone in for electing a slate of officers because they are fed up with the kind of formal organizations usually set up in Sri Lanka. These groups last for a short while and then lapse into inactivity, at most serving as stepping-stones for the politically ambitious. The simple expedient of having a single farmer-representative is judged adequate and appropriate for improving water management.

One long channel (G5) crosses administrative boundaries, with its head in unit 33 and its tail in unit 32. Each unit was settled with households from different parts of Sri Lanka, so no social links bridge the hydrological divisions. The farmers' solution has been to have four representatives, two from the head and two from the tail. Previously there were lots of fights between the two colonies, farmers said, but since the organizations were formed last season, there has been cooperation. The four representatives have worked out a schedule of water rotation among all the field channels served by G5.

Why have head-end farmers been willing to help tail-enders? I ask. There is a feeling that "others must also live from water," an IO answers. "Farmers may be selfish, but they should not be selfish if this hurts others," another says. I am given also a concrete reason. Before,

1. There is controversy in the literature on this matter of whether water user organizations, or local organizations generally, should be single purpose or multifunctional (e.g., Tendler 1976; Cernea 1984). The most defensible general conclusion seems to be for groups to start with a top priority then diversify into other areas as members feel both the need for this and confidence that they can handle these other tasks (Esman and Uphoff 1984, 139–41).

tail-enders often came at night to close upstream field channels so that water would flow to their fields. Head-enders thus had to stay up all night to guard their intakes. "With an agreed-upon rotation, everyone now can get some sleep, which is much appreciated." We find moral and practical considerations reinforcing each other.

Farmers have moved beyond our initial plan to organize first at the field channel level and then at the D-channel level. They have established an informal Area Council made up of representatives of the twenty-two farmer groups around Gonagolla, nineteen within the IOs' area plus three that farmers started on nearby channels. The council meets each month, with local officials invited to attend and discuss matters of common interest. If officials can meet with all the farmer-representatives at one time, they do not have to visit each separate area, which helps them. Farmer-representatives set the agenda, and a farmer presides. This is a fine idea, but something we had not expected to see so soon. It is now 9:30 P.M., and the IOs who do not live in this house have to travel some distance to where they stay. So we end the discussion and drive to the circuit bungalow near Ampare for the night.

Engineers' Views

Next morning we meet the new deputy director for Ampare, Senthinathan, who now serves as project manager. He expresses a much more favorable attitude toward farmer participation and the IO program than did his predecessor. This reflects both his own experience and attitudes and the positive impact that IOs and farmer organization activities have been having in the area.

We also meet the technical assistant in Ampare who handles water distribution within the Left Bank system. The irrigation engineer (IE) responsible for Gal Oya has reported that the number of petitions he receives from farmers about water problems has declined since the IO program started, and I want to know if this report is true. (He also said that he now takes more seriously the petitions he receives, but I don't feel I can pursue this.) The TA says he believes there are fewer complaints coming in. Cooperation among farmers and better working relations with the ID staff are reducing the number of problems that need to be presented to headquarters in Ampare. This, I think, should encourage senior ID staff to cooperate with and sustain the farmer organization program.

Second Meeting: Converting Opponents and
Instituting a Bottom-up Approach

Driving with Sena and me this Friday morning to meet the second group of IOs are Munasinghe, our IO supervisor, and Doug Merrey, the anthropologist advising the ID in Ampare on institutional development as a member of the USAID-funded expatriate consulting team. As we ride, Mune tells us that two of the farmers who were previously "our program's biggest enemies" have now become "our biggest supporters." Waratenne, the UB9 yaya palaka who gave Gunasena so much trouble last summer (referred to in chapter 3), has said publicly that after initially thinking the IO program was a "useless scheme," he is now convinced that farmer organization is the only way to get their problems solved. Waratenne made his recantation with a humorous statement: "At first I thought the IO fellows had just come to watch our village girls bathe in the channels, but now I know they are serious about their work." He told farmers that they should keep politics out of their organizations. "Politics is the enemy of water management," he declared, though he himself is a party organizer in his area.

The other farmer who opposed the program is even more powerful. Kalumahattea controls forty acres, was the right-hand man of the former member of Parliament here, and has been chairman of the village council as well as president of the Multi-purpose Cooperative Society. Mune says he was behind the harassment of Anula last June. Now he speaks glowingly of the IOs and says they should be kept in the area until the organizations are firmly established.

We arrive at the school in Uhana where last June we met with these same IOs from late afternoon until well after dark. This time we have bright morning sun. Hemachandra, who worked with me as a research assistant in 1979 when I was doing village studies in two other districts of Sri Lanka, is coordinator for one of the two teams assembled here. In response to my question, "What have you learned since I was here in June?" he makes a lucid statement of their approach as it has evolved in the field:

> When the program started, there was no connection of farmer to farmer and no connection of farmers to the government departments. The farmers had many problems, so the first thing IOs needed to do was get to know the farmers and their problems.
>
> Then we tried to create a sense of unity among farmers by various techniques. The first effort was to get channels cleaned and repaired by shra-

madana to demonstrate the value of group action. Then we got farmers to work out rotational schedules for sharing water. This was based on informal organization, with no slate of officers, only persons who volunteered or were chosen to oversee the work.

The next step was to get more discussions going among farmers, to get them to identify their problems and to undertake solutions, such as repairing weak bunds along channels to stop flooding. This showed what they could do on their own.

But some problems need outside assistance, so we started meeting with local-level officials to explain the group approach and to solicit their cooperation, which was almost always forthcoming. We arranged meetings between the farmer-representatives and officials of the Irrigation, Agrarian Services, and other departments.

Action was taken on some problems identified by farmers that the officers could deal with using their authority and resources, such as repairing gates or providing sprayers. Once farmers began helping themselves, JPs and other government personnel often began to work better and harder.

Still, certain problems can be resolved only with the help of district-level officials, such as changing water delivery schedules or getting more fertilizer delivered to Agrarian Service Centres for sale to farmers. Farmer-representatives and local officers could meet with the TA, and if he could not resolve their problem, they could go to the irrigation engineer or deputy director in Ampare. If joint efforts have been made by farmers and officials at lower levels and were not sufficient, this creates some expectation that persons at higher levels should assist.

Hemachandra has described a "bottom-up" approach to development that is almost textbook perfect. Only this is not just theory; it has evolved in practice over the past year. It is stated better than we could do in their training program, and it maps out a strategy for utilizing both local and government resources more effectively.

The IOs begin by spending time with farmers individually, gaining their friendship and confidence, then meeting with them in small groups, and finally in larger gatherings with all farmers in a given command area. They ask: What are your problems, and how can you solve these as individuals? Quite naturally, the IOs say, the need for collective action is recognized and followed up by the group.

When I ask what problems they are encountering, the same one mentioned last night is raised: What to do about tenants? In this area they estimate that as many as half the cultivators are farming land on lease or mortgage. Such arrangements are illegal but are not rooted out because some of the land is now controlled by government officials who employ as laborers the original allottee or his children. Some allottees hve been reduced by debt to the status of beggars working on their own land, says one IO. Who should come to the farmer organization meetings? The allottee? The de facto owner? The cultivator, who may be still a third person? The government's policy is not to recognize mortgaging or leasing, even though it exists, so if we have a formal membership requirement, only original allottees can be members. This would make for confusion, since allottees often are not actually handling farm or water operations.

In one area, Anula reports there are serious problems caused by illicit liquor (what Americans would call "moonshine"). Many older farmers are frequently drunk, and this leads to quarrels among neighbors, to loss of money and even land, and to poor agricultural practices. Younger farmers have been the first to join the new organization on one of her channels and have made it a "temperance union" alongside its water-management efforts. They are trying to persuade older farmers to give up drinking, and the situation is improving somewhat. This function is one we did not foresee for water management groups, but improving social relations may be a prerequisite for improving irrigation. This shows the need to give farmers flexibility in shaping their organizations.

I ask what problems there may have been with the rotations introduced last dry season. Siriwardena (Siri), the training coordinator, describes how they were resolved in UB16, a long distributary channel. There are too many farmers (forty-five) on this channel for a single group, so they divided it into three sections, according to the water situation and how close together people live—the head with twelve farmers, the middle with eight, and the tail with twenty-five. Then representatives were chosen for each of the three sections. Since the worst water problems arise with the tail, representatives from the head and middle meet with all the tail-end farmers to work out a rotation. (Two farmers meeting with twenty-five must put a lot of social pressure on the two, I think.) Head and middle farmers have agreed to close their pipes on the first day of each five-day issue, sending all water to the tail. On the second day, middle farmers open their pipes, and on the third day, head farmers open theirs; on the fourth and fifth days water goes to whoever needs it.

At the start this was not very successful, and there was a lot of confusion owing to unpredictable deliveries from the main system. The ID did not inform farmers of changes in its five days on, five days off schedule. But by the end of the season, the arrangement operated at about 75 percent effectiveness. The organization has proposed saving water in this maha season by reducing the twenty-five days of continuous flow given for land preparation, but the TA was not able to change the schedule.

A small storage tank halfway down the channel for watering livestock had filled up with weeds and silt, reducing throughflow. It has been cleaned by shramadana, and one section of the D-channel that overflowed every rainy season, washing out part of the channel bund, was been improved. The Irrigation Department used to spend 5,000 to 7,000 rupees to repair it each year. Farmers got the ID to give them two large concrete culverts and transported them from the ID storehouse in their own bullock carts. They installed the culverts with their own labor and repaired the road. Next Sunday, Siri says, they will be doing more shramadana work, and he invites me to come and see for myself. I say I will try.

The farmers on UB6 have been very active, another IO reports. After IOs came to speak with them, they formed an agricultural association with six elected officers, more formal than the others. The organization met every two weeks to do shramadana and discuss local problems, but the farmers do not have serious water constraints because they get water directly from Uhana branch canal. They are actually wasting water. They decided this maha season to take water for only three days out of the five they are entitled to. The gate that controls their D-channel is broken, however, and cannot be raised or lowered by its metal stem. The ID used to leave it propped open continuously with a stone, but now the farmer-representative, when he takes his daily bath in the Uhana canal, operates the gate according to their water-saving schedule. "Going 'bathing' in the canal to open and close gates is something that ID personnel did not like to do," an IO tells us.

There have been some problems on the adjacent channel, UB7. All farmers agreed to a rotation serving the tail end first, but they changed it to head end first "because head-enders couldn't bear to see water passing them by," says the IO working with them. The channel is full of silt, with three to four feet of sand accumulated over a half-mile stretch at the head, which greatly reduces the flow of water to the tail. Farmers could remove this, but some say (correctly) that cleaning this stretch is the ID's responsiblity. Moreover, they say (with anger directed towards the ID) that if farmers do the work, officials will claim they did it themselves and will file false claims to get paid for work the farmers did.

Getting farmers to agree to clean the channel is slowed because head-enders benefit from the siltation. More water flows into their fields when it cannot run downstream. The IO thinks cooperation can be mobilized, but so far his efforts have been stalemated by a combination of "rational" and "spiteful" considerations. I am encouraged when an IO tells us that farmers have already cleaned the D-channel from UB7.3 down to the tail voluntarily, even though much of this work should have been done by the ID. Farmers say this is the first good cleaning the channel has had in at least twenty years. Also, some farmer groups on UB7 have started holding classes with the agricultural extension agent to improve their agricultural methods. So there are signs of progress even here. This was reported as a "problem" channel in June.

I ask Anula whether she is still having problems with the ID officials who were opposing her efforts last June. She says they are cooperative now. They were put up to their troublemaking by the most influential farmer in the area, who has since changed completely and is now calling the IOs "saviors" for the farmers. Mune says this is one of the farmers he told us about already. How Kalumahattea was won over is quite a story.

The IOs discussed in their team meetings the problem Anula was having and decided to work indirectly themselves. The team coordinator, K. B. Dissanayake, was assigned to get to know the man personally. This turned out to be difficult because the farmer refused at first even to acknowledge K. B.'s presence when the IO approached his house, even when he was sitting on the veranda reading a newspaper. K. B. persisted and kept coming back. After several visits the man finally spoke to him, and they began getting acquainted. K. B. smiles and almost blushes as Anula narrates how he won the rich farmer over. K. B. has a charming personality, but he is also very intelligent.

Before approaching the man, whose name means literally "dark gentleman," K. B. talked with other farmers to find out what he was like. It appeared some flattery was in order. K. B. asked Kalumahattea how it was that he was so successful while other farmers were not, since all had come to Gal Oya about the same time and got the same size of land allotment. The rich farmer responded that others had acquired no scientific knowledge of agriculture and got poor yields. Also, they drank liquor and squandered their money. When they got sick or went into debt for other reasons, they mortgaged their allotments and became serfs on their own land. (He should know about this, because he has gotten control over many allotments himself as a liquor seller and moneylender.)

K. B. suggested that cooperative action among farmers could deal with each of these problems, spreading knowledge of better agricultural practices, stopping drunkenness, and assisting members who became ill. Kalumahattea agreed that organization was needed to help farmers help each other, even if he himself had risen above needing such help. He is now one of our biggest boosters and is a farmer-representative on UB7, K. B. adds.

I question whether he is not just politically motivated, figuring that if he can't beat the farmer organizations he will join them—and take them over. This is possible, comes the answer, but so far he has acted in good faith. At a recent meeting of farmer-reps he attended where no IO was present, Kalumahattea proposed postponing the meeting until an IO could come, "since farmers still don't fully understand the concepts behind farmer organization." He could have tried to dominate or direct the meeting himself, K. B. points out. I suggest that maybe this shows an opposite danger, that farmers will become "dependent" on IOs. But they assure me this is not happening; many times farmers meet on their own and take action, though they still want IOs to attend.

Noon has come and passed, and we must be going. We take our leave, saying we will meet them all tomorrow morning at the Inginiyagala circuit bungalow, just below the main reservoir where all the IOs will gather for a day of discussion with Sena, Doug, and me.

Third Meeting: "It's Hard to Be Selfish in Public"

We drive to the house near UB9 where the two women IOs in Gunasena's group live. Asoka and Sanda have cooked a splendid meal of rice and curries. We start talking over lunch about their experience with water rotation and water saving. Most of the efforts started last summer continued through the season. For the past three months, the farmer-representatives from all UB9 field channels have held monthly meetings, starting an informal D-channel organization, the next step in our strategy. IOs have discussed with them the value of trying to solve farmer problems collectively saying, "Previously the powerful got most of what was available, and others could get only what was left over; all should get their fair share through farmer organization."

Such egalitarian principles sound fine, but I have to ask how "the powerful" accept such an argument and whether they try to obstruct the organization.

Sanda replies that in a meeting of farmers, the one or two who have to give up something for the common good do not object and go along. "They may be selfish in private, but it is hard to be selfish in public." This observation is so simple, obvious, and important that it almost stuns me. She apparently thinks I am looking skeptical, however, and adds as an explanation, "Pressures for fairness can be built up through discussions among farmers." I ask whether this behavior can be perpetuated. The IOs say that farmers understand how scarce water has become as population and cultivated area have expanded and water supply appears to be reduced. This is a good answer, but I hope to myself that farmers will not be forced to find out whether there is some point beyond which water scarcity brings out dog-eat-dog behavior. So far the reallocations of water being made are benefiting some quite visibly without demonstrably hurting others, apart from the inconvenience of having to close their pipes or give voluntary labor. There can be some risk to head-enders of crop loss if their fields have not been kept saturated, but we are basically getting more value from the same scarce resources, a positive-sum situation.

I am curious to know about the change of heart in the yaya palaka who was opposing our program last June, accusing Gunasena of selling marijuana and of plotting revolution through the IO program. Waratenne initially posed some threat to the program, but the IOs say that of late he has repeatedly demonstrated his good faith and personally provided the refreshments for a recent shramadana, having previously opposed such work. I consider it significant that he has offered to help form an organization for those who cultivate in the drainage areas of UB9, since they are unrepresented and are some of the poorest people in the area. Quite a change.

What about cooperation with local officials? Virtually all are now very helpful, although a district agricultural officer did not show up for a meeting he had scheduled with farmers because, he said, he could not get an official car for transport. This contrasts with the support of the new DDI, Senthinathan. I am told that he recently drove all night from Colombo to keep an appointment he had made to meet IOs in Ampare the next morning.

How soon do the IOs think they can withdraw from the area, leaving FRs and officials to carry on the program themselves? Farmers are still sometimes a little worried about meeting officials, one IO says, but a lot of progress has been made in this direction. The real test for the organization will come in the next yala season, when water will again be very scarce. Assuming that the organizations perform well in sharing and saving water, then IOs can move on to new areas, perhaps leaving one member of the team behind as liaison and troubleshooter to cover an area now served by five or six IOs.

How long might it take them to organize farmer groups in new areas now that IOs have gained experience in this first area? Opinions differ whether one season or two seasons (six versus twelve months) would be needed. The conclusion is that, given their techniques and confidence now, they might be successful in one season but should be allowed to remain an additional season if progress is not rapid enough. Much will depend on how responsive officials continue to be. So far, IOs have had good cooperation from them, and most of the time they need only get farmer-reps and officials started talking with each other. We take leave of this group because we still have two more groups to meet. We will see them again tomorrow morning at Inginiyagala.

Fourth Meeting: Preventive Maintenance

Ariyapala's group meets us in an empty room, not currently rented out, in a row of one-story structures, including shops and dwellings, along the Uhana branch canal. Across the canal a vigorous volleyball game is going on from the time we arrive, about 5:00 P.M., until it is too dark to play. With work and school over for the day, volleyball absorbs the remaining energy of village youths in many parts of Gal Oya.

This IO group has been working at the head of the system where rehabilitation work is starting. They have had to organize meetings between the ID and farmers to discuss problems on the respective channels. Since the UB1A meeting (reported at the end of chap. 2), it has been difficult to get those farmers together again, they say. Farmers are skeptical that work will actually be done on their channel, and they get plenty of water in any case. I tell the IOs they need not to worry about organization for these farmers, only seven in number, since their offtake can be controlled once physical rehabilitation is completed. They will hurt only themselves if they mismanage water. The IOs agree, adding that these farmers are fairly rich anyway, some owning tractors and even rice mills.

Other channels seem more inclined to cooperate. The biggest problem the IOs face here is the long gap between the redesign meetings with farmers and the commencement of construction work. There was 100 percent cooperation from farmers at the outset, but since little rehabilitation has been done, some farmers say it is a big farce. Farmer activities are not stalled, however. Ratna describes how, on her channel, farmers decided to repair two potential breaches in a channel bund wall before they got worse and gave way. They discussed going to the ID to get some tractors and wagons for hauling dirt and stones but then decided to do the work entirely on their own, using their own carts.

Most significant for our program, all this was done without the IO's involvement. We congratulate Ratna on this work, so IOs understand that they do not have to play a central role to get credit. Even some farmers who did not own land in the area but who used the road came and participated in the shramadana, she says, and some who had previously refused to join in voluntary work came this time. Ratna's report from UB1 is encouraging not only because farmers took initiative on their own and did not seek ID assistance (which they could reasonably have requested), but because this was done where we had not expected or predicted it, in head-end areas where water supply is reasonably good.[2] It is now quite dark, and we still have another group to visit. We head off in our van down the road along Uhana branch canal toward Udayagiri temple.

Fifth Meeting: Progress at the Head End

Before we go to meet the IOs, who are sitting outside the temple's lecture hall under a bare light bulb, the Buddhist priest here wants to show us his newly built temple. It stands on a spot where archaeologists and historians think there was a temple over two thousand years ago. Around us in the scant light, we can see old stone columns and slabs, possible evidence that this is one of the oldest sites for a Buddhist temple in Sri Lanka. The priest, who has devoted the past forty-five years of his life to raising funds and supervising the reconstruction of

2. Our group had hypothesized that farmers' willingness to take responsiblity for water management would resemble an inverted U-shaped curve, with farmers least willing where water was relatively abundant or absolutely scarce and having most incentive to participate in management where it was relatively scarce (see Uphoff, Wickramasinghe, and Wijayaratna 1990).

a fine new temple, shows it to us with justifiable (but un-Buddhist) pride. Sena tells me the priest has been very helpful to the program, always making meeting space available for the IOs whenever they need it.

The IO group here is headed by Chuti (whom readers met at the beginning of chap. 3). She and the others greet us warmly even though it is already past 8:00 and they have been waiting for some time. Chuti jokingly tells me she has grown a few inches since we last met. Certainly she has grown more confident and mature, a pleasure to talk with and listen to. This group has an incredibly complex channel system to organize. There are a dozen field channels, and even more sub field channels for the six of them—over eight hundred acres plus encroached areas.

We begin with a discussion of the head field channel. In Colombo, I read a report from this team that UB2.1 farmers were unenthusiastic about farmer organizations because they had no major water problems, but in recent months they too have started an organization and begun water management work. What about this? "At first the farmers were unreceptive, and this was discouraging. But we concentrated on field channels farther down the D-channel and got positive responses there. Finally, when UB2.1 farmers could see others engaging in shramadana and water rotation, they got interested themselves." They now give up one day's water out of every five for downstream farmers which represents a "victory" for the program, showing that head-enders too can be engaged in water management efforts if approached in a reasonable (and patient) way.

On UB2.1.1, the ID's desilting work lowered the water level in the channel so that four allotments no longer can get water to their fields. The ID agreed to build a check structure across the channel to raise the water level and help these fields, but the work supervisor and laborers did not install it high enough. (An IO says farmers insist the workers were drunk at the time, but this cannot be verified; I note this as a reflection of current attitudes.) Farmers decided to remedy the situation by themselves and used their own carts to transport stones. A temporary check has been put in to improve on the new one, which can be properly constructed when the rehabilitation work is done.

I ask about the woman in a purple sari who attended the UB2.2.3 design meeting last June. Chuti says she is a cultivator who has a particularly hard life. Her husband abandoned her some years ago with twelve children to support, having mortgaged two of their four acres to get going-away money. With the help of her children she barely manages, but she does participate in the farmer organization. Munasinghe tells us with a smile that as district land officer he annulled the sale, since it was strictly illegal, "so the woman should get back her full allotment." A small but happy accomplishment of the program.

In this group of six IOs there are three women, making it a good group to ask about women's participation in the farmer organization. Is it sufficient? Could and should it be increased? We do a quick survey and estimate that roughly 8 percent of households in the area are female headed. The IOs say there is about one woman member per organization, so if the average size is fifteen members, there is no marked disproportion. One of the farmer groups in Chuti's area proposed that a woman be its farmer-representative, but she declined. In Sita's area the daughter of an allottee is acting as secretary to the group, which needs someone skilled in writing. The consensus is that no particular problems exist for women's participation in the organizations.

It is past 9:30, and we need to adjourn. As we walk back to the van, there are nervous jokes about whether snakes would bite people on hallowed temple ground. We are glad the IOs now have bicycle lights operated by generators, though they light up our way only dimly. We take the women IOs back to the houses where they stay while the men depart on their bikes. It is a long drive back to Ampare.

The Director Comes on Inspection: Tense Moments and a Surprise

Back at the circuit bungalow, we find that the director of irrigation has arrived. He is having a late evening meeting with ID staff upstairs on the veranda. His voice is so loud that I can overhear him expressing his skepticism about the IO program and about a role for farmers in water management. He sounds much as he did in our tense meeting at ID headquarters in Colombo a year ago. At 11:00, before we go to bed, he and I have a brief discussion that is friendly enough, but reflects the anticipated divergence of views.

Our conversation is resumed the next morning at breakfast. He decides he has time this afternoon (Saturday) to meet the IOs, since they will all be gathered at the Inginiyagala circuit bungalow. By moving his schedule up, he might be able to return to Colombo a little earlier. He is tired from all the traveling he has been doing.

This may account in part for his surliness. One of the reasons he made the trip to Ampare is to review the IO program and form his own impressions. He tells me that almost everyone he has spoken with about it is "pro," but he is inclined for the sake of argument to remain "con." I say he is entitled to be skeptical until he is satisfied there is sufficient evidence of effectiveness, since the resources and responsibility for the program are really his department's. Our agreement was that ARTI-Cornell would have two years to develop a successful field-tested approach to farmer organization. After one year, we think there is already ample basis for the ID to institutionalize the program, but we would like another year for further testing and modification.

As we part after breakfast, the director says, "I don't know where all this participation business is going to lead. If we take it too seriously, banks will have to ask their depositors what color to paint their lobbies!" I hope he is joking and try to point out that depositors at least have some choice about which bank they will patronize; irrigating farmers depend on a monopoly provider of water for their livelihood. But he doesn't want to hear any argument and breaks off the discussion with a wave of his hand. He will join us this afternoon after spending most of the day inspecting rehabilitation work. He heads to Ampare while Sena and I drive with Doug to Inginiyagala for an all-day session with IOs.

The bus that was to pick up the organizers arrives two hours late at the circuit bungalow—typical luck. We manage to get enough chairs and benches arranged in a circle on the veranda so we can all sit there, looking up at the massive reservoir dam five hundred yards away. I begin by reading from a recent Ministry of Lands publication that speaks approvingly of the IO program and its progress to date, but this does not eliminate the insecurity felt by IOs who know the ID is not yet committed to the program.

We tell them that the director of irrigation himself will be coming this afternoon to meet with them. I say that he is frankly skeptical about the need for our program and even of the need for any organized farmer participation in water management. But I tell them, recalling my grueling meeting with him a year ago, that he is fair-minded. With enough evidence and logic, he can change his opinion. They should be careful to be very factual, we say, and avoid any exaggeration. "If you say that farmer organizations now keep cattle off channel bunds, you had better be sure of this, because the director just might ask you to take him around to see for himself."

The IO program hangs in the balance today, but we try to minimize anxiety by getting on with our business—a combination of debriefing, group problem solving, and informal training plus camaraderie. Many of the issues taken up are ones raised already in our separate group discussions, but with more background we can focus on specific problems and possible solutions. The bungalow keeper has a large lunch ready for us at noon. In the afternoon, the TA who distributes water to the Left Bank comes to discuss scheduling for the maha season and the difficulties he has keeping to schedule. IOs need to know this, because they must explain problems to the farmers.

A little after 4:00 P.M., the director arrives with Senthinathan, several other ID staff, and the USAID project manager, Ken Lyvers. We rearrange the chairs so everyone can still sit within our circle. The most important hour or two of the IO program is about to begin. Chuti has been chosen by the IOs to give a short welcoming speech, which she does with her usual charm and eloquence. Hemachandra then describes "the IO approach." We had asked him to repeat the explanation he gave Thursday morning, but this time he goes into too much detail, and his statement lacks the clarity and thrust of the day before.

There is some restlessness in the group. We have not planned a schedule of presentations to the director, and there is now a long awkward pause. Sena and I thought Munasinghe would act as chairman, since we judged it inappropriate for either of us to play such a role. But Mune is sitting back just as we are. This will have to be the IOs' show. We have helped them this far, and they have responded magnificently. Now they have to present themselves and their work to the director. We should not be making the case for them, since they, not we, will have to carry on the work if he accepts them.

The hesitation at least makes it clear that we have not set up a rehearsed presentation for the director. I hope he appreciates this lack of coaching. One would never plan such an awkward silence deliberately—it is excruciating—but maybe it will work out for the best. Mune finally comes in with a few comments and asks the IOs to speak up. Ariyadasa stands and recounts the farmer activities on UB1 that Ratna described to us yesterday. He is tall, with a Lincolnesque beard, and speaks with intensity and feeling.

Ari tells how farmers identified two places where the bund was in danger of breaching, and how they decided on their own to repair it without seeking ID help. The atmosphere has been fairly formal, even tense, up to this point, and Sena and I have been exchanging private worried glances. But as Ari talks I see the director starting to sit forward in his chair, listening intently, no longer aloof. When Ari is finished, the director recalls that when he was the irrigation engineer here in Ampare, almost twenty years ago, one of his greatest disappointments was that farmers took so little responsibility for maintaining the system. "I don't know how often this kind of thing occurs, but if what you have just described is happening now, this is a welcome thing." The atmosphere is suddenly more relaxed. The director is now talking with the IOs instead of inspecting them.

IO after IO gets up to discuss what farmers have been doing to save water, clean channels, repair bunds and roads, and work with ID and other officials. This last point, it turns out, is crucial to the director. Somebody has told him that IOs have been criticizing the Irrigation Department, and he thinks it comes in for enough criticism without having the IOs, whose salary the department pays, also knocking it. It is evident, however, that the IOs have been working collaboratively with ID staff and have been promoting farmers' cooperation with the department.

As the meeting goes on, I am impressed by how articulate and evidently knowledgeable the IOs are. It is normal in a group of several dozen to have a few persons who are good speakers and do most of the talking for the group. But already more than a dozen have spoken, and all have been clear and persuasive. There seems to be no end of capable persons in this group of thirty.

After two hours, the director rises and says that the hour is getting late and he has been "on circuit" for four days straight. He says he will speak in English so that his meaning is perfectly clear. (He has been speaking in Sinhala, but Tamil is his mother tongue.) This causes me some apprehension, because it may mean he wants to make some fine distinctions or to insist on some important qualifications in his support. He thanks the IOs for their discussion and says he has learned a lot from them: "Frankly, I have been skeptical about this program, about the need for it, and about its feasibility. But I am encouraged by what I have been hearing. I think something new, something useful is happening here, and I am prepared to give full support to the program. I have consulted my senior staff and they say something useful is going on here."

The director now surprises us by saying that he is prepared to take the IOs into the Irrigation Department straightaway. He knows there may be some difficulties, and he appreciates that the IOs will need to discuss this among themselves, but he hopes he will get an answer and a favorable one from them. There is stunned silence.

This is more than any of us dared hope for. There is applause from the group as he sits down. One of the IOs nicely thanks the director for his words of appreciation and promises that the IOs will consider the offer. (Such an offer seemed so unlikely that they have no group position worked out in advance.) I hope that this failure to accept his "proposition of marriage" on the spot will not cool the director's new-found ardor. This is something that the IOs, in consulation with ARTI, will have to decide for themselves.

The ID visitors go inside to have a soft drink and a snack prepared by the bungalow keeper, while Sena and I meet with the coordinators of the IO groups to discuss plans for the next day. There is elation among the IOs, but the director's "inspection" of our program is only half over. He would like to spend the next morning observing field activities. It so happens that farmers on four different channels, without knowledge of the director's impending visit, have planned shra-madanas tomorrow morning. We tell the director he can choose whichever work sites he wants to visit. He asks to see the ones that are most accessible, since he is tired from traveling. We also want him to meet farmer-representatives in the Gonagolla area, since their organizations are the most advanced.

The IOs who leave in the ID bus won't get home until 9:00 or 10:00 in the evening, but everyone is pleased the meeting has gone so well. We say good-bye to those we won't see tomorrow and return to the circuit bungalow near Ampare, where for the next three hours we have wide-ranging and now more relaxed discussions over drinks with the ID engineers. The director acknowledges that they can no longer maintain, as he insisted in our meeting a year ago, that the ID manages water adequately. He recalls that when he joined the department thirty years ago, his senior engineer told him: "The ID calculates the right amount of water to be issued from the reservoir, and once it leaves the sluice, our job is done." Those days are past, the director says. Most of the discussion, however, turns on less serious matters until after midnight, when one engineer, a friend of the director for many years, pleads tearfully with him to be transferred out of Ampare because of continuing fears about ethnic violence. Things have been peaceful in the project area since last August, but apprehension about futher outbursts continues.

Field Visit: "We Wanted to Drown Ourselves"

The next morning, we get off to a late start. It is difficult to assemble half a dozen persons early on a Sunday morning. The director will have time to visit only one location, so I offer to go with Hemachandra to see the farmers' work on UB16, while the others go to UB2.9. Hema and I head off in an ID jeep, driving along the road paralleling the Uhana branch canal. It is rough road, barely a vehicle's width in one place.

We turn off onto a cart track that runs along the edge of UB16 D-channel. I can see half a dozen places where the road has been recently patched, as much as half its width, because the road and the repair are different colors. The road would have been impassable to vehicles before, but now it is quite usable. This work was done by the villagers as part of their weekly shramadana, Hemachrandra tells me.

When we arrive near where the farmers are working today, we leave the jeep and walk up a path. It widens and I see twenty persons, including women and young boys, hacking away with hoes and machetes, widening and leveling the path into a road. They are working with an energy seldom seen in Sri Lanka, especially in the direct heat of the sun. Several men are moving a large stone out of the way with iron bars. We stop and speak to them. Two farmer-representatives are there working with the rest.

We walk further to another group of about a dozen persons chopping out undergrowth. Ten of the participants have come from a neighboring village to help with this road, which connects to their area. They say the path was laid out almost thirty years ago, when Gal Oya was first settled, but it has not been touched since then. Now they are making it passable for bullock carts and perhaps even jeeps.

One of the farmer leaders, Neeles, asks if I would like to see some of the other work they have done, and I say I would. He heads back the way we have come, walking briskly, showing none of the deference to me as a foreigner that is common in villages. This is a good sign of confidence and independence. I see the stock tank that was cleaned and the newly installed culverts the IOs told me about. Neeles says proudly that they have been working every Sunday to improve their area, as I can see. By now we are surrounded by a dozen persons, including two other farmer-reps and three of the IOs working in this area. The ID jeep driver who brought me acts as an interpreter and is probably surprised by what he hears.

Neeles is obviously the person others look to for leadership. Looking at him, one would judge him the simplest and poorest of farmers, but few people I have met convey ideas and purpose with more forcefulness. The threadbare sarong, the bare feet and chest, the unkempt hair will mislead anyone who only looks and does not listen. Neeles says they will continue this work every week until their village is thoroughly improved. Evidence of their determination is all around me. He says that with channel cleaning and rotation last season they were able to get water to all allotments on UB16 for the first time. They have also identified fifty-four acres that can be cultivated if a few changes are made in the channel design during rehabilitation and if some more water can be issued to the area. They will present this plan at the redesign meeting.

They have already discussed where and how to reduce damage to the channel bunds. They would build a buffalo wallow on the edge of the stock watering pond and prohibit any other access for buffaloes. Could this be enforced? The response is, "Of course." Neeles says they know that people who use the channel for bathing knock dirt and stones into it. I have heard engineers complain of this many times, assuming people do so out of ignorance or laziness. The people here know it is something to be avoided and controlled, but there has to be group action. Neeles points to where they would put in concrete steps for bathing access. He says use would be restricted to that area. I wish ID engineers could hear these suggestions on how to protect the channel coming from farmers themselves.

This display of initiative prompts me to ask why their group planning and work started only recently. Neeles credits the IOs for helping them see how they could solve their problems by group action. This is a gratifying answer, but much as I like it, I press the point: "However useful the IOs' coming may have been, they brought no new resources to the community. It is the farmers' own efforts that are being used, and farmers had these resources for a long time without employing them in this way. What has changed?"

Neeles answers in an oblique but revealing way, with a dramatic metaphor of the kind one finds often in colloquial Sinhala speech. He says that before June, when the IO program started here, farmers were treated very badly whenever they went to the Irrigation Department or to other officials as individuals. "We sometimes were so humiliated we felt like jumping into the canal and drowning themselves. Now, however, when farmer-representatives meet with officials, we are well received and treated with respect. This gives us great encouragement."

This is a consideration I have overlooked before. I should pay more attention to the contribution that lower-level officials have been making to the progress of the program. One of the incentives for farmers to work together and make contributions from their own resources to solve local problems is the cooperation they are now getting from local officials. Being treated with respect can be tremendously motivating to farmers who are used to being looked down on by their social and educational "superiors." They are now being listened to and responded to, often for the first time in their lives. I resolve to give more credit to officials in future discussion of the program.

Will the farmer organizations continue if the IOs leave? Neeles answers yes. "But we should continue to have some IO support," he adds. I tell him this is our plan. Can farmers who have already formed organizations help to spread them to other areas? Yes, but it will be good to have some IO assistance in this, he responds.

We need to be leaving to join the others, but Neeles says someone is bringing fresh coconuts for us to drink. These are a real treat, much preferable to bottled drinks. But farmers who are close enough to a store usually bring something bottled for refreshment, not knowing that coconuts are more appreciated. I promise to visit UB16 again, to see the progress they have made. They are proud of what they have done, but one does not get the feeling they are doing it for others; rather, they are doing it for themselves. The jeep traverses the bund road back to the Uhana main road, where we drop several IOs by their bicycles and proceed to Gonagolla.

Farmer-Representatives Meet the Director

When we reach the TA's office near Gonagolla, the director and his staff are already meeting with a dozen farmer-representatives, hastily called together that morning by the IOs. As I enter, an old farmer is speaking. He is white haired, slightly stooped, and humbly dressed, but his gentle voice is clear and confident as he describes the changes in water management introduced "since June." He cites earlier problems in getting water and tells how it is now shared. Other farmers, not quite as well spoken, add to and confirm this report.

One farmer assures the director that there will be no more breaking of gates. "The ID can put all the padlocks it likes on channel gates, but that won't solve the problem of water theft unless there is farmer understanding and cooperation." He makes a pun, saying that "the canals are full of *ibba*," using a Sinhala word that can mean either "padlock" or "turtle." This suggests with a touch of humor that without social control over individuals' behavior, padlocks will continue to be broken off the gates and chucked in the canals.

After spending an hour with the farmers the director takes his leave, having a long drive back to Colombo before him. The oldest farmer,

Ratnayake, rises and thanks the director for coming to speak with them. He says they know tht farmers in other schemes have serious water problems similar to their own, and he asks whether the Irrigation Department can spread the IO program to other areas too. His request is so perfect an endorsement of our program as to be embarrassing. With a half-laugh, Sena asks whether the IOs coached him to say this. The farmer appears surprised by the suggestion, insisting, "This is my idea." Sena smoothes over the situation by saying he was just joking.

The director and his ID and USAID traveling companions walk outside the TA's office and get into their cars. Out of earshot of the farmers, he berates the TA and work supervisor for having removed and not replaced D-channel gates in this area. He is unhappy that farmers who want to save water for others downstream have to use tree stumps or rocks to block off channel flow.

Doug Merrey, who was with them during the morning visit to UB2.9, tells us the shramadana there was impressive. There had been no advance notice, yet over fifty farmers were out doing repair work on the channel. The director groused a bit about the "technical standard" not being very high (the TA had not shown up to supervise the work as requested), but he was visibly impressed that the refreshments served had been paid for by the farmers themselves. Usually some rich person or someone politically ambitious provides the refreshments to gain favor. That each farmer contributed three rupees for tea and *kiribath* (Sri Lankan milk-rice, a special treat) impressed the director as a sign of the farmers' independence and involvement.

Communalism, Politics, and Other Issues

Sena, Doug, and I go back inside to speak with the farmer-representatives at greater length. One of the issues I want to pursue is cooperation between Sinhalese and Tamils in water management. From the start of the project this has been a real concern, because there is considerable communal tension within the country, and violent clashes broke out here last summer. When I ask whether the area council they have set up here can later include Tamil representatives from the tail areas, the farmer-reps become agitated. Several stand up and lean in my direction as they all talk at once. Sena says they are insisting there should be no ethnic difficulties between farmers. One says with evident pride that they donated water to the Tamil area last season, and this was a "great victory." I know that some of these farmer-reps personally protected Tamil ID staff here during the disturbances last August, so I do not pursue the matter, though I may want to probe a bit more later.

One farmer-representative starts telling me that they now clean the field channels regularly by themselves. If they have a problem they cannot solve, they take it to the TA or another official. (This is a recurrent theme I am glad to hear from farmers—self-help as the first step, supported by government efforts as needed.) I ask whether the program can continue after the IOs leave. Farmers respond by pointing out that it was their idea to set up the Area Council to meet and work with officials. There is no way I can test their sincerity and capability, but they express confidence and willingness to keep up the work already begun.

Can they tell me more about their experience with saving and sharing water? One tall farmer, Punchibanda, speaks with a directness and detail I wish he had shared with the director of irrigation. He says there are twenty-four farmers on his channel, and in some seasons tail-enders failed to get any water. He himself has the second pipe from the head: "I have lived here for twenty-nine years, and never once did I close my pipe, that is, before last June. I used to take water without regard for tail enders. But now things are different. Since I was elected farmer-representative, I see that all farmers get water. With the amount of water available, if we share it among themselves, all can get enough, and some can even be donated to others downstream."

He says that he doesn't take decisions on his own but consults others. If one farmer won't close his pipe as had been agreed, "I ask him to close it, and if he still refuses, I close it myself, because I know that the other farmers back me." Punchibanda says he even did some night patrolling of the channel during the dry season to be sure only eligible fields were drawing water. What if farmers do not cooperate? "We tell them to close their pipes or else." The threat is not specified. Most of the breaking of gates was done by encroachers. "Before, farmers didn't care to protect the gates, but now we are united, and we must protect even the nuts and bolts." He concludes, "I hope this program prospers a thousand times," and sits down.

I ask about payment of farmer organization officers, at the D-channel or branch canal level, if not at field channel level. The consensus is that nobody should be paid, that they should all be "honorary." I pursue the matter of financing the organizations and their activities. How much responsibility would the farmers be willing to take through their organizations, and how might this be financed?

One farmer suggests, with others nodding agreement, that they would be willing to maintain distributary as well as field channels if this responsibility were clearly given and if the ID would provide some mechanical assistance and technical advice for tasks beyond farmers' means.

I ask whether they think it necessary for D-channel organizations to have some kind of dues or tax. Most feel this assessment would be a good idea, and one farmer suggests the money could be used to pay for shramadana refreshments. Others reject this, saying, "The whole spirit of shramadana will be lost if this is done." The majority favor making personal contributions to cover the cost of refreshments, keeping organizational funds for things like replacing the bolts on a gate if they are removed.

Another farmer suggests that when there is a pest attack, it is important for all farmers to spray quickly to contain the outbreak. Now, invariably some farmers do not have the cash on hand to pay for spraying, and the effectiveness of others' spraying is thus reduced. If there is a group fund, farmers could draw on it to spray the whole area immediately and then could reimburse the fund at the end of the season.

Thinking about all the examples of abuse of such funds throughout South Asia, Doug asks, "Won't these funds be liable to misuse?" Farmers say the money would not be given to one man but would be deposited in the bank, where at least three signatures of officers could be required to use it. "We won't give the money out haphazardly," says one farmer-representative.

What about party politics? I ask. Won't partisan differences disrupt the organizations? One farmer says emphatically, "It is all right to have politics in the home, but in our organizations there should be no politics." Won't this position be undercut by politicians? I ask. The response is, "Then we will have to remove any politicians who try to interfere." Can they do this? "We will be strong enough," is the reply. "If politicians promise to help all of us with our problems, that is okay."

But will individuals and groups resist blandishments from parties or politicians when they offer special favors in return for support? "Those days are gone," answers one farmer, receiving emphatic nods all around. "Politicians should do things not just for individuals or groups but for the community as a whole," says another farmer. "If a politician has helped the community, he need not ask for our vote. He will get it."

Knowing how pervasive partisan politics has been in Sri Lanka, Doug observes that this would be quite a change, and farmers agree. One farmer says there was a time when the people were misled by politicians. "But now farmers cannot be misled." Seeing the skepticism on our faces, he adds, "We too were misled," and most farmers laugh. Their self-criticism win some credence, but I cannot resist asking how widely this view may be shared by farmers elsewhere. They think it is widespread by now. Expressing the negative self-image found among so many here, old Ratnayake says, "We Gal Oya farmers are the least intelligent people in Sri Lanka, and if we can now see this, surely others can see it as well." All I can say in response is that they are not unintelligent at all.

Having probed for cleavages along party lines, I ask again about communal divisions. Can there be one area organization for all farmers, both Sinhalese and Tamils, or should there be two, one for each? I add "because of the language problem," to put the question euphemistically. There is no need to have two, is the reply. Old Ratnayake says in Sinhala so simple and sweet that I need no translation: "There are no Sinhalese farmers and no Tamil farmers—only farmers."

We have no prejudice, someone adds. I remind them that there was violence against Tamils right here in Gonagolla just five months ago. Several farmers stand up and speak agitatedly. "A few rogues made all the trouble. They get drunk, gamble, and look for opportunities to loot. They will use any pretext to benefit themselves and make trouble." They seem aggrieved by this blot on their community's reputation. Violence is disapproved among good Buddhists, though that has not always deterred its use in the past.

One farmer tells us how three days ago, on *Thai Pongal*, a Hindu holiday, because of the rumors of possible ethnic conflict, a number of farmers went the day before to the Tamil fishermen who operate on Navakiri reservoir nearby and suggested they remain at home the next day, to give no opportunity to troublemakers. The farmers offered to watch over the fishermen's boats and nets to make sure nothing happened to them. As it turned out, there were no clashes in Ampare district or elsewhere, but this evidence of Sinhalese willingness to protect Tamils from injury and loss of property impresses me, and I tell them so. (It is perhaps just as remarkable for farmers to take an interest in the well-being of fishermen.)

Another farmer says that during the communal conflicts last August, when they guarded the homes of the TA and work supervisor, there were also seven hundred to eight hundred cattle owned by Tamils grazing in the area around Gonagolla. These were watched over by Sinhalese farmers when Tamils fled, and all were handed back when the troubles died down. (It would have been easy for the cattle to be sold or slaughtered.) I express my appreciation of this, and another farmer says there has been some cooperation between Sinhalese and Tamil farmers for years. Sinhalese farmers get cattle from Tamils for plowing, and Tamils get their seed paddy from Sinhalese. At a time when ethnic chauvinists on both sides are stirring up communal feelings, I wish such acts of generosity, even courage, might be more widely known. "Farmers are united," says old Ratnayake. "Now we do things on our own." He repeats his suggestions that this program of farmer organization be extended to other schemes.

We need to leave, since we too must make the long trip back to Colombo. They invite us to stay for lunch, but we decline the invitation with apologies. The drive back takes eight or nine hours, and it is already 1:00 P.M. We promise to come again, and we try to show our appreciation by shaking hands and exchanging smiles all round the big circle gathered about us. Sena, Doug, and I have much to discuss on the long trip back.

Reflections

As I write this report, I repeat to myself what I thought so often during the visit, when I heard and saw the things described here—this is too good to be true. We must guard against euphoria or complacency. One of our articles of faith has been that there are bound to be problems and difficulties, even failures, in any effort to promote change. Introducing participatory development where it has been absent is liable to meet resistance and many setbacks. From the beginning of their training, we told the IOs to expect problems and to be prepared to cope with and learn from failures. We asked for honesty and self-criticism as standard operating procedure, and we still do.

We have the start of a very interesting experiment in behavioral change. The organizations were intended to change farmers' behavior—so they would use water more efficiently, enforce discipline, carry out rotational deliveries, and do proper maintenance. This has been achieved more quickly and easily than expected, and we are waiting for some collapse to show us that our earlier caution and skepticism about how rapidly farmers would accept new practices and responsibilities were correct. What is more surprising is the amount of change

we are seeing on the part of officials. No longer need farmers feel humiliated in the way Neeles on UB16 described to me. Within the Irrigation Department there appear to be changes, such as attending more
seriously to farmer petitions.

The role of local officials in making the program effective has been
critical. The "bottom-up" approach makes sense to them if farmers
first take the initiative to solve what problems they can by themselves
and then seek assistance with problems beyond their local capability.
The farmer organizations and representatives are making it easier for
ID staff to do their work properly and the good examples set by IOs
have encouraged other local officials to take their responsibilities more
seriously, to get to know the farmers personally, and to become more
active. The government is not paying these officials more salary to do
the 10, 20, even 50 percent more work with farmers that they are doing now. This payoff from the IOs' effort cannot be measured, but it is
a gain in government performance that in real terms might well justify
the cost of the IO program quite apart from any production increases
through better water management.

One of the most interesting things observed was the "conversion" of
two powerful local leaders to the program and the principles it stands
for—group action, equitable outcomes, and broad participation. I
would not have predicted this, at least not so quickly or dramatically.
One can say that these influential farmers decided to join what they
could not beat. But how did the IOs get enough farmer solidarity
welded so quickly that, despite opposition, there was something these
opponents felt constrained to join? We must watch to see that they
don't capture the organizations, of course.[3] Sanda's explanation for
the change is that these people, like others, act selfishly as long as the
matter is "private" but that they find it difficult to persist in selfish
actions once a problem is taken up "in public."

We have already lost several IOs because they had a chance to take
more permanent, if less exciting, employment elsewhere. Ariyadasa,
who helped break the ice in our meeting with the director of irrigation,
is leaving to become a teacher near Ampare, but he has asked to continue IO work part time or on weekends, since he thinks teaching will
not take all of his time. A couple of IO marriages are planned, and two
maternity leaves are coming up, but the IO teams feel confident they
can reassign and reschedule work so that progress is not impaired by

3. The literature on subordination of local organizations to elite interests is reviewed in
Esman and Uphoff (1984, 187–93, 209–18).

the leaves. When we assess the IO cadre, we find only two persons we thought would perform well who have not worked out as we expected, whereas half a dozen IOs from whom we did not expect much have turned out to be effective, reliable organizers. We are sorry to have lost several IOs to other jobs, but we think we can recruit and train similarly capable persons in the next batch. The cadre of IOs is now capable of helping train new recruits through apprenticeships.

When I report on our experiences to a national seminar on water management held at ARTI in Colombo after our return from the field, someone asks how I account for the changes described. I answer that I am not sure. Although I maintain a certain optimism about human beings, I try to preserve a realistic, even pessimistic view of motivation, to guard against naïveté and disappointment. While we would like to see generosity, we have not planned on it. I confide that I myself am surprised by many of the outcomes so far. We are seeing demonstrations of cooperation and generosity—within farming communities, between ethnic groups, on the part of officials, and between officials and farmers—that are unexpected.

Many of the elements of strategy embodied in the original IO program design, though subject to modification as experience has grown, appear to have been sound. After all, we were building on successful epxerience elsewhere in Asia. Confidence invested in the IOs and in the farmers is being repaid with interest. In my conclusion, I observe that what we have seen in Gal Oya cannot be understood and explained without some reference to the very "social" but very "unscientific" influence of friendship. I say I don't know how to deal with this as a social scientist, but the friendship within the ARTI group, within the IO cadre, among farmers, among these three groups, and also increasingly with ID staff seems to be part of the mobilization of energy and ideas that has had such an impact.

The next round of decisions and efforts will need to be tackled with the same combination of realism about human motivation and confidence in human potential that got the program this far. It is encouraging that our ARTI-Cornell efforts can now proceed with closer cooperation with the Irrigation Department.

A Farmer Victory and Some Setbacks

Unfortunately, the project steering committee blocked the transfer of the IO program to the Irrigation Department, largely out of private fears in the ministry that the program would be "co-opted." The director of irrigation, having made what he thought was a generous offer to take the IOs in and being still somewhat ambivalent about the program, was not disposed to fight for it. Also, after some discussion the IOs concluded that they were not eager to push for a transfer. They thought (hoped would be a better word) that they might yet become permanent staff with ARTI, something ARTI never saw or suggested as possible. Also, joining the ID staff would entail some cut in IOs' salary, because as temporary employees some fringe benefits were incorporated into the salary they received through ARTI. On the other hand, getting permanent appointments would be of great value, so the salary level itself was not decisive. Perhaps most important was IOs' apprehension about entering a hitherto exclusively "technical" department. They wondered what their future might be as nontechnical personnel. At Inginiyagala the director had only said he would take in *these* IOs, not any more, making the position of the cadre within the department uncertain.

In retrospect it was probably a mistake for the IO program not to be transplanted into the department after this first year, making continuing efforts to improve the situation from within. ARTI and Cornell, however, did not want to push the IOs into the ID, since it was their lives and their futures that were involved. In March, the project steering committee created an Institutional Development Unit to be located in the ID's office in Ampare and tentatively also a parent unit at headquarters in Colombo. But the latter never materialized, since complicated maneuvering began soon afterward over whether to increase the ID's capacity for improving water management or to create a whole new branch, division, or department for this function. At the end of March the IOs' contracts were extended for another year just as they were about to expire. The delay in issuing new contracts contributed to a decline in IO morale and initiative.

The 1982 yala season turned out to be even more water short than 1981, with the reservoir's supply 15 percent below that of the previous year. The Irrigation Department authorized Left Bank farmers to plant only 5,000 acres of paddy, whereas in previous water-short years the limit had been set at 12,000 (still only a fraction of the total Left Bank area). About eighty farmer-representatives met with the Government Agent on February 19 and made a strong case for increasing their authorized area. They promised to increase their water use efficiency and asked that the allocation made to the Right Bank for growing sugarcane be reduced. More livelihoods were at stake on the Left Bank, and the sugar plantation could manage with half as much water if it irrigated not just during the day but also at night, as farmers had to do when water was limited.

The GA praised the farmers for their organized efforts and self-discipline and said he would seek authorization for 12,000 acres for the Left Bank by reducing the Right Bank's water quota. He also agreed to have farmer-representatives sit on the District Agriculture Committee (DAC), made up of district heads of government departments, members of parliament, and other high officials who formulate policies for the Gal Oya area. The four representatives chosen by farmers included Narangoda from Gonagolla, who facilitated water saving for Tamils on G1, and the two former opponents of the program from the Uhana area. The choices were balanced by area, age, and party affiliations.

Farmers were jubilant over this victory, finally getting a long-sought reallocation of water from sugar to rice production. This boosted the popularity of the farmer organizations. So did their getting their own spokesman on the DAC, literally rubbing shoulders with top officials. That 1982 was going to be an election year may have helped the GA and the District Minister cast their lot with farmers rather than the sugar corporation.

There were some apparent setbacks with the farmer organizations at the same time, however. Munasinghe wrote to us at Cornell to say: "Do not be surprised to hear that we cannot be very happy about the performance of farmer associations. I was expecting these to be very active organizations by now. But my sincere feeling is that these have become weaker than at your last visit. We have identified various reasons, and I suppose we can rectify the mistakes and make the associations into stronger bodies before long." Mune added on a more positive note that he was "the happiest man" when the government agent agreed to increase the area authorized for planting on the Left Bank to 12,000 acres.[4] He reported also that "farmers are very happy that their representatives are now in the DAC." Munasinghe's views were echoed in a letter from the IO Hemachandra:

I think we have taken a long step ahead by getting four members represented on the DAC. During the past five hundred years of our history

4. In fact, Left Bank farmers planted between 17,000 and 18,000 acres in the yala season, 50 percent more than authorized. By very careful management of the scarce water, aided by some fortuitous rains in July, farmers got practically a normal harvest in this very water-short year, an additional victory for the program.

such a thing has not happened. During the past two or two and a half months, most of the IOs have been disappointed and less encouraged in their work. Planning and implementation of new tactics and participation generally came down. So many reasons influenced this. At last we discussed them and found the reasons for the discouragement in our work. We discussed this with ARTI. So that we could again proceed with the program in the correct way, we arranged a program for farmer-representatives' training and divisional-level officers' training.

Hemachandra reported on a meeting March 24 with the District Minister, who had asked to meet the IOs as a group. The DM was "very satisifed to hear our discussion," Hemachandra wrote, and Wijay in a telephone call said similarly that the meeting "went very well . . . beyond our expectations." ARTI had been apprehensive about possible politicization of the program, but the District Minister gave no hint of this, endorsing their work after hearing IOs' reports. So there were positive as well as disappointing signs. On a personal note, on March 31 the first IO marriage occurred, as Gunasena and Anula got married in Ampare, with Munasinghe and the Government Agent as attending witnesses.

At the project steering committee meeting referred to above, it was agreed that ARTI would recruit and train a second batch of IOs to extend the organizing effort farther down the system. This time it was not possible to get permission from the Job Bank to advertise publicly for the positions, since a National Youth Service had been established to register and place unemployed university graduates. It sent a hundred reasonably suitable candidates for ARTI to interview, but as usual the selection and appointment process were slowed at many points.

June 1982 Visit: New IOs and Further Progress

On arriving back in Sri Lanka, I was able to participate in training the new group of IOs at the Inginiyagala circuit bungalow. Their enthusiasm would have given impetus to the program if it needed a push, but in fact none was needed. Farmers were once again carrying out rotations and making efforts to save water for tail-enders in this water-short dry season. In Gonagolla, the technical assistant had turned over to farmer-representatives the keys to operate the gates of G1, G3, and G5, a token of trust. When farmers in that area organized a shramadana campaign to clean the section of canal that flowed through the town, the jala palaka and the yaya palaka both got into the canal and worked alongside forty-five farmers from seven field channel groups. Farmers in units 32 and 33 continued to share G5 water, no longer having to lose sleep at night to steal or guard water. In other areas, IOs reported similar progress.

Why didn't farmers undertake such cooperative efforts before? I asked an IO couple recently married. "Earlier farmers were not worried about others, only about themselves," was the response. After discussions with IOs, after training programs, and after hearing from tail-end farmers about their water problems, farmers were more willing to cooperate. Was the changed behavior due more to different farmer attitudes or to the organizations that created a structure for decision making among farmers? The answer was that both were necessary. I pressed them to assess the relative importance of each, and Anula suggested that changed attitudes were 75 percent of the solution, with 25 percent due to the newfound unity among farmers. Her husband Jayalath said that when farmers from the tail of UB7 had previously asked head-enders for water, the latter wouldn't listen, but now they did. He pointed out that on UB6A, where there was no organization, water saving was nevertheless being undertaken.

In my field notes, I jotted down that maybe I needed to reconsider my "structuralist" orientation. I had been regarding roles, incentives, and sanctions as more significant than ideas, seeing the latter as essentially derivative from the former. Now IOs were making me see ideas as having some autonomous influence on behavior. In that same conversation, Anula and Jayalath stressed the importance of the friendship they had forged with farmers as a key to the changed behavior, establishing trust and a sense of mutual obligation.

The most interesting conversation during that visit to Gal Oya was with Kalumahattea, the influential farmer on UB7 who had initially opposed the program but was now a farmer-representative himself. He received me more graciously than he had received K. B. at first. He was very critical of Irrigation Department personnel, though full of praise for the new deputy director. When I asked him a leading question about the dangers of politicization of the farmer organizations, though himself a party organizer and former parliamentary campaign manager, he responded fiercely: "Politics is a cancer for farmer organizations. It is very difficult to eradicate, but we will have to manage this. Politics is more likely to emerge in organizations if the ID is not doing its work properly. Then people go to the politicians. If the department

gives farmers full support, there is no need to go to the politicians."
When I asked him whether there could be cooperation between
Sinhalese and Tamil farmers, he echoed the statement I heard in
Gonagolla last January: "As farmers, we are all one group." Asked
about the role of IOs, he said, "They should work as advisers in a certain way, becoming friends of farmers," supporting Anula's statement
that morning.

During the visit I was able to meet the District Minister, P.
Dayaratne, in his home one evening. He said he had urged the Minister
of Lands to make the IO program permanent and extend it elsewhere,
and the minister had agreed to this in principle. The DM repeated his
commitment to keeping the farmer organizations nonpolitical, saying
he had chosen not to meet with the farmer-representatives himself so
nobody could accuse the program of any partisanship. Instead, meetings with farmer-representatives were held by the Government Agent,
though the DM interacted with FRs on the District Agriculture Committee. He expressed satisfaction with the contribution that farmers
made to DAC deliberations.

There was also a very satisfactory informal meeting with about
twenty farmer-reps at the circuit bungalow. I reviewed with them
my understanding of the program and its progress, seeking feedback, which was almost all positive. The young farmer-rep they asked
to give the vote of thanks at the end of our five-hour meeting concluded with the extravagant pledge, "We farmer-representatives are
prepared to go even to hell if it will help the farmers." This brought no
smiles, only earnest nods of agreement. It became clearer to me why
the program was making so much progress. It had the push and direction of some marvelously committed people, brought forward by
the IOs' efforts.

Before we left Ampare, we invited the Government Agent to join
us for supper at the circuit bungalow, and we talked until almost
midnight. He was extremely supportive of the program and of revising
various government rules and roles to give farmers a larger voice in
agricultural management (though this could not be done without
approval from Colombo). Munasinghe told me later that in a meeting
with personnel from various government departments the GA had
told them bluntly, "If you are not prepared to work with the farmer
organizations, I am prepared to have you transferred out of this district tomorrow." Unfortunately, this may not have been much of a
threat for most officials, since Ampare was still not considered a very
desirable post.

A Scandal

One thing I left out of my June 1982 trip report because it was so sensitive can be mentioned now that we are further away from the event, which for a while threatened the program's continuation. We knew when we decided to have both men and women IOs, initially all unmarried, that we ran some risks of having the program discredited in conservative rural communities if there were any "indiscretions" within the cadre. In the training program, Sena was explicit in warning new IOs how important it was to respect traditional norms.

For a year we saw no cause for concern, but eventually an illicit liaison between two IOs came to our attention. The woman IO involved was from the area, and a few months after returning to Gal Oya she had married a local man and had a child. To make matters worse, her father was a farmer-representative. When gossip started to spread, her husband threatened to kill the male IO involved if he was found with the woman again, so ARTI reassigned him to Colombo to do record keeping in the office.

Despite several warnings and the transfer, one weekend when Munasinghe had come to Colombo to meet with ARTI staff, the fellow took a bus back to Uhana and met the woman at Munasinghe's house, to which they had a key, as did many IOs. The woman's bicycle was not well hidden and attracted the attention of a watchman, who saw the couple together in our supervisor's bedroom.

The news came back to Colombo almost as fast as it probably spread around the town. The two were suspended from the program by telegram pending investigation (getting an affidavit from the watchman). Although such conduct might not be grounds for dismissal in the United States, there was no question that this constituted misconduct within the program. They had been warned about this because of the discredit it would bring upon all IOs.

The other IOs were outraged to think that their hard work and sacrifices could be undermined by colleagues in this way. They supported the dismissal, and some blamed Munasinghe for being so lax with IOs that the couple could come and go in his house at will. (This informality and friendship I considered a virtue rather than a fault.) I happened to be in Ampare when the all-IO meeting called to discuss this matter took place. After several agonizing hours, it was decided that dismissal was all that could be done, hoping the affair would blow over.

In fact, the program did not even feel a breeze. IOs heard no reproaches or even talk about the incident. It was as if a big stone had been thrown into a pond but there was no splash, not even a ripple. In retrospect we understood that the popularity of the program among farmers, and officials too, had protected it. IOs' reputation was so substantial that farmers apparently shrugged off the story as reflecting badly only on the individuals involved, not on the whole cadre—who could have been tarred with the same brush had someone wanted to be malicious. The IOs as a group passed this unplanned and unwanted test of their standing in the community—a test too painful to impose deliberately—with surprising and gratifying ease. It showed us that the goals and performance of the program were widely accepted by farmers and officials.

5

Moving Farther Down
in the System (1983)

This visit in January 1983 is the fourth time since the IOs were first
fielded in Gal Oya that I have been able to visit them to observe their
progress and problems. With me on this week's visit are Lucky, now
head of ARTI's Water Management Group, since Wijay is beginning
his Ph.D. studies in the United States; Sena and Hammond; Vidanapathirana (Vida for short), who has taken Wijay's place as the group's
economist; and Jeff Brewer, an anthropologist just arrived to work
with ARTI as a Cornell consultant. I am eager to learn how the batch
of IOs recruited and trained last summer is doing.

After breakfast and before leaving Ampare, we stop at the ARTI office in town where the socioeconomic research started in 1979 is administered. It is a bungalow across from that of the Government
Agent. Three field investigators are already there at 8:45, tranferring
onto master forms the data from questionnaires to be taken back to
Colombo. Farmer record-keeping is now in its seventh season, though
the sample has had to be scaled back from over five hundred households to half that owing to financial constraints. Investigators' cover-
age of half the colony units in the Left Bank has given ARTI valuable
knowledge for dealing with the problems of farmer organization. The
walls of the office bear maps and charts showing the location of monitoring points, trends in yield, and various hydrological and other data.
Unfortunately, few are up to date.

Agreement is reached that the data will be ready by next Wednesday,
and we pile into the van to drive to Weeragoda, where most of the new
IOs have been working since September. We know that expanding any
effort to a larger area is difficult even if it is fairly successful in its pilot
area, and I will focus on this problem, which is like shifting from first

gear into second. Hammond will concentrate on the progress of rehabilitation work. Jeff and Vida are new to the IO program and will be
getting acquainted with the field situation. For Lucky and Sena, this is
one of their many visits for supervision and learning.

In a New Area: Shramadana on M8

We drive through the Uhana area where "old" IOs are working to
get to the Weeragoda area. At the first house where a new IO is living,
we are told he is in the field. We stop at two more houses and
each time get the same answer. We have not set up any schedule in
advance, so I wonder whether we will spend our whole morning just
driving around. Then we spot two IOs near a small junction and stop.
They have just been looking in on a shramadana campaign nearby.
After drinking a glass of tea with them at a small shop, we walk over
to the work. Thirty farmers are clearing a field channel that comes
off M8 distributary channel. Because it is so long, the top mile and a
half functions as a D-channel and is therefore the responsibility of
the Irrigation Department, but the ID has done little work on it for
many years.

There are serious problems with getting water to the tail end, so this
group work was planned at a farmer meeting just last night. That is
why no technical assistant from the Irrigation Department is here advising them. I comment that the silt being excavated and thrown on the
banks of the channel is likely to wash back in. A farmer explains that
this is just a quick job to get water to the tail. (We learn later that some
fairly influential persons have land there, which may have helped to get
such speedy mobilization.)

We walk along the channel, past a stone marker mutely demarcating
the ID's jurisdiction down to that point, talking with the IOs about
their experience so far. They seem quite knowledgeable about the
channel and the people on it. They started under very difficult conditions last September. There was no water left in the main reservoir,
and water could not even be issued for domestic use. Because of bureaucratic delays, the IOs had no bicycles for some weeks and had to
get around on foot, over areas much larger than the first batch were
given. The water shortage made it difficult to find lodging because
even households with enough space and food were unable to provide
water and thus declined to rent rooms to IOs. It therefore took them
some time to get settled.

Supervision was also reduced, since they are more distant from Ampare. When the rains finally came, torrentially, Munasinghe could not visit them because his motorcycle could not get through the mud. That diminished the guidance they got. The experienced IO who had started as their team's coordinator resigned to take a teaching position, which worsened the supervisory gap. The work has gone reasonably well, however. Being on their own has not adversely affected the IOs' work.

I ask Indra whether she has had any problems doing IO work, and without hesitation she says no. We leave her and the other IO at the junction where we met them and head farther down the system to M16, where water problems are more severe.

M16: Expressions of Satisfaction

We find Kusumawathie at the house where she stays. We are invited into a neat sitting room, with Japanese calendars on the wall and a Singer sewing machine. Kusum is working with forty-five farmers on thirty-four allotments along four field channels off M16 distributary channel. We will meet the other two IOs in her group later. There is to be a meeting at 3:00 this afternoon for all farmers on M16.6, arranged at their request. They have no explicit organization yet, but the farmers wanted to get together and informed the IOs of this. Farmers on M16.4 have already started an informal organization.

I ask what problems they have had. Kusum says that at first farmers thought the IOs were just research assistants gathering data, "not doing anything." (Some may have met ARTI staff doing baseline studies or monitoring.) The IOs had to explain their role and tell how it would lead to some action. Then some farmers thought the IOs were political appointees sent out before the election, so they had to clear up this misconception.

Has she has any problems as a woman IO? No. In fact, farmers probably trust female IOs more than male IOs, for whatever reason, she says. There is only one problem. "We have to walk long distances here, since the women still don't have bicycles." Munasinghe whispers that ARTI cannot purchase the available women's models because they were made in China or Japan (USAID funds cannot be spent on goods from these countries). Men's bicycles, made in India, have already been purchased. Lucky promises to do what he can to take care of this problem.

Does she like the work? Very much, she replies. I ask what she likes about it, and she says that the interaction with the farmers is very satisfying, especially when advice is listened to. How long would she like to do such work? Her answer to this is less direct, "As long as we are here." Seeing my puzzlement, she adds that she feels she is providing some useful service because the people here have such a difficult time just making a living. Farmers realize that if they work together, it is for their own benefit. We thank Kusum for her hospitality and go back to the van to drive farther down the system.

M18: Starting Work in a Tamil Area

It is noontime, and we find one of the two IOs who work on this D-channel in the room they share. Jaufer is a Muslim whose first language is Tamil, though he also speaks Sinhala and fairly good English. We have already lost four of the six Tamil-speaking IOs trained in the second batch because they were able to get more permanent employment.

The IOs here have the most challenging assignment, because irrigation water has seldom reached this far from the reservoir. They arrived during the drought and had great difficulty getting drinking water and housing (plus they had no bicycles for the first month due to bureaucratic delays). One can hardly blame those IOs who left to take other jobs. The colonization officer working in this area let these IOs move into part of his government bungalow. Jaufer and Subramaniam appear well settled in. I wonder who pasted the picture of Queen Elizabeth and Prince Philip up on a wall next to one of a Hindu god. The two fellows have a list of local officials tacked up on the opposite wall.

The area these IOs are working in is very large, almost nine hundred acres. Orignally four IOs were assigned here, but Jaufer says he and Subramaniam can probably manage the work by themselves if given enough time. Four groups have already formed spontaneously and elected farmer-representatives. Farmers on some other channels are now also seeking to form groups ahead of schedule.

The problem we face is that farmers see the organizations primarily as means for making demands to get water. It is not clear how much more water may be available to the middle and tail end of the system or how soon, even with rehabilitation and farmer cooperation at the head. If expectations are created that setting up a group will get farmers more water, the whole organizing effort can be crushed if water does not come. At the moment, a large issue of water is being made for the maha season, partly in response to farmers' demands but voiced through political rather than organizational channels. Farmers expect that if organized they can get water for the yala season too.

Will it be possible to sustain farmer organization here even if there is no water in the dry season? Jaufer thinks so. They need for unity is widely recognized among farmers, as is the need to clean channels before the maha season. True, some farmers ask, What is the use of organizing if there is no water? But others say they have water problems to solve even in maha.

Could Tamil farmers here cooperate with Sinhalese farmers at the head of the system in Uhana? Jeff asks. These farmers think so, says Jaufer, though he adds that the language problem may make for difficulties. This may be a diplomatic way of expressing some reservations. We all have some, not knowing how much cooperation can be institutionalized when language and cultural differences coincide with the inevitable clashes of interest between head-enders and tail-enders.

We ask Jaufer how he likes the work. While he has one of the most difficult tasks in the whole program, he also has one of the sunniest personalities. He replies with a smile, this time in English, "This is a good experience." What has been the most surprising thing so far? Jaufer thinks a bit and says, "The farmers' attitude that the irrigation system belongs to the government, not to them." He tries to get the farmers together to solve their own problems and to consider the system as their own, telling them "otherwise there is no future." Farmers realize there is no point in quarreling about water.

We leave in the van as Jaufer pedals off to observe a planned water rotation in his organizing area. It is well after noon when we stop for lunch at Central Camp, a cluster of shops that got its name when the irrigation system was being constructed. In front of one roadside restaurant, an aged cook is making *rotis* on a makeshift grill, whirling the dough balls like a pizza chef to make them round and flat for frying. Either plain or with curried fillings, these make a good lunch.

Farmer Meeting: No Action Yet

We now look for the IOs working on M5 and are told by the wife of one of them that there is a meeting with farmers at the community center, half a mile down the road. We find Wimalasena standing by a small blackboard talking to about twenty farmers plus many young onlookers. Two other IOs are seated, unfortunately at the front of the meeting, while the farmers sit on benches as in a school. We will comment on the undesirable social dynamics of this seating arrangement when we meet all the IOs next Tuesday.

As the meeting progresses, Wimalasena speaks ardently, practically lecturing the farmers, and we become uneasy and exchange frowns. This is not what we thought we advised in training, though Wimalasena exudes goodwill and sincerity as he talks about how farmer problems can be tackled and solved by working together. As he talks, a government jeep with a deafening loudspeaker stops nearby to publicize a lottery, drowning out the meeting for a minute before driving on. Wimalasena, looking suddenly tired, sits down.

There are a few moments of silence. An old farmer gets up and starts to talk about a channel bund on M5.7 that washes away each season. "You have come to see it," he says to Wimalasena, "and we appreciate that. If all of us get together, we can take action to fix it. Problems can be solved if we can all be like brothers." He talks about some of his problems with fellow farmers. "Somebody came one day and blocked the pipe to my field so it got no water." He laughs to suggest he is not angry. Maybe the farmer who did this is sitting here. "I won't use bad words as maybe these will come back to me. But why didn't the person come to me so we could solve this amicably?" Another farmer gets up and talks about problems of getting seed paddy and of being given short weight when it is available. He suggests farmers should all buy their seed together to ensure the right amount and quality. Several other points are made, and the conversation starts to drift.

One IO suggests they elect a farmer-representative, but one of the young farmers proposes they first get together and do shramadana on their channel. This is the approach we have recommended, to start with informal work and to have leadership emerge in the process of group interaction. The IO may be pushing too hard to get an organization started. Sensing this, he withdraws his suggestion and offers to join in any shramadana the farmers agree to. One farmer says they should do the shramadana now, while there is water in the channel and they can fix the levels properly.

Farmers confer among themselves, but no concrete proposal is forthcoming. Since no farmer is taking leadership, the IOs thank them for coming and say they will meet with the group again. The meeting dissolves. Sena and Lucky huddle with the IOs to discuss how they think the meeting could have been conducted more appropriately, while Munasinghe meets IOs individually to settle up travel claims and reimbursements. While this is going on, a number of youngsters gather around Jeff and me to try out their English shyly.

Returning to Ampare: IOs Everywhere

It is now late Saturday afternoon. Wimalasena and Punchibanda head off on their bicycles, and we take Asoka in the van to drop her in Uhana. About a mile down the road we meet two new IOs bicycling along. They are returning from a shramadana campaign on M5.4.1 and tell us about it. A third IO, Dissanayake, comes along on foot. He is Sinhalese but knows enough Tamil to work in the area where we will be going tomorrow. We give him a ride into town.

After we drop the IOs in Uhana and drive along the main road, we meet Jayalath returning from Ampare. He has been visiting his wife Anula, also an IO, in the hospital there. They are expecting a child very soon. A few miles farther on we meet our program's first coordinator, K. B., walking along the road. He left for a more permanent post as a teacher but has come back to visit his IO friends and Mune over the weekend. He says he would gladly rejoin the program if it is made permanent, because he likes this work better than teaching. I can't help asking Munasinghe whether all these meetings of IOs along the road are by chance. He insists, "Absolutely nothing has been arranged for you. We had no fixed schedule today." Which is true.

That evening at the circuit bungalow, we are joined for supper by the deputy director of irrigation for Ampare. Among other things, we discuss Senthi's recent decision to break and remove a newly built measuring structure on UB16. Farmers on that channel had insisted when their area was being surveyed for rehabilitation that the existing twelve inch pipe offtake from the branch canal was not large enough to serve their area. But ID engineers insisted their calculations showed it was correct and left the pipe the same size.

Senthi says it is now agreed by everyone that the area does get too little water. He has been impressed with the patience and persuasiveness of the farmer-representatives, including our friend Neeles, who have petitioned for a larger-diameter pipe. Senthi says the ID will install one, but this cannot be done until the end of the season. In the meantime the ID is hammering out the measuring weir it installed so it does not impede flow into UB16. Such responsiveness to farmer needs is a visible sign of good faith from the ID side. We hope it is fully appreciated by the farmers.

I am gratified to hear Senthi say that it would be good if the IO program could be introduced in other schemes a year or two in advance of rehabilitation efforts, so that farmer organization could be established before the new work begins. This is one of the clearest votes of confidence we have had so far. The DDI before him wanted nothing to do with the program.

As we head back toward Weeragoda the next morning, we pass workmen chiseling out the measuring weir at UB16's gate, just as Senthi told us. We drive down the bund road along UB16 to see for ourselves the water problems at the end. As we are returning, we meet an ID jeep coming. In it is the new chief irrigation engineer, Manoharan (another Mano), making an inspection to see whether water is now getting to the tail. We have a friendly talk, and I am amazed to find a CIE in the field at 9:30 on a Sunday morning to check on farmers' problems. I hope this sense of personal responsibility will become the mode for all ID staff, though I hope also that the system can be made to operate well enough with farmer-representatives as trusted intermediaries that such inspections are unnecessary.

Sinhalese-Tamil Cooperation on M5.4.4

We drive a long way down the branch canal and then turn off along M5. It is itself practically a branch canal, serving about 2,500 acres. The road gets very bumpy by the time we approach where groups 4 and 5 are based. Because of drinking water shortages when they first came, the IOs could not find accommodations in any private homes, so the manager of the cooperative gave them permission to use part of his warehouse as sleeping and living quarters (there are only men in this group).

This area is most difficult to work in for a number of reasons: it is a border area between Sinhalese and Tamils; it is on the boundary between Ampare and Batticaloa districts and thus an administrative no-man's-land; and few officials spend time in the area. The people here tell the IOs the only official they ever see is the *Grama Sevaka,* a village-level administrative officer. There are running battles in this area over water because it is usually so scarce. Ethnic differences make the situation "very dangerous." The number of Tamils in units 16, 17, and 25 has shrunk in recent years by 50 percent or more. Many moved out after the 1981 violence.

As we sit on benches under a mango tree talking with IOs, Suma-nasena walks up. He has been with a shramadana campaign on M5.4.4 not far away and invites us to come and observe it for ourselves. This subfield channel is more than a kilometer long and serves twenty-seven farmers, fifteen of them Tamil, twelve Sinhalese. When we reach the head of the channel, we see water flowing heavily into it, as the first issue of the maha season has just reached this far down the M5 distributary. As we walk along the channel, I notice two young-sters blocking up pipes watering head-end fields. Through Sena I ask whether these are their own fields. I suspect not, since they are in their early teens, and I am surprised that they feel free to stop the water flow. They reply that the fields are not theirs but that these have had enough water already. I ask, won't the owners complain? They say the owners are with the work group downstream and will not object. "Everyone is supposed to get a share of the water but no more." Such a level of trust is impressive, especially here.

As we proceed it is clear that a thorough job has been done in re-cutting the channel walls and new turnouts to fields. The flow of water is strong and apparently unprecedented. Several concrete structures appear in danger of washing out as the water rushes through cracks around them, but Hammond assures me there is not much danger of serious damage.

Finally we reach the work group. As evidence that this channel has not been cleared for a long time (several decades), they are hacking out with axes the roots of a large mango tree that have grown through the channel, blocking it a few hundred yards from the end. Most of the group are taking a tea break, so the timing of our visit is good. Two elderly women offer me boiled cassava somewhat apologetically. This is the food of the poor in Sri Lanka, but I take it gladly, since I grew to like it years ago in Ghana. They follow with pieces of *jaggery*, local unrefined sugar eaten in lumps as a sweet, and then unsugared tea in battered old glasses. By their menu I know this is the least prosperous group of farmers I have visited, but it is all good fare. The whole group gathers under the tree, resting and taking refreshments.

This three-day shramadana has involved all the families on the channel. The ethnic mix of the group makes it even more inspiring to see a thirty-year-old channel brought back to its intended capacity. By happenstance, the first flow of water for the season is just reaching the section of the channel where they are working as we arrive. It is exciting to watch the farmers see their effort rewarded by an abundant flow.

One of the young farmers who seems to be a budding leader in the group wears a sun hat with "Disco 82" emblazoned on the crown. He is of the IOs' generation, but unlike them wears a sarong (not pants) and no shoes. There is easy interaction between the IOs and farmers as they huddle in discussion. I reflect on our earlier fears that university graduates might not be able to bridge the social status gap with poorly educated farmers. We also were concerned whether IOs' modern dress and manner would estrange them from farmers, but we have learned that IOs' attitude and demeanor, more than outward appearances, make for acceptance and cooperation (or the opposite).

Before we leave, I ask the farmers whether they think they can continue to work together, both Sinhalese and Tamils, to solve their problems of water management. The response might best be translated as "Why not?" As we walk back to rejoin the others, Sumane tells me more about this area. As recently as eighteen months ago there was violence between the communities, with some Tamil homes burned and their occupants chased out. Before that in 1957 and 1977 there were also clashes. Given this legacy, it is not surprising that there was previously no cooperation to clean the channel and rotate water deliveries. Who would have expected that so soon after their arrival the IOs would have gotten this much joint action? The timing of water delivery from the main system could not have been better to reward cooperative efforts, of course. Sumanasena is visibly pleased with what has been done here the past three days, and he deserves to be very satisfied.

M5: The Most Adverse Situation

The IOs have reassembled at the cooperative warehouse to meet with us. They tell us that problems arise because none of the IOs speak Tamil. (ARTI tried to recruit more Tamil speakers for this batch but was not successful.) One IO tells how he tried to explain to one household in his limited Tamil that IOs wanted to help farmers with water problems, and the woman of the house brought him a glass of water. This occasions good-natured laughter all around.

For many weeks the IOs were hampered by not having bicycles, but that is no longer a constraint. They describe another problem, however, that turns out to be complex. There is a lack of cooperation not only between ethnic groups but also within the Sinhalese community. Sinhala settlers seldom go to each others' houses. They don't even give messages to one another when IOs ask them to. They are divided by caste, political, and economic differences. There are some large hold-ings (thirty to forty acres) put together by "outsiders" when Tamil set-tlers fled the area. These absentee owners do not cooperate with anybody and do not support farmer organization, having rented out their land under tenancy arrangements that are illegal, as is their con-trolling the land.

As an indicator of caste divisions in the area, when IOs finally organized a shramadana campaign, they found that low-caste par-ticipants did not join in drinking tea the first day. Only when the IOs insisted did they take any the second day. They still do not sit on benches with the higher-caste farmers when group discussions are held. Such divisions will certainly impede efforts to unite farmers.

The IOs report intense competitiveness and individualism among farmers. Water conflicts are common, with bribes or political patron-age used to get extra water or to install illegal pipes. About three-fourths of the Tamil cultivators here now live outside the area, many having withdrawn to reside in their ancestral homes nearer the coast. One-quarter of the Sinhalese live elsewhere. This is obviously an un-attractive area, difficult sociologically quite apart from its water prob-lems, which are severe.

I ask if farmers in this area had previously heard about the IO pro-gram. "Some in unit 17 knew," says one IO: "But still, many thought we were political activists at first. Some even called us Che Guevaras. It didn't take long before the people understood that we had come to deal with water problems, but then they expected us to give them wa-ter and to operate channel gates, which wasn't our responsibility." Hammond comments that this group has the most challenging task of any. He knows the whole Left Bank system from his fieldwork here, and he finds the M5 area "impossible" hydrologically as well as "abandoned" administratively.

At this point the IOs bring up some personal problems. The roof of the warehouse is leaking, and when rains come, their things get soaked. Also, they need a pump to repair bicycle tires. "This is not like in a town where you can go somewhere to get your tube fixed." We agree they should be given pumps, though we do not know if govern-ment rules will allow this. They have no lights for their bikes yet, and that too is a reasonable request. Right now they have just one flash-light, belonging to one of the IOs, that is shared between the two groups. Also, they would like to have racks for their bikes to carry notebooks and other things. They have to bicycle at least three miles to the nearest shop to buy food supplies. Lucky says he will look into these matters, and we take our leave.

As we drive away, Sena comments that this group contains some IOs who were most outspoken during their training and who were thus viewed as the radicals of the lot. "They are still very socially committed," he observes, "but they are now more thoughtful, more tactical in their approaches." Sarath, he recalls, had the reputation of being very quick-tempered, a fighter who could be easily sparked off. "We were a little concerned about him. But he has changed a lot." How? "He is more considered, more careful in his statements for having seen how people live and what it takes to change their situation." Some sobering observations as we head back to Uhana for lunch.

Meeting First-Batch IOs

The coordinators for groups that cover the first organizing area are gathering at Munasinghe's house in Uhana this afternoon. Sumanadasa from Gonagolla is one of the first to arrive. He says some of the momentum there has been lost. The two organizations that started spontaneously outside the IO organizing area are no longer functioning. One collapsed because a large landowner controlling thirty to forty acres (illegally) kept coming at night to close the head-end pipes and take water to his fields at the tail. The group was not able to stop this. Also, the number of IOs in Gonagolla has fallen from five to two, and they cannot work so actively with all the groups anymore.

The meeting was called by the IOs themselves to improve their own communication and coordination of efforts, but I ask for some time to review the situation. The most difficult area is UB2. Sita and Chuti say it has twenty-five groups now, but because of the changes in the assignment of IOs to different areas, some groups have fallen behind. The four groups at the tail have lapsed into inactivity, but this may be because they have had four different IOs working with them the past nineteen months due to personnel turnover. (One IO had to be fired for neglecting his work and borrowing money from farmers.) These channels have a higher incidence of tenancy, which exacerbates any problems of organizing farmers, but that is another matter. Sita and Chuti count up the number of effective organizations and conclude that sixteen of the twenty-five are working well. Most of these are toward the head end and are more willing to conserve water, so the failure of some tail-end groups to function hurts mostly themselves and not others.

As we review the situation, it appears that where the ID has consulted with farmers fully and done good rehabilitation work, the organizations have progressed satisfactorily. Where the physical work has been poor or promised improvements have not been made, or major problems remain uncorrected, there is dissatisfaction and farmers withdraw. Can the organizations manage water in the next dry season? Sita and Chuti confer quietly in Sinhala and think a bit. "When it comes to irrigation and water management, there are no problems. Farmers are very cooperative in distributing water within channels, and farmer-representatives on UB2 get together to share water among channels." They are confident that farmers there can handle most of their water problems.

Only on UB2.10 do they anticipate considerable waste of water, depriving UB2.9 of some of its supply. They explain that UB2.10's turnout was installed too low, so it draws too much water and leaves the upstream channel with less volume and less head. It is impressive that IOs have such a grasp of the technical aspects of irrigation, and I am glad that the problem they report is more technical than organizational. I am also pleased that the IOs are not hesitant to speak about problems. Sena and I regard it as a sign of maturity that after reviewing all the D-channel areas, the IOs suggest we excuse ourselves to let them go ahead with their planned meeting.

M5 Again: Support for Our Initial Hypothesis

We drive back down along Uhana and Mandur canals to meet three IOs organizing farmers at the head of M5 distributary. We sit in a school building about eighty feet long with half walls and a thatched roof held up by rough-hewn timbers. When we tell them we plan to meet all the IOs on Tuesday, they huddle in some consternation. All have shramadana campaigns or farmer meetings already scheduled for that day, so none of them can join us.

One of the IOs has just come from assisting in a shramadana nearby. Thirty-three farmers have been clearing the lower half of a two-mile channel that has not been so thoroughly cleaned since it was built in 1952. It was the farmers' own idea to start at the tail and work upward, says the IO, since nine allotments at the tail had not been getting water even in maha. The rest will be done next Thursday.

Is good leadership coming forward? I ask. Anula says yes, mostly young farmers, but some older ones also. In the Uhana area where she worked before, farmers could be organized after some time and discussion, but here they come forward more willingly and propose shramadana, for example. This supports our hypothesis that water user associations would be easier to form in the middle of the system, where water is scarce but not too scarce.[1]

Hammond asks whether farmers will be as willing to cooperate in the upcoming yala season when water is scarcer. Anula says that this maha season some farmers are already closing field channel turnouts when they have enough water, and water stealing is not happening as before. I ask how they know, and she says that farmers tell them. Also, obstructing the flow of water at night has been reduced. "Farmers know that a lot of channels downstream are not getting water, and they want to help get water to them."

This sounds too good. I ask what problems remain. The IOs say that most farmers haven't been able to stop a minority from taking water illegally from the channel for their garden plots, by using siphons or cutting bunds, when tail-end farmers who are entitled to the water are not yet getting enough. Also, some have extra unauthorized pipes to their fields. Will tackling this problem damage farmer cooperation? One IO says some farmers have threatened to remove the pipes themselves, but since certain pipes have been in place for ten or even twenty years, a decision to remove or authorize pipes should be left to the ID when it rehabilitates the channels.

We ask these new IOs about their experience with the work so far. "It is fine as long as we are healthy." What is the problem? Both got malaria when they first arrived in the field and had to go to the hospital for a while. Fortunately, chloroquine can control malaria fairly quickly, but even a brief bout is miserable and debilitating. I greatly regret that we have not been able to issue mosquito nets to the IOs. Government regulations do not authorize giving nets to staff at their salary level. We drop the IOs at the main channel road and pick up two others who are waiting to go with us to the next meeting.

1. This is discussed in Uphoff, Wickramasinghe, and Wijayaratna (1990).

Farther Down Mandur Canal: CIA Spies? Another Test

The IOs assembled in a farmhouse cover M6, M9, M10, M11, M12–15, and M17. Their areas are really most spread out. I ask about the response of farmers here. On M6 there are fewer water problems and less response, says one. "Otherwise it is pretty good, though we face some problems of political pressure." I ask for some elaboration.

The Rural Development Society in this area, being aligned with the governing party, has taken the initiative to conduct shramadanas to clean field channels. This is fine, but it preempts the technique IOs have usually used to get organizations started. The area is quite politicized, since parts of it used to be loyal to the opposition party. Recently there have been many "crossings over" to the ruling party, complete with public recantations at big ceremonies. Most farmers do not like this trend, but nobody objects publicly because no one wants to be seen as opposing the government. IOs have been invited to speak at these events and were told what to say. One IO who spoke says he hurried over these points and talked mostly about the need to build up self-reliant groups.

This situation creates real problems. The IOs should not be "against" collective action even if it is organized with political motivation, and they should not get themselves at loggerheads with influential local figures. Nor should they get themselves or the farmer organizations entangled in partisan politics. The group has discussed this issue and concluded that for now they had best bide their time. "These energies will get dissipated," says one IO, "and partisan efforts will slacken."

An IO says that things got fairly tense last October, shortly before the national election, when one of the District Minister's campaign speakers at a party rally in Ampare publicly claimed credit for our farmer organization program for the government and the ruling party. The next day, at an opposition party rally, a speaker criticized our program, saying it was intended to impose a water tax on farmers. He went on to say that the IO program was an American creation and that the IOs could be *agents of the CIA*. Oh, no!

Wijay happened to be in Ampare at the time, so he met with all the IOs and they talked for hours about what to do. They decided that rather than respond with a denial, they would go on with their activities, concentrating on technical tasks to avoid any appearance of being political organizers. To everyone's relief, the attack gained no credence or momentum and was quickly forgotten in the jostling of the campaign.

This indicates that the program is popular enough and well enough accepted among farmers that resentment can't be whipped up on partisan grounds or by emotional appeals. The charge of being a CIA agent has been like dynamite in this part of the world. If there were unspoken resentment toward the program or even a lack of sympathy, more could have been made of the attack.

I ask about farmer response to the organizing efforts so far. IOs say that cooperation has been particularly good where water problems are worst, for example, on M12. (This inverse relationship is fortunate.) Where they have fewer problems, as on M10, it has not been as quick or strong. Hammond discusses with IOs the progress and problems in physical rehabilitation of the system. It is getting dark. The few bulbs in the sitting room permit us to see one another, but it is time to leave, since the household may have its own plans for Saturday evening. We thank the family whose house this is and take leave of the IOs, driving back to Ampare along the now familiar canal road.

Expressions of Official Support

Monday is spent with staff of the Irrigation Department in Ampare focusing on rehabilitation of the system. We need to know about this in order to facilitate farmer input into the plans and actual work. We also want to improve relations between technical staff and water users, trying to counter stereotypes and attitudes from either side that may impede cooperation. I am struck that Irrigation Department staff now seem to take the IOs and their work very much for granted.

At 10:00 A.M. we are able to meet for almost an hour with the Government Agent. Having become a solid supporter of the IO program since it got started in the field, he expresses satisfaction with the IOs and with the progress of farmer organizations. The most specific outcome of our meeting is the GA's suggestion that we draft a memorandum to the District Agriculture Committee, describing the three-tier structure of farmer organization that has evolved at field channel, distributary channel, and branch channel levels, so it can be formally recognized by the DAC.

The groups so far have no legal standing, though some D-channel organizations have been registered with the ID to be eligible to do contract work. We have declined previous suggestions to formalize the organizations but now feel more confident that the emerging pattern and scope of organization is viable, being understood and supported by farmers. ARTI will prepare a memorandum for the February DAC meeting, and the GA invites Lucky or Sena to attend if possible.

Day-Long Meeting with IOs: A Monkey Dance?

Tuesday morning we head back in our van to the school near the M5 gate. IOs from both batches are to meet us there, though some cannot come because of program duties already planned. As we drive past M3, we see a large number of farmers digging out that channel. We stop and find out this is their third day of shramadana. The ID has reclassified this D-channel as a field channel, making its maintenance now their responsibility even though it serves 115 acres, more than twice the previous cutoff point for farmer responsibility. Farmers have accepted this task because the channel was not adequately maintained before. They seem to be doing a thorough job, wanting to get water to their fields quickly. The ID has repaired one broken structure on the channel, so it is cooperating in the effort.

At the school, most of the IOs are assembled. A light rain is falling, and the school benches and chairs are arranged in a circle that avoids the occasional leaks through the thatched roof. I am curious to see how the organizers will organize themselves. When we are all seated, Ariyapala, who has been selected to succeed K. B. as the program coordinator, welcomes us and starts through an agenda worked out by the team coordinators. Given its length, we will have difficulty covering all this.

Ariyapala wonders out loud how to handle the first big item—reviewing progress and problems to date. We decide to start with the new area, since it has received the least attention previously. Premachandra, our Weeragoda coordinator, begins. He recounts the difficulties of the second batch—the yala drought conditions, the lack even of drinking water, the difficult time finding accommodations, the lack of bicycles for some months. He reproaches ARTI for lack of planning on transport, and Lucky acknowledges that the institute's performance was not satisfactory. He starts to offer an explanation but stops, saying the criticism is correct.

Premachandra continues, saying that there is considerable progress now. Probably the main difficulty at present is establishing contact with government officers in the area. As we heard on M5, they come infrequently. Also, because the program had no coordinator for several months after K. B. resigned, little was done to establish links with officials at district level.

I am impressed with Premachandra's discussion, soft-spoken but clear and confident. When I saw him previously in Gonagolla, he was always reticent, and I did not regard him as a leader. But Munasinghe says that without Prema's strenuous efforts to get the work in Weeragoda started under very adverse conditions last fall, the program would have stalled.

Premachandra says that the frequent turnover of IOs and the reshuffling of groups has created discontinuity in IOs' contact with farmers. He wants a more systematic way of filling vacancies, but mostly they need more IOs. He ends by saying that there are a lot of problems, but he is confident they have made much progress already in Weeragoda. The rain has stopped now, and with the sun coming out it is getting hot and more humid.

Jayalath speaks next about Uhana, prefacing his remarks by saying they now have to do their work with fewer IOs. They no longer meet farmers individually but instead work with farmer-representatives and groups of FRs. (This is an appropriate shift in focus.) There have been some changes in farmer-representatives made by the farmers themselves, which is good, because it means that farmers take the program seriously. But this creates more work for the IOs, who have to spend a lot of time with the new FRs.

The rehabilitation program still presents farmers and IOs with some problems, since there is confusion about who will do what on D-channels and field channels. Farmers are unhappy that private contractors hire "outsiders" for work that farmers could do. In this water-short year, they seek any possible employment. After taking this matter up with the TA, it was decided that farmer organizations should prepare a list of local persons who would be hired, without regard to politics. Contractors are apparently satisfied because they now get laborers who take the work more seriously than persons who come with letters of recommendation from the MP.

On one D-channel, the farmer organization elected a person to watch over the quality of construction work on behalf of farmers. "Unfortunately," the IO says with a smile, "this man took the task too seriously. It went to his head and he became a nuisance. He went around measuring everything, getting in the way, and making unwarranted suggestions." This man was relieved of the responsibility, which was given instead to the farmer-rep, who has handled the task with more tact.

On UB2, farmers pointed out to the ID that substandard materials had been used in constructing one weir. The TA, on inspection, found the allegation was true and decided the weir should be rebuilt at ID expense. On UB15, another IO says, farmers complained that the structure was inferior, but the TA did not believe them. Then there was a spell of heavy rain, and it began to wash away. This convinced the TA, who ordered the contractor to reconstruct it. A similar incident is reported from UB1. This seems a good time for a tea break. Three IOs who left a few minutes ago return with two large enamel pitchers full of hot tea, which is poured into several dozen small glasses that are passed around.

After the break, we hear about the situation in Gonagolla. There, all the groups clean and maintain their channels, rotating water deliveries if there is need. "Originally farmers didn't see much use in an organization," Sumanadasa says, "but now they are trying to coordinate and improve inputs for better production. Some lethargy is setting in, however, since the organizations are not getting full cooperation from all officials. Area Council meetings are no longer held every month, as the number of officials participating has dropped off and many of the decisions taken are not implemented." This reflects the machinations of the divisional officer of Agrarian Services, who for unknown reasons is resisting the program. Farmers decided that regular meetings should be revived, but still many of the relevant officers do not cooperate with the organizations.

One of the new IOs stand up and makes a short speech: "We must realize that there are two forces in Sri Lanka—the force of the officers and the force of the farmers—and they are bound to come into conflict. We all know that the force of the officers is stronger. I was dubious from the outset that we could create viable independent farmer organizations. Farmers will put more faith in political associations that can give them jobs and can pressure officers. Farmer organizations can be expected to fail."

This elicits an active discussion of whether farmer organizations can succeed in the face of party organizations and government bureaucracy. Many apprehensions are expressed. Is self-help a tenable philosophy and strategy under Sri Lankan conditions? After the skeptical IO restates his challenge, Ratnasiri, one of the most articulate IOs from the first batch, responds. He recounts the successes IOs had in the first area, despite initial opposition from political figures and certain officials. He argues that a self-reliant approach is in fact the only way Sri Lanka can achieve rural development.

The program has suffered a setback in Gonagolla, I suggest, but that doesn't mean that all efforts to organize farmers are doomed to fail. Rather, it should prompt us to develop a better approach to problem solving, identifying priority problems and not letting up on solving them as long as they remain obstacles. The discussion has gotten rather intense, and it is good that the ARTI van has driven up with forty-five lunch packets—rice and curry meals wrapped in plastic and then newspaper; Sri Lankan take-out food. We adjourn for lunch, and a few IOs organize the distribution of packets and the washing of hands from pitchers of water brought from a nearby house. No knives and forks here.

When we reconvene after lunch, the first topic is improving communication. Perhaps people are not so focused or motivated after lunch, because the discussion drifts rather inconclusively. There is a discussion of how the IO unit office in Ampare will be set up and of training programs for farmer-reps. It is midafternoon, and I watch boys playing *elle,* the Sri Lankan version of baseball, by the school with sticks and a rubber ball. Their shouts of excitement occasionally distract our discussion, but nobody seems to be bothered.

It is time for an afternoon tea break, to lengthen attention spans again. We have already been meeting for most of seven hours, and everyone needs diversion. As Sena and I sit together discussing the meeting so far, somewhat disappointed, two very old farmers come walking softly into the schoolroom. They look around curiously and then shuffle over to us and start talking. They say they are returning from a shramadana on their field channel and wanted to meet us. Their channel, off M3, is two miles long, and together with other farmers they have cleared six hundred feet today. Such work must be difficult for them because of their advanced age. Their faces are weather-beaten, and their long white hair is tied in knots at the back of their heads. Apparently planning a leisurely discussion, they sit down and pull out wads of betel leaves from the waistbands of their faded, worn sarongs, offering some to Sena and me. We accept out of courtesy, but we pass up the lime that makes chewing betel very astringent (and turns the mouth bright red).

I ask when their field channel was last cleaned, and they say this is the first time it was ever done thoroughly. Before the IOs came, they say, there was no possibility of cooperation. Sena asks if they are original settlers, and they flash proud smiles: "We came to unit 18 on November 12, 1952." They remember the exact date. They came from an upcountry district that has a very different climate, and adjusting to Gal Oya was difficult. "There was not drinking water available at the time, mosquitos were terrible, and people suffered often from malaria."

Sena asks them if they are sorry they moved to Gal Oya. The answer is evasive: "Now things are fine." What about their yields of paddy? "At first we got only five or six bushels, because of lack of water and fertilizer. But now when both are available we harvest thirty-five to forty bushels per acre. The water situation is improving." One comments, matter-of-factly, not morbidly, "We are sorry that we are about to die just when things are getting better." The other farmer, embarrassed by this, compliments the IO: "There was no cooperation before Punchibanda came."

What things should farmers cooperate on? I ask. They can get together to improve the roads, to clear land, to level fields, to remove stumps from uncultivated highland, says one. The other adds that they can repair houses and replace thatch on roofs. Obviously they have many ideas of what could be done by collective action to improve their lives.

One of the IOs brings two glasses of tea to us as we talk, and we ask him to give it to the farmers. At first they seem embarrassed, but then they accept the glasses and drink the tea slowly, visibly pleased. The IO goes off to find more glasses to bring more tea. Most others have finished their tea and are waiting to resume the meeting. When the farmers finish their tea, they hand the glasses back to the IO. An animated discussion ensues between them and Sena. He turns to me and says with a smile: "They want to do a dance for us. They want to thank us for our hospitality and for the IO program, and they would like to entertain us." How should we respond to such an offer? It might be difficult to avoid smiles or even laughs if these handsome but aged men try to dance before the group. My concern is heightened when Sena says, "They call their dance 'The Monkey Dance.'"

We don't want to offend them by turning down their offer, but neither do we want any embarrassment. In his most gracious manner Sena thanks them for their offer, explaining that we need to resume our meeting. It won't take long, they protest, but mildly. They see the IOs are reassembled and amble away with a wave and a smile, hitching up their sarongs and drawing forth more of their precious betel leaves to chew.

We have to move through the rest of the agenda more quickly. The need for more Tamil-speaking IOs is brought up. Lucky explains that we hope to recruit and train a batch of thirty by July. The next two agenda items take much longer. The sun is going down, so it must be after 6:00. The last issue to be considered concerns the IOs' "future," as they refer to their prospects for permanent employment. Lucky says the Minister and Secretary of Lands have agreed to a permanent cadre of about 20 percent of the IOs who have joined the service. This appears to be the proportion who should be encouraged to make a career of this work and who would be needed to train, supervise, and coordinate a continuing number of IOs taken on contract (say for two years) to carry out the direct organizing work in the field. This means not all current IOs can expect permanent jobs, but there are at least prospects for those who are most effective and enthusiastic.

It has been a long day. We could use more time for stocktaking and informal training. But it is evident the work in the field is going reasonably well, owing in large part to the initiative and resourcefulness of the IOs. Even those who have not impressed us by speaking up are getting generally good results. The van heads off to the east, packed with IOs needing a lift to their homes downstream. Sena, Lucky, Hammond, Mune, and I start walking westward along the canal road. Since we have been sitting all day, stretching our legs feels good. It is almost an hour before the van catches up with us and drives us back to Ampare.

Discussions with the Deputy Director

Wednesday we meet Senthinathan in his Ampare office. Most of our discussion turns on problems of system redesign and operation. We discuss the question of farmer participation in quality control for the new construction, and he is quite agreeable. He says he has already instructed his TAs to cooperate with farmers and to be responsive to their suggestions and criticisms. When Senthi joins us for supper that evening, he tells us that this is the first time Uhana farmers have left all gates off the branch canal closed when the ID shut them to send water downstream to Weeragoda and Mandur. Senthi says when he visited the tail end of the Left Bank system, he told Tamil farmers there that they should thank those in Uhana for cooperating by not taking excessive water at the head end.

The DDI also gives the IOs credit for this, but we know that farmer organizations have not yet been put to the most severe test, which will come when there is absolute, not just relative scarcity—when water become crucial to people's livelihood. Upstream farmers have been willing to help out tail-enders on their own channel and farther down the system, but so far this has not entailed any jeopardy to their crop.

How cooperation will fare in zero-sum situations we do not know. We are moving toward a structure of consultation and cooperation that could produce greater benefit from any given amount of water. Such a positive-sum capacity may become valued enough that conflictual situations will be negotiated among those affected to maintain the organizational structure. With IO initiative and with administrative support such as Senthi now gives, we can hope for further movement toward institutional development.

New IOs and Staff

After my January field trip, ARTI recruited and trained fifteen more IOs in February, filling vacancies created by resignations of IOs who found more permanent jobs elsewhere. Because it was shorthanded, ARTI treated this group as half a batch and gave it less formal training. Indeed, much of the introduction to concepts and techniques was given by experienced IOs. Then the new recruits were assigned to IO teams as apprentice organizers without the month of supervised field training given to the first two batches.

This group evidently was quite motivated by the two weeks of training. In the tradition of previous self-managed training programs, the IOs organized their own going-away party, for which they sent out an impassioned invitation. Wijay, by now at Cornell, received a copy and translated it for me, insisting that he could not do justice to the poetic and powerful Sinhala it was written in:

> We'll disperse tomorrow;
> We are going to a zone where some people are sweating blood to build a
>   nation, and we will live with them.
> The friendship that has grown among us over the past few days causes
>   our hearts to feel sorrow, as we are dispersing tomorrow.
> *But* we endure it for the sake of greater friendship in the future.
> Colleagues, we invite you to join us in our farewell event.
>
> February 25, 1983

Their appreciation of friendship as a motivating force was unmistakable.

Plans for a full batch of Tamil-speaking IOs did not proceed as smoothly as we had expected in January, but ARTI and Cornell prepared themselves for the task. Recognizing the invaluable contribution Sena had made to the program as an all-purpose training consultant, we hired as a Tamil counterpart Nalini Kasynathan, a university lecturer who walked into ARTI one day.[2] She had grown up in Batticaloa, below the tail end of Gal Oya. Her father was a member of Parliament from there for many years, so she had good connections. After a cyclone devastated Batticaloa in 1978, she formed and headed a voluntary organization to mobilize assistance, so she had an activist disposition to go with her academic qualifications.

About the same time, Jeff advertised for a Tamil-speaking research assistant, and the most qualified applicant was a young woman who had recently graduated from a university in Tamil Nadu, India, with top honors in agriculture. Shortly after Jeff hired Bhawani, ARTI appointed her as a research and training officer to help with the planned Tamil program, because she was more capable than any other applicants. She and another new R&TO, Rajendra (Raj), were both very energetic additions to the Water Management Group.

Farmers Break Structures

On March 23, half a dozen newly built measuring weirs on UB9, UB10, and UB11 were broken by someone during the night to increase the flow of water into these channels. After all the IOs' efforts to get farmers to cooperate with the ID and not break structures, this looked like a real setback. Yet in an unexpected way this incident ended up strengthening relations between farmers and engineers.

Farmers in Uhana had been told at a meeting in February that the first date for water issue to start land preparation for the yala season would be March 15. The District Agriculture Committee later decided the date would be March 20, to give the ID more time for rehabilitation work in the canals, but this was not widely publicized. When by

---

2. Mrs. Kasynathan had come across a copy of my June 1981 trip report while interviewing an engineer near Batticaloa, and she wanted to know if ARTI had any more such reports. She was starting some research on the distribution of water in Gal Oya, both as a geographer interested in spatial relationships and as a Sri Lankan concerned with the causes of ethnic tension. She wanted to determine to what extent location and supply difficulties accounted for the inequalities of distribution rather than ethnic factors, which were usually pointed to as an explanation.

March 23 water was still not reaching the tail of some Uhana channels and it appeared that the measuring devices were impeding the flow, six weirs were demolished.[3]

Farmer-representatives insisted they had no role in the damage, but they admitted the breakage should not have occurred. They said that farmers would pay to repair the structures once sufficient flow was given. Such an offer was unprecedented and a vindication of the farmer organization program in Sena's view. But the story was more complicated. The irrigation engineer in Ampare, to his credit, when he saw that not enough water was flowing through some of the weirs, had told farmers they could remove them. So at least some of the breakage was "authorized." More significant, the DDI acknowledged that water shortage on these channels were at least partly due to ID miscalculations. (To save on construction costs, the department had built one fewer cross-regulator than designed for the Uhana canal.) So he declined the farmers' offer to pay. For farmers to offer to pay to repair broken structures was something new, but so was engineers' admitting technical mistakes publicly. This episode could have led to recriminations between the two groups; instead, the exchange of mea culpas contributed to better relations.

Looking for a Home

With ID engineers now pleased with IOs' work, we thought it should finally be possible to get the program established on a permanent basis. Everybody in government circles was talking about improving water management, but administrators were assuming that IOs would be available to continue the kind of work being done in Gal Oya without taking the actions necessary to maintain the cadre. ARTI could not get the Ministry of Lands to focus on creating an administrative base for the IOs. Sena called me in Ithaca to report that we might lose half of them in May when hundreds of teaching positions would be opened up for graduates.

3. At least some of the weirs were certainly obstructing the flow and supply of water. Quite by chance, our part-time engineering consultant, Kariyawasam, had made some measurements of flows in channels just the day before the incident. He found the flows on UB9.6 and M5.4 were as much as 75 percent less than designed volume, at least partly because the crests of weirs downstream had been built as much as thirty inches too high. ARTI "classified" Kari's report to keep it inhouse, so that the ID would not think his investigation prompted farmers to break the structures.

That would be a disaster for the program, since many of our most experienced IOs would leave in such an exodus. There was talk in the Ministry about using IOs in a World Bank project to rehabilitate half a dozen major schemes in the north, and USAID was thinking about a follow-up project that would use IOs in other schemes. There were also feelers from the Mahaweli Authority about trying the IO approach in some of its systems. ARTI, with its limited administrative capacity for this kind of task, was hard-pressed to keep the program alive (funds for it came from the Irrigation Department, to be sure). Its two-year commitment was completed. The number of IOs was continually diminishing and might drop precipitously, so the program could die a natural death. Since ARTI wanted the embryonic program to be transplanted, not aborted, however, it carried on.

June 1983 Visit: IO Cadre in Jeopardy and Another Test

My return visit to Gal Oya at midyear was under the long shadow of losing half or maybe most of our IOs. Government delay in announcing six thousand new permanent teaching appointments gave the program a reprieve but no commutation of sentence. Our newest IOs had been put into vacancies in the Uhana, Gonagolla, and Weeragoda areas and were helping to extend the program to several thousand more acres downstream.

Gonagolla farmers were, as before, taking important initiatives. They had run into difficulties getting credit for agricultural inputs from the Agrarian Service Centre because almost all had defaulted on previous loans. So through their organizations they worked out an agreement with the Centre by which anyone who deposited 1,000 rupees as security could get a 2,000 rupee loan. Farmers deposited 157,000 rupees to get double that value of seeds and fertilizer for the yala season. Further, they made a deal with the local cooperative, which also regarded most farmers as "uncreditworthy." Each farmer-representative could vouch for up to five fellow farmers, who would then get 2,000 rupees' worth of seeds and fertilizer on credit—what banking institutions call "character-based lending."

After bureaucratic delays, the Gonagolla farmer organization finally got enough improved seeds and fertilizer from the Agrarian Service Centre to plant a forty-acre model farm to demonstrate modern production techniques. (This was something the ASC should have been glad to assist, but wasn't.) The farm produced eighty bushels per acre at harvest time, 60 percent above the area's average, and the 19,000 rupee loan from the Centre was repaid. Unfortunately, the monthly Area Council meetings continued to be irregular or poorly attended, largely because of the farmers' disappointment, which we heard about in January, that officials did not participate more actively and help implement joint decisions.

In Weeragoda, on the other hand, we found that newly started monthly meetings were both regular and well attended, by officials as well as farmer-reps. On M5, the most difficult Left Bank area hydrologically and socially, there were more indications of progress. The Sinhalese and Tamil farmers on M5.4.4 whom I visited in January had recleaned their channel before this season. About 70 percent of the lower one thousand acres of M5's command area, which had not gotten water in the yala season before, were being cultivated this year, thanks to a complex rotation system worked out among over thirty farmer groups. Caste differences had not been eliminated, but in general farmer cooperation was solving most water problems. "The big change is that force is seldom used in water disputes now," said one IO.

Such progress made IOs proud of their contributions to the program, but most had already received or expected letters of appointment to teaching positions. At a meeting with all IOs, one of them said, "We would all prefer to stay with the program if we could get permanent appointments." Nobody disagreed. IOs pleaded for ARTI to get the government to make it possible for them to continue in this work.

A further test of the program occurred during my visit. A rumor was circulating that IOs were getting farmers to do rehabilitation work on a voluntary, unpaid basis in order to pocket the money that should have been paid to farmers for this work. Allegations of corrupt practices are worrisome in Sri Lanka because most are credible. Protestations of innocence, even if correct, may not save a reputation. We found out later that the rumor had been started, out of spite, by our former foe-turned-friend, Waratenne. He was unhappy that he had been replaced as a member of the District Agriculture Committee (farmers wanted such responsibility to rotate). The program had previously withstood a sex scandal and an accusation of spying, and this rumor of corruption likewise died quickly, without evident damage. Not only did farmers trust the IOs but they valued the program too much to let it be tarnished by malicious tales. This test of the program's credibility encouraged us, though at the time it caused real consternation.

An Irrigation Department Turnaround

On the trip back from Ampare, we returned via Batticaloa to visit Manoharadas, now chief irrigation engineer there after leaving Gal Oya at the time of the 1981 communal riots. He told us, among other things, that the director of irrigation, though publicly skeptical, even critical, of farmer participation and interdisciplinary irrigation management, was now supporting these changes within the privacy of internal departmental discussions.[4]

This became evident when we returned to Colombo and had a meeting with the director, set up by Herb Blank, who had taken over as USAID's project manager for Gal Oya after Ken Lyvers left. The director announced that the Irrigation Department had prepared a proposal for reorganization that would interest us. It would establish an Irrigation Management Branch which included a Sociology Division responsible for farmer organization. Since the head of this branch would be chosen from among its heads of divisions, this made it possible, at least on paper, for a sociologist to be the additional director of irrigation. Most likely the additional director would be the head of the Engineering Division, but even to accept the possibility of a sociologist in such a high position of authority was quite a step for the Irrigation Department.

For us the most significant part of the proposal was a diagram showing a multitiered system of farmer organization as part of the new proposed management structure. There would be field channel representatives for each area of about fifty acres, much as we had established in Gal Oya. These would form committees at the D-channel level, with Sub-Project Committees corresponding to our Area Councils. A Project Committee would be chaired by an engineer but

---

4. This information was given in confidence at the time, but since the director died two years later, his private contribution to the process of bureaucratic reorientation should be noted. It reminds us that things are not always what they seem. Ponrajah's curmudgeonly opposition to participatory irrigation management may have reflected his internal conflicts between continuing in familiar technocratic ways approved by his peers and adopting more contemporary practices urged by "outsiders."

would have a farmer majority. The director told us that this plan had the support of all his deputy directors.[5]

Unfortunately, when he subsequently took the proposal to his small circle of *senior* deputy directors, they rejected it, apparently out of sheer conservatism. When Ponrajah presented it to the ministry, it was given not as an official proposal from his department, but as his personal suggestion. The secretary and minister, distrusting him because of his public belligerence toward change, decided to proceed with their own reorganization plan to improve water management. This aggrieved the director, who felt his proposal was never even properly acknowledged. Adding injury to insult, the first floor of the ID's headquarters in Colombo was taken over by the fledgling Irrigation Management Division (IMD), which operated formally under the ministry rather than under the department.

The question of a home for the IO cadre got hung up between the department and ministry as they jockeyed for preeminence in water management. (There was no question that the ID would retain responsibility for design and construction.) This cold war froze our program's progress. We could not get any decision that would forestall an exodus of IOs, and *two-thirds of the cadre left the program*. Disappointed but without hesitating, ARTI started planning right away to train a new batch of thirty in September.

To complicate matters, ARTI's board of directors would not agree to what we thought was a reasonable compromise with Lucky, as devoted and innovative a member of the group as Wijay had been, to whom UNICEF in Colombo made a handsome offer. When the board declined to make any accommodation, Lucky left to take up the very attractive UNICEF position.

Communal Violence and Continuing Work

In my June 1983 trip report, in discussing problems and dangers facing the program I wrote: "I was struck this time by how often people

---

5. I learned later that he had pressed a few reluctant deputy directors to concur. The proposal was drafted by a committee of DDs that included our engineer friends Senthi and Kumarasamy. That these two would approve such changes was not surprising, but another member of the committee was Senthi's predecessor as DDI in Ampare, who had been an early antagonist of our program (chap. 2). Foes may not be intractable or unalterable, we learned from this.

expressed concern with rising ethnic tension. The increasing terrorism by separatists, and the many calls for retaliation and suppression could boil over into widespread violence that would surely shake Gal Oya" (p. 80). Four weeks after I wrote this, what was feared happened. The whole world read and heard about the crazy and cruel violence that knifed through Sri Lankan society at the end of July 1983. Fortunately there were no reports of violence in the Gal Oya area. A personal consolation was that none of the Tamils working with our Water Mangement Group was injured, though all suffered property damage.[6]

Amazingly, work resumed fairly quickly. My Cornell colleague Randy Barker arrived on August 8 to work with ARTI for several weeks, as previously planned. He made a field trip to Ampare with Jeff on August 12. Sena informed me by phone on August 16 that plans were going ahead to train two new batches of IOs in September, one Tamil and one Sinhalese, though not together as we had planned. The Tamil batch would be smaller than decided earlier and made up of persons only from Ampare and Batticaloa districts to avoid suspicion of "outsiders."

Newspapers carried almost no accounts on Sri Lanka after the sensational and tragic events of midsummer. Finally I got a letter from Jeff in October reporting that the new IOs seemed to be working well. "Bhawani and Nalini are both happy about the Tamil group. They appear to be enthusiastic, and all are surprised to find how enthusiastic the farmers are." There were enough apprehensions at official levels about appointing young Tamils as "organizers" on behalf of the government that ARTI finally designated them "field investigators" with the nominal task of gathering data on farmer organizations. Since there were no such organizations, however, the new recuits would have to set these up to study them. This was a fiction acceptable to the department and ministry as long as ARTI took responsibility.

Raj wrote about the same time, saying that things had "turned for the better" since his previous letter: "The new batch of [Sinhalese] IOs are doing excellently. According to Mr. Munasinghe, they are doing much better than expected. There was a small delay in their first pay because the appointment letter from the Ministry didn't come in time.

---

6. Bhawani hid with her aged parents in an empty water tank while their home below them was burned to the ground. The new home of the ID's deputy director for water management, Kumarasamy, was also burned down. He and his family hid in the jungle for two days before they could reach the safety of a USAID adviser's home. Nalini's and Shyamala's houses in Colombo were looted. ARTI staff were pressed into emergency service administering one of the refugee camps in Colombo.

Because of this, most of the IOs had a lot of hardships, but in spite of all these problems they didn't neglect their duty." This problem of delayed payment caused by administrative complications or neglect had become practically a ritual of initiation for all IOs. When the salaries were finally received, however, they were almost double the previous level, possibly reflecting the higher value placed on the IOs work.

While the program was regaining momentum organizationally, the agricultural situation was deteriorating. Jeff called at the beginning of December to say the water position had become dreadful. This was going to be the worst maha season in forty years, with the reservoir containing only 51,000 feet in storage at the beginning of the season instead of 100,000. My note to the Sri Lanka group at Cornell, passing on Jeff's report, observed that our activities in Gal Oya had, paradoxically, fared well under conditions of water shortage, "but this may be a situation with water too scarce for cooperative solutions to work." This caused apprehension for my next visit to the field, since we had not only the consequences of man against man to contend with but also the adversity of nature.

6

The Start of a Tamil Program and Sinhalese Farmer Initiatives (1984)

This visit in January 1984 makes the sixth time I have journeyed to Gal Oya with colleagues to observe the progress of the institutional organizer program. With me are Sena and Nalini, our training consultants; Bhawani, who is supervising the new Tamil IO program; Nandaratna, our new research assistant hired to help with translation and documentation; and Hammond, Ph.D. recently completed, come to observe and assist with technical aspects of the program.

It is 6:00 on a Friday afternoon when we drive into the yard in front of Munasinghe's house. He is there with three IOs from the Gonagolla area. Soon D. S. Ranasinghe, now the IO program coordinator, arrives on his motorcycle. We have mixed feelings about this means of transport, but the program now covers 15,000 acres, making bicycles impractical. D. S. was not one of the most energetic IOs at first, but after the exodus of two-thirds of the IO cadre in July, he was given the coordinator's responsibilities and has assumed them very effectively. It is a pleasure to see him performing at a level we had not expected.

I go over with D. S. and Munasinghe which of the "old" IOs are still in the program. Only twelve remain, but Sena has told me the new IOs are at least as good as the earlier ones. This I want to assess for myself. We chat a bit in the yard and then drive on to the circuit bungalow near Ampare, where we spend the evening.

Bakiella: Entering Tamil Territory

The next morning, Hammond stays behind to talk with ID officials in Ampare while we drive to a small community in the Tamil-speaking

143

lower reaches of the Left Bank system. We have a number of young Tamil men working in this area, appointed officially as "field investigators" rather than as IOs. This will be my first meeting with them.

We pull into Bakiella a little late because our vehicle got a flat tire en route. Three of the investigators are waiting at a local tea stall to meet us. They lead us to a sitting area covered by corrugated tin sheets. It is clean swept and has sixteen chairs, for the eleven of them and the five of us. Bhawani starts the discussion in Tamil.

The group coordinator, who has worked previously for ARTI as a field investigator, begins in fairly good English. He says almost all of the farmers in the area where he is working are resident; we had anticipated a large proportion of absentees, which would make organizing more difficult. "Farmers are welcoming the IO program. At first they did not know what it was, but day by day they are coming to know it." The next IO, who is working toward the tail of Gonagolla branch canal, says farmers there have very little water and are enthusiastic about anything that might change this. Another IO describes the situation on Vellaveli channels V12, V18, V22, and V23. Water seldom flows below V8, so they are eager for improvement of the system's performance.

The next organizer says proudly, evoking a few snickers from the others, that he earned his degree at the university in philosophy. He and two others are working along Silikody branch canal, including two channels that have both Sinhalese and Tamil farmers. Still other Tamil IOs are working farther down Silikody and Gonagolla. I ask whether farmers they have spoken with know about the farmer organizations started at the head of Gonagolla in 1981. "They have not heard," is the response, reflecting the lack of communication between the two ethnic groups.

Sena asks about the farmers' reaction to the IOs when they first went around. At first farmers thought we had come to take away their food stamps, says one IO, but now they often give us nice meals. Bhawani whispers to me that this is a welcome change; when they started work here, several wrote that even the householders from whom they rented rooms were feeding them only grated coconut and cooked pumpkin, no meat or fish. That they are now given more appetizing food is a sign of acceptance.

Nalini asks about farmer participation in the redesign meetings. One IO describes how they contact the engineer to get a date and then communicate this information to farmers. They ask each farmer they talk to individually to sign a sheet saying he has been notified. We debate whether this procedure is a good idea, since it may seem coercive. The IOs say it was done to avoid complaints by anyone who might later claim he was not notified of the meeting. We finally agree it is acceptable if the IOs explain that the purpose is to make sure everyone is notified. In the house we hear a chicken squawking wildly. It suddenly flies through the window behind us, flapping its wings in our faces as it makes its escape.

We talk about the meetings themselves. The ID engineers do not always come as promised. Reasons are given later (illness, lack of transport, etc.), but when no word is sent at the time, the IOs are put in an embarrassing position. At one meeting an engineer new in his job and apparently feeling insecure did not want to get out of his jeep when he arrived. Farmers gathered around, and the discussion proceeded with him sitting in his vehicle. When they needed to inspect problems along the channel, the engineer stayed in his jeep and drove along it with farmers trailing behind. Finally he had to get out, and the meeting became more satisfactory. Now, the IOs say, he is more willing to meet farmers face to face.

Nalini asks for any instances where farmers' suggestions helped the engineers do their jobs better, and several are reported. On M18.11, there were fifty acres that did not get water. Farmers proposed that the tail-end area be served directly by a new channel. Right-of-way was granted by head-end farmers at the redesign meeting, and when the engineer checked the levels he found the new channel connection was feasible.

What percentage of farmers come to redesign meetings? 90 to 95 percent. Who does not come? Part-time farmers and persons cultivating subdivided plots. Nalini asks, How do group versus individual interests get resolved when in conflict? Tail-enders tend to stick together as a group, says one IO, whereas head-enders tend to be individualistic, so the engineer tends to decide in favor of the former.

I spend a little time going over the some of the history of the program. We invite as many of the workers as can come tomorrow morning to join us in Ampare for a meeting of all IOs. Tomorrow is a special Tamil holiday, Thai Pongal, and it is also a Sunday. Moreover, they are not even officially IOs, so I stress that they are not required to come but are welcome to join us for discussion and lunch. Some say they will. After taking pictures of the group, we drive back to Uhana for lunch with Munasinghe and Hammond at the home of a farmer's wife who does custom cooking. With her five children and a cat and dog hanging around, things are a little chaotic, but the rice and curry are excellent.

DAC Farmer-Representatives Take Initiative

We have been told that the four farmer-representatives chosen by their fellow FRs to attend District Agriculture Committee meetings are getting together this afternoon at the Uhana school to propose agenda items for the next DAC meeting. The Weeragoda representative, Ratnayake, who is also a "native doctor" (herbalist), is already there. He shows us a letter he got this morning saying that the January DAC meeting, scheduled for the nineteenth, is being postponed until February 6. In a way this represents progress; last June farmers-members were not being informed about meeting dates, let alone changes.

He pulls out a list of farmer proposals for the DAC agenda and begins to go over them with Sena and me: "1. Farmers are willing to pay back overdue loans, so as to become creditworthy again, if the government will agree to reduce the interest rate on overdue loans from 12 percent to 3 percent and to accept payment in ten installments, over five years, instead of requiring that it be paid in one lump sum." This is unusual—farmers initiating a plan for repayment of overdue loans. But now that their water supply is more assured, the value of improved agricultural inputs is increased and they have some incentive to regain access to credit from government banks—private loans bear interest as high as 25 percent per month.

"2. The first water issue for the Left Bank, announced last October to start January 15, should be postponed because there has been enough rain this month and the fields do not need water now." The water saved in this maha season can be used in the coming yala season, Ratnayake explains. How often have farmers asked *not* to have issued to them water they are entitled to? The monsoon rains, devastatingly overdue in early December, have now started and have even fallen in excessive amounts.

"3. The water tax the government has proposed to start in April should be reduced, so as not to make it a burden on the farmers." In Colombo I heard this referred to as a service charge, but here farmers already refer to it in Sinhala as a tax (*badu*).

"4. The government's decision to issue ownership certificates to farmers in Gal Oya as in other major settlement schemes should be expedited." Settlers were never given title to the land they were allotted. It was thought this would prevent mortgaging, sale, or subdivision, but it did not. So the new policy is to recognize farmers as owners rather than as allottees. They know about this plan and are eager for it to be implemented.

"5. Promised action to prevent herders from the coastal areas from bringing their cattle to the Left Bank area should be taken as soon as possible. Units 17, 18, and 26 are badly affected by this influx." Farmers have been criticized by ID officials for grazing their cattle on channel bunds, since this destroys the banks and causes erosion. But farmers insist they have no alternative, since outsiders have taken over all available pasture area.

The list is signed. "Weeragoda Farmers Organization." There was no such organization last June. Ratnayake says they have regular meetings now of all the farmer-reps in the area. Representatives from the other areas have come while we talk, two from Uhana and two "alternates" in place of Narangoda from Gonagolla. I have met all of them before, and we converse a bit as we pull school chairs into a circle for the meeting.

Who will act as chairman? The farmers nod back and forth, and Ratnayake from Weeragoda moves into the chair behind the school desk, which signifies such responsibility. It was the Weeragoda farmers who suggested these get-togethers in advance of the DAC meetings. Ratnayake explains the postponement of the next DAC but suggests they go ahead anyway. Sena and I are impressed with the seriousness and confidence with which the representatives get up in turn to read their proposals.

Piyadasa reads the Gonagolla suggestions first: "1. A number of the newly built structures are deteriorating fast, and some action should be taken by the ID quickly to rescue them or soon they will be destroyed." He is probably referring to the causeways along main canals that washed out in the heavy rains in late December. This has been a real embarrassment to the ID, since about a dozen large and very visible structures collapsed. Whether it was due to poor design or faulty construction is in dispute. Quite possibly there was some corner cutting by contractors. Rebuilding awaits some resolution of this question. Farmers, having had engineers' expertise held up to them for so many years, may be pointing to this failure to taunt the ID, or they may be expressing genuine concern over a large and unfortunate waste of resources.

"2. Paid rehabilitation work on channels should be entrusted to farmer organizations." This has happened in a few places in Uhana, where field channel groups got themselves registered with the ID. It is something the ID leadership in Ampare has agreed to in principle, but staff at lower echelons impede it because they can profit by giving or getting rehabilitation contracts privately.

"3. The scheduled water release for January 15 should be canceled." (This was on the Weeragoda list too.)

"4. What happened to earlier resolutions of the DAC about farmer problems? Can we have a report on the present status of previous resolutions?" (This needs no elaboration.)

The chairman Ratnayake next stands and reads the Weeragoda list, which he has already gone over with us. There is discussion particularly of the water tax. Then a rather young fellow, perhaps in his early twenties, reads the Uhana list: "1. Steps should be taken to make sure that the Paddy Marketing Board can buy all the new crop." PMB facilities for buying and storing paddy are full. Farmers are afraid that when the maha harvest comes in they will not be able to sell their produce for the guaranteed price of 3.08 rupees per kilo and will have to turn to private traders, who pay a little as 2 rupees. Farmers then will have too little income from their crop to pay back the loans they have taken for seeds, fertilizer, and chemicals, due at the end of the season. This will make them "uncreditworthy" again. They want to avoid having to make distress sales to private buyers, who they suspect are in collusion with PMB staff to reduce that agency's purchases and thereby force farmers to sell privately.

"2. A lot of the paddy crop has been destroyed by the floods in December, and farmers are requesting compensation for their losses." Sena asks whether they had crop insurance, since Sri Lanka is one of the few developing countries to have such a program. They describe a bureaucratic "Catch-22" of conflicting deadlines for paying premiums, making claims, and so forth. Farmer-representatives say that at the last DAC meeting, before the floods, they asked to have the insurance extended by one month because of late planting, but nothing was done. Now they stand to get no compensation for their crop losses even though they have paid the premiums for many years.

"3. The DAC should approve giving farmers back our guns." What for? we ask. One FR explains that some time ago, the government required them to turn in their guns. But now so much damage is being done to the crop by wild boars that they need their guns back.

Uhana's list impresses me less than the other two, perhaps because it is less directly concerned with water management issues. But then, water problems are less severe at the head of the system.

I ask whether there is some inconsistency between their requesting not to pay the new "water tax," on the one hand, and asking for many services, on the other. Where will the money come from if farmers do not contribute to the government? Ratnayake quickly responds that they are objecting not to the tax as such but rather to the amount: two-hundred rupees per acre is too much. Also, they want to be sure that the money collected is kept and used in the district, so they are not paying for improvements elsewhere where people do not contribute anything. (They know about "free riding" even if they don't use this terminology.) Also, they want to know whether the tax will be reduced if a farmer is not given adequate water. Their acceptance of the tax at least in principle is something I had not expected so readily.

We do not want to interfere with their agenda discussions, and we have to attend another meeting in Weeragoda anyway, so we take our leave. As we drive along the Uhana branch canal, we see several of the collapsed causeways that the farmers complained about. They are not a pretty sight. I hope faulty construction is more to blame than the design.

Weeragoda: A Rival Farmer Organization

We are to meet nine of the new IOs who have been working in this area since October. Their coordinator resigned last week to take a teaching job, and Munasinghe has told us the group plans to elect a new coordinator. Sena and I discuss whether this is a good idea. We accept it in principle, since it represents the kind of self-management we have been encouraging, but if the IO cadre is to become a regular part of the government service, it may not be possible to sustain such a practice.

When we arrive at the house where the IOs are waiting for us, we learn the deed is done. Gamage has just been elected. We congratulate him, and the meeting begins. We review everyone's assignments. One of the women is working with over two hundred farmers, three times the number we started with in 1981.

What are the problems they face? They talk first of the progress so far. There are eighty farmer organizations at field channel level, though no D-channel organizations exist yet. They say farmers have asked for these groups. An informal D-channel organization now exists for M5, the most difficult one. Quickly other organizations are named: M1, M6, M8, M12, M16.

Field channel organizations are asking for legal recognition, says one IO, because they have started collecting their own funds. One group has agreed that its members will put up 100 rupees per acre for irrigation and agricultural improvement. But because the group is not officially recognized, it cannot open a bank account to deposit the funds. This is a welcome "problem."

I ask about shramadana work, and the estimate is that only between 70 and 80 percent of the channels were cleaned this year. Until torrential rains fell late in December, there was a drought and farmers did not expect to get any irrigation water this season. Now they are catching up on maintenance work. One IO says that according to the plan they drew up for the last quarter of 1983, there were to be eighty-two shramadanas, and most of these have been done. This is a surprise. Sena and I ask to know more about this plan. It included targets for forming organizations, numbers of farmers for IOs to meet individually, numbers of training sessions, and numbers of meetings.

How was this planning done? Sena asks. Gamage explains to us that when the IO program was in the doldrums last fall and needed revitalization, ARTI staff got the idea of a planning exercise. It does seem to have helped pull the program together, but Sena and I have misgivings about the "target" orientation thereby created.

Gamage goes over problems and constraints that he sees. Some organizations are trying to mobilize funds by taking contracts to earn money by their labor. Some have a regular savings program, and some get the payment that farmers are no longer making to their yaya palakas, a source of potential conflict. Since the organizations cannot open bank accounts, the elected treasurer keeps the money himself— not a satisfactory arrangement if large sums are raised, as the farmers intend to do.

Savings funds are expected to be able to extend emergency loans to farmer members. They will also permit bulk purchases of agrochemicals, reducing farmers' costs. The fund can also pay incidental costs like FRs' travel expenses, postage and stationery, and field trips. A meeting is scheduled for January 25 to discuss rules and regulations for more formal farmer organization. Farmers have many ideas for diversified activities.

Now something very interesting comes up. There is a rival farmer organization in the Weeragoda area. A teacher who seems transparently politically motivated has started it, with the help of a former IO (from the second batch) who has been appointed as a teacher in the same school. Imitation is the sincerest form of flattery, but we wonder what to make of this.

Gamage tells us more about the organization. When it started, it criticized our farmer organizations "for shielding ID people" and for not being critical enough of government officials. But now it hopes to have government assistance and takes a different line, saying that it wants to work with government officials; that it seeks a pension scheme for farmers; and that there should be a "farmer" elected from this area to Parliament.

Several of our farmer-reps went to an initial organizing meeting and were even elected as officers. This sounds worrisome, but the IOs are sure this organization will go nowhere. The FRs who attended its meetings have come to the IOs and been very critical of the new organization. They say they will not attend any more meetings because they see through its political motivation.

One IO tells how he recently planned a meeting with farmers at a local school at 4:00 P.M. When he arrived, he found a meeting of the rival organization in progress. He did not go in, but the farmer-rep walked out of the meeting, and several other farmers did too. They went to a nearby temple to hold their own meeting. The leader of the new organization, however, publicly complained the next day that his meeting "had been broken up by a thug." IOs have a good laugh at this.

One of the women IOs says our farmer organizations have been criticized by some opposition party members for being involved with outsiders. Someone said that our program is just intended to help foreigners write dissertations and will end when the research is finished. (Such knowledge of the outside world, even if quite incorrect, is remarkable.)

Hammond expresses his concern that the farmer organizations may be getting too diffuse, beyond their principal task of water management, and that this may make them inviting political targets. Sarath and Gamage respond that activities with the ID are still the main focus. The monthly meetings include ID officials, and the TA for this area has agreed to meet with farmer-reps on the second Wednesday of every month to go over water management problems. At this meeting they go over every channel and prepare a report that goes to the irrigation engineer for Gal Oya. The date was decided with the TA so that he can attend all meetings.

Hammond says he is pleased to hear this, since it means the existing farmer organizations should have little to fear from the new one. One IO reports that the farmer-representative who was elected the new organization's president for unit 26 has already resigned, saying it is useless. The afternoon is practically over, and the IOs will have to travel home in the dark if we don't end soon. We say good-bye, and they head off in different directions on their bicycles while we drive back to Ampare, a long trip.

Meeting All IOs: "Tour d'Ampare" and a Phantom Cashew Plantation

Next morning after breakfast, the ID jeep that is supposed to pick us up is late, so we set off on foot from the circuit bungalow toward Ampare, where we will meet all the IOs at the central school. As we walk along, Sena tells us how, three months ago, a first-year student at the Technical Institute we are passing was killed by a wild elephant while walking here at night. We are glad it is daylight. Halfway into town, the jeep meets us and takes us to the school.

About two-thirds of the IOs are already gathered, and more are walking in from the main road. One IO tells me they have organized a special event for next Wednesday, a holiday. With farmer-reps, they have planned a bicycle tour from Uhana to the main reservoir at Inginiyagala, a trip of twenty-five miles. They expect about 150 farmers from units 22, 23, 24, and 27 to participate. IOs want farmers to become more conscious that their irrigation system is served by a fixed amount of water, though with huge recent rains the reservoir is at one of its highest levels ever. They also want to build up camaraderie among farmers. (When I tell Mano, the irrigation engineer for Ampare, about this plan, he says he will join this "Tour d'Ampare" himself.) Bicycle touring is very popular in Sri Lanka.

By 8:55 enough IOs have arrived for D. S. to call the meeting to order. He is more at ease in his chairman's role than either of his predecessors were. After introductions all around, he gives a progress report, which begins with a discussion of problems and difficulties. He recounts things we are all too aware of, such as the decline of the IO cadre to just twenty-one last July. With the arrival of the new batch, they have been able to cover the phase 1 and 2 areas properly again and to extend the program to new phase 3 areas (see map).

In fact, now we are ready to take up the whole Left Bank area, he says. There has been enough publicity given to the program through the media and word of mouth that there can be an organized system to produce more progress in the future. They have discussed with farmers how the organizing effort should be completed in 1984, and they hope to hold a conference with farmer-reps from the entire area by the end of the year.

The main cause for this rapid progress, D. S. says, is the IOs' activeness. This sounds uncharacteristically boastful, but he explains that IOs are now covering areas three times larger than those previously handled. This represents an acceleration for which the hard-working IOs deserve much credit. He says that there has been some more attrition in the IO ranks since October, and we need to find a way to resolve the program's status—whether IOs have a "future" in it. He winds up a concise ten-minute presentation. If all this is correct, and I have no reason to think it is not, the program has actually progressed in the past six months despite some blows that could have been lethal.

Jayalath stands to give the report from Uhana, where five IOs are covering an area where twenty-two worked before. In addition, three new IOs are organizing LB14, 15, 16, 19, and 20, with another three covering LB22, a three-mile channel serving over one thousand acres. The new IOs are starting off with areas two to three times larger than in 1981. He says that relations between farmers and officials are good, even with the divisional officer of Agrarian Services at Gonagolla, who previously put obstacles in the program's way. When I ask how this improvement came about, Jayalath says he organized a *pirith* (a Buddhist propitiation ceremony) at the DO's Agrarian Service Centre to which farmers and IOs were invited. This seems to have created amity with the DO and was a good initiative by Jayalath, who has not previously impressed us as being particularly acute.

The farmer organization on UB11 has done rehabilitation work under contract with the ID. The farmer-rep had to visit the ID and other government offices more than a dozen times to get the contract awarded and the work paid for, but this group's success is an example for other groups and has showed them they should not get discouraged, says Jayalath. I am impressed by their perseverance. Jayalath tells about the planned bicycle tour. It will test IOs' and farmers' organizing capacity and build up cordial relations, he says.

Jinadasa now stands to give the report for Gonagolla, where only two of the first fourteen IOs were left after July. This attrition did not give the farmers a very good impression, he says, but the new IOs were able to persuade farmers of their sincerity. "Now we have changed the unsatisfactory situation."

The District Agriculture Committee is giving more importance to its farmer members, Jina says, which was not the case previously. He reports on a recent DAC meeting where a big controversy arose. Punchibanda, the farmer-rep from Gonagolla, asked where farmers could pasture their cattle so as not to have to tether them on the canal banks, causing damage. The Government Agent asked whether the cattle could be taken to a jungle area near Navakiri reservoir. At this the district manager for the Agricultural Development Authority (ADA) intervened to say that this area would not be suitable for pasture because it was already being developed as a cashew plantation.

Punchibanda objected, "There is no cashew plantation there." This angered the ADA official, who insisted he had reports that showed the plantation would soon begin production. The farmer-rep, knowing the area better than the official, stood his ground, repeating that there was no cashew plantation in that place. Now an investigation is being made, which has embarrassed the official. Last week he told a meeting of cultivation officers that "the farmer organizations are going around criticizing government officers." It is too bad we have antagonized this man, but nothing can be done about it now.

Jina goes over the assignments of IOs and says all are making good progress. Since they are getting into areas where farmers speak Tamil, IOs have to work through farmers who speak both Sinhala and Tamil. (Sena explains to me that the Tamil IOs are working separately from their Sinhalese counterparts because the two batches are administered separately, which is a pity.)

Jina says that organizing farmers is much easier in the Tamil areas. He says, "All an IO has to do is give a call and all the farmers come." (The IOs laugh at the Sinhala expression he uses: literally, "If the IO just goes *huuuu*, all the farmers come.") He concludes: "The main problem is continuing loss of IOs. Four out of sixteen IOs have left Gonagolla for permanent jobs since October. The remaining IOs have expanded the area they must cover to fill in the gaps, but this makes the work more difficult, and two more may go soon. There are limits on what the remaining IOs can do." Jina sits, and D. S. thanks him for his report. The coordinator for the Tamil group arrives and is introduced. He sits next to me at the front of the schoolroom and whispers to me that he left Bakiella at 7:00 this morning but it took almost three hours to get here by bus.

Gamage is now introduced as the new area coordinator for Weeragoda, and he reports on progress there, essentially saying what we were told yesterday afternoon. He adds that farmer training activities are proceeding well, with the assistance of Ratnasiri, the training coordinator. Three classes have been conducted, and more are planned. Three more of the Tamil "IOs" arrive. Chairs are rearranged, and several IOs crowd together to make room for them to sit down. The Tamil visitors are introduced, and there is some awkwardness about whether they should also give a report, since they are not officially IOs. Sena makes a short statement in Sinhala on work in the Tamil area, and we move on.

Next Ratnasiri gives a report as IO training coordinator. He starts going into too much detail, and Sena asks him to be more general. Some impressive numbers are given of courses held and farmers trained, but we are wary of purely quantitative measures. What is the program doing to evaluate its training efforts? I ask. How do you know whether you are meeting farmer needs? Ratnasiri has not been very organized in his presentation up to this point, but he responds precisely and coherently. "First, we ask farmer participants after each program to get feedback directly. Second, we consult the farmer-representatives afterward, because they know the needs of farmers. Then, consulting the IOs who attended the sessions is a third way. So many times IOs have pointed out our weaknesses in the training program." I am encouraged by this self-critical statement. D. S. thanks him for his report and announces a tea break.

To satisfy myself whether there really is a systematic effort to evaluate training, I go over to Ratnasiri and congratulate him on the report, asking what changes have been made in the training program on the basis of these evaluations. He says they now include some discussion of land problems—for example, how farmers can solve land disputes among themselves to save time and court costs. This seems a useful addition.

What has the debriefing of farmer-representatives taught them about deficiencies in the training program? He says they learned that the explanation of water measurement given by the TA in their last course was not adequate. He pulls out a notebook to see what other notes he has taken from the debriefings. Though I can't read his scrawl in Sinhala, he is being systematic in his work. I thank him for showing me the book, and he says in broken English, "Our training program is very dynamic," by which I think he means it is changing all the time, a good way to conceive of it.

Because we have not arranged to have tea brought in, the IOs must walk to tea stalls a few blocks from the school. The break lasts half an hour. At 11:15 D. S. says, "Let's begin." For the next hour, we discuss progress and problems with the physical rehabilitation. Some of the work has been of inferior quality. Some things that the ID said would be done were not. Also, few contracts have been given to farmer organizations to do some of the work themselves as previously promised. Hammond asks: "What is being done by the IOs to help bring the ID and farmers together? It is easy to blame the ID for many things. This is the first time such a complex rehabilitation effort is being undertaken in Sri Lanka, and there are bound to be mistakes. IOs need to work more closely with the engineers."

A number of IOs interrupt as they hear this being translated: "We are being misunderstood. There is plenty of contact." One says there are meetings every month between farmers and ID staff. "Unfortunately, the ID keeps making promises that are not kept, which makes our position difficult." Another IO says, "These monthly meetings may not be enough. Perhaps we should meet as a group to discuss what more can be done."

After some extended discussion led by Hammond of how to improve water use efficiency within D-channel areas, it is time to adjourn for lunch. P. B. Ratnayake gives the vote of thanks, summarizing the discussion. The vehicle ferries us in groups back to the circuit bungalow, where a huge lunch is ready. Mountains of rice, along with meat, fish, and vegetable curries, disappear in remarkably little time. Sena wonders what the IOs get to eat in the homes they stay in.

When the meal is finished we go out in front of the bungalow to take pictures. The IOs assemble in their respective groups and smile obligingly for the snaps. It is after 4:00, and the bus ride back to their communities will take the rest of the afternoon for most of them. Because there is no electricity in Ampare today, our van has not been able to refuel, so we cannot even give them rides back to the bus stop. They seem not to mind and set off on foot in a long straggle of congenial groups.

That evening Manoharan, the irrigation engineer for Ampare, joins us for dinner. We are glad for his company and for the opportunity to talk about how things are going in the system. But we know he has come to talk not with us so much as with Bhawani, to whom he has recently become engaged. Nalini and Sena helped arrange the match between the two families, who initially resisted the proposal. We are pleased for both of them, even though their marriage may mean that

ARTI will lose a very capable staff member if Bhawani has to resign because she cannot remain based in Colombo.

Progress in a Most Difficult Area: Revising a Judgment

Monday morning at 8:00 we go to the ID offices in Ampare and drop in at the IO unit office behind the main headquarters. This office has been expanded to twice the size it was last June, and the maps and charts on the walls are more numerous and informative. The area now being covered by IOs is shown on a large map, and there are some nicely illustrated charts for use in farmer training programs.

We talk with several engineers and with the head of the American consultant team assisting the ID. Warren Leatham has made a big difference in getting the rehabilitation work on track, and his experience in other countries has made him sympathetic to farmer participation. Senthinathan is out of the country for three weeks, so we cannot see him on this visit. At 9:00 we head for Gonagolla to drop in on a farmer training session previously scheduled in a primary school. We find that Ratnasiri has things well in hand, so we say we will come back to meet with the farmer-representatives later in the day, and we drive to a school beyond Gonagolla, where nine IOs are waiting for us. Sixteen were assigned to this area last October, but four have left already. Originally the IOs here operated in three groups, but with the attrition they have regrouped into two.

Both Sena and Mune have told me that Jinadasa is doing a fine job as area coordinator, directing a fresh batch of IOs working in difficult circumstances, keeping morale high, making adjustments in assignments, and so on. This is all the more appreciated because at the end of the first training program we did not rank Jina high enough to appoint him as an IO. He was brought into the program only because a vacancy occurred when Ranasinghe Perera was appointed a research and training officer at ARTI. Perhaps because he got a late start and suffered from the opposite of a halo effect, we never regarded him very highly.

How wrong we were. We can see now that what he may have lacked in flair or eloquence he has compensated for by steadiness, dedication, good nature, and a problem-solving determination. Because he graduated from the university too long ago to be considered for a teaching appointment (this is also the case for D. S., our program coordinator), this work has become his career, and he has poured himself into it commendably. It gives us encouragement for the whole program to see how Jina and D. S. have blossomed in leadership roles. I also reproach myself privately for underestimating what persons may be capable of accomplishing when given the opportunity.

The IOs go around and tell me which areas they are working in. I ask whether they are using any of the farmer-reps in the organizing effort. Yes, Jina says, half a dozen (whom he names) have been visiting farmers to "create enthusiasm" where groups are weak—for example, on some channels on G3. The FRs went house to house to revive the groups. This is appropriate because IOs had earlier invested considerable effort on G3. Especially since they are spread so thinly now, it is good that farmers will take some responsibility for remedial efforts.

Sarath Silva has been working in Rajagalatenna for less than four months. Because their first IOs had left, farmers were afraid the new IOs would also vanish, so he had to spend two months going house to house. Last month the farmers started meeting in groups at field channel level. Just yesterday, he says, four organizations were formed by farmers. Shramadana work has been done on LB35, 36, 37, 38, and 39, but not on LB34 so far. Fourteen of the expected twenty-three field channel groups are already formed.

Tilak reports on work farther down the Left Bank main canal and along Silikody branch canal. There is a complicated situation on LB40, where the four field channels upstream have Sinhalese farmers, and thirty-two Tamil farmers cultivate at the tail of the distributary channel. The four field channel groups have cooperated for the first time to rotate water among themselves and also to send water to the tail. This had never happened in twenty-three years, Tilak says. "The Tamil group is the best group we have." The prospects of getting water after so many years should be a strong incentive for group activity.

Farmers on LB41 have asked why the IOs are not yet coming to organize them. That distributary is silted up and needs much work. Tilak says he simply has not had time to do any work there. But there are plans for a massive shramadana on LB41 later this month, with farmers from LB40 and the Silikody channels coming to help. It was the farmers' own idea, he says. They have set a target of getting one-hundred farmers to participate in the channel repair work and will definitely get at least seventy-five. This is as pleasant a surprise as the ethnic cooperation on LB40. I have not heard of such mobilization of labor to help clear a D-channel that is not the farmers' own. Tilak explains that this area has in the past been a stronghold for the JVP, a "leftist" party. At first the idea of farmer organizations was criticized by some farmers as not radical enough, but the JVP leader has now become a farmer-rep, and he says this is "a good way of going forward, that it can solve major problems."

Next we review experience in the middle reaches of Gonagolla in some detail. Many of the problems and breakthroughs sound familiar. Sena congratulates the group on their fine work and expresses special appreciation to Jina and Batagoda, the coordinators. They and the other IOs respond that we should not forget the two IOs who worked in Tamil areas previously and have since left the program, or the three IOs who are not here today to report on their activity themselves. I am impressed that the IOs are so eager to give credit to others. This may reflect the kind of leadership Jina and Batagoda have given, forming a set of individuals into a mutually supportive team. Batagoda says: "This is not a group to take any step backward. Though we may leave tomorrow, while we are here, we will work with every bit of strength we have." He says this to praise the group, but it reminds us that we are likely to lose still more of the IOs from the program.

We go outside the school to take pictures. The IOs gladly pose together. Sena suggests I get a snap of the women riding their bikes, which have become something of a symbol of the IO program. Nanda and Soma laughingly wheel off and stay ahead of our vehicle all the way over the rough road to Gonagolla. They are wearing their saris, and I am a little apprehensive about this. Sure enough, one woman's sari snags in her bicycle chain as we reach the town, and she looks at us with chagrin as we drive past. She catches up with us by the time we drop Jina off near his house, and we wave good-bye to all the IOs as we head toward Uhana for lunch.

Founding Fathers on UB9

The farmer-representatives for UB9 field channels are meeting this afternoon to draw up a constitution for their D-channel organization. They have elected officers but are unsure about the next step. P. B. wants us to meet with them, but we are reluctant, thinking we might interfere in the farmers' self-management. Our curiosity wins out, since we want to see what rationale farmers offer for this next step in organizational development.

With a little difficulty we find the house where they are meeting. Waratenne, one of our original antagonists but now a supporter, is there as the new president, along with a young and dynamic secretary, Jayasundere, and six other representatives. We crowd into the sitting room on chairs. All sign an attendance book that the secretary passes around. There have been several meetings already. Hammond asks if they have regular meetings and is told they come together whenever there are problems.

Waratenne, dressed in white tunic and sarong, stands and assumes a rather formal posture to open the meeting, "organized by the representatives of UB9," he says. He calls on Gunaratne, a farmer-rep whom I have not met before but who quickly impresses us, to extend a welcome. He says that they had invited the IOs to come to this meeting but they are happy also to have us as visitors. It is not clear whether he is just being polite. The secretary stands and reads the minutes of the last meeting.[1]

The chairman asks if there are any corrections. He says that the nine-inch pipe serving field channel 9.1 is too big and should be reduced to help UB9 tail-enders. This should be corrected in the minutes. (I wonder how often farmers have asked to reduce their offtake.) Gunaratne rises and says that though some facts were not included, those that are in the minutes are important from the viewpoint of farmers, and so it is a correct report. With no other comments, the minutes are approved.

1. It may of interest to know what such minutes said: "The meeting was called to order on December 2 at 3:30. The agenda was approved, and the chairman welcomed members. The TA and IO participated. 9.1 representative had a complaint that three farmers on his channel need pipes but only one could be provided, according to the TA. In such situations, farmers should get help from their organizations [presumably by rotating or adjusting water deliveries].

"9.2 representative said that farmers can't be organized if given no facilities in organizing farmers [this is unclear], but everything is okay on 9.2 itself and he thanked the TA for coming.

"9.3 representative said they have the longest field channel and less water flow, which could be augmented with a causeway across the channel. TA said to present this request in writing. 9.4 is okay.

"On 9.5 it was reported that rehabilitation work has stopped halfway. The rest should be completed. TA agreed.

"9.7 reported that water is not reaching the tail. [Gunaratne interjects for our benefit that this tail-end always has water trouble. There is actually enough water because it spills over the bunds of the D-channel and causes damage.] The farmers decided to request the ID to build up the bunds.

"9.8 has culverts that are too small to drain the overflow, and on 9.9 and 9.10 other difficulties were reported. The TA promised to attend to these matters.

"The problems on 9.1, 9.8, and 9.10 were identified as the priorities to be dealt with. In proposing the vote of thanks, the secretary said that they were doing the best they could for the organization, but there are still some problems from the side of the officials, including the lack of a legal constitution. The meeting was adjourned."

Waratenne consults his notes and calls for the next item. He explains that the main purpose of this meeting is to prepare a constitution. Another farmer-rep arrives. P. B. leaves to get a chair and comes back without one, so they share his chair. This is very significant and indicates close relations between IOs and farmers, since sitting on the same level implies social equality. (Lower-caste persons are supposed to sit at a lower level than any seated high-caste person, or to stand.)

"At the last meeting, we discussed the aims of farmer organization at field channel level," Waratenne says. "Now we are considering organization at the distributary level." Rain starts and makes much noise on the tin roof. "The main thing keeping us from working effectively is not having a constitution. Will we get one from the government, or will we have to prepare one ourselves? This is very important work. Farmers are the most important people in any country. At present we are not able to do any work at the government level because farmers are not accepted by officials. According to my estimate, eighty percent of farmers in the Gal Oya scheme are still not getting a chance to put their problems before officers." The rain comes down harder, and the noise gets louder. He goes on: "The IOs are not going to stay for a long time. We must prepare a strategy with the help of all. Until recently we faced many problems, but we couldn't get assistance from anyone, certainly not from officials. The IOs are very good, but they are not getting enough response from officials either. For small matters we can't go to the IOs all the time anyway. We want to run our organization the right way, for which we need a constitution. If we prepare a constitution, where do we go to get it registered? The ID? The Government Agent? Once we have a recognized institution, with letterhead and rubber stamps, we can go to any office." Waratenne asks me whether the Tennessee Valley Authority in the United States, possibly a model for Gal Oya valley, has farmer organizations. He wants me to say something.

I am impressed that this Sri Lankan farmer with maybe six years of education knows about the TVA. I say that farmer organizations are different in the United States, not going into how much difference it makes when average farm size is several acres rather than several hundred (or thousand) acres. "You should not put your faith in letterheads and stamps, but rather you should express your unity of purpose and ideas in your own constitution." What are the problems that can and should be dealt with at the D-channel level? I ask.

The secretary picks up the discussion, saying that in the maha season there is sometimes too much water in the D-channel. "Recently the

channel was about to overflow and breach, but it was built up by the
farmers themselves to save it. When we told the TA about the problem,
he said he would talk to the engineers about it. When an engineer came
to the Area Council meeting with farmer-reps, he promised to build
the bunds higher, but this has not been done. We know he has a large
area to serve, but something needs to be done so no breaches occur in
the future." Hammond says that there should be no problem of too
much water in the D-channel during the coming dry season, but
Waratenne says that when farmers at the head close their pipes to send
water down to the tail, the level in the channel gets fairly high even in
yala. He adds proudly: "Farmers have asked *not* to have a water issue
this month. This year, farmers sowed their maha crop with the rains in
November, instead of waiting for a water issue from the reservoir for
plowing the land. Then when they needed a special issue from the reservoir in December because the rains had stopped, they got one from
the ID." Hammond says this is a great improvement over the way the
system was operated when he first came here four years ago. "The ID
is listening to farmers' suggestions now, and this is a benefit from having farmer organizations." The farmer-reps nod their heads in agreement and say there is more cooperation.

Sena says perhaps they can start their effort to draft a constitution
by discussing their goals, but Waratenne asks again, "If we draft a constitution, who will approve it?" Sena says, the Government Agent.
"Once it is approved, you can have a bank account, take contracts to
do rehabilitation and maintenance work, and so on." Gunaratne says
they were under the impression that they could get a draft constitution
from higher levels of government. This prompts Sena to point out how
most previous rural organizations have been institutions "from Colombo": "As such, they are controlled by government officers. If farmers want this to be their own organization, they should not look to the
government to get their constitution."

The first part of any constitution, I say, is its statement of purpose.
"What are your purposes?" Gunaratne rises and says: to ensure that
all farmers get water when they need it and in adequate amounts. The
secretary adds: to have a say in decision making, not only at this level
but at higher levels. Waratenne states solemnly: to safeguard and protect the canal structures. (Sena and I exchange glances; he was behind
some of the breaking of measuring structures last March. This may be
a hypocritical suggestion, but like most people, he may be inclined to
hold up ideal behavior as the public norm.)

Gunaratne suggests: to increase the income and standard of living of people through better water management and to help reduce the cost of maintenance of channels. The ID is bearing a lot of cost now, he says, and this could be reduced. I ask if this means taking care of the D-channel as well as field channels, and he says yes. This is important, because the ID has legal and financial responsibility now for D-channels. These farmers are willing to accept more responsibility than they have had before.

Waratenne speaks again: to deal with "crises." Sometimes there can be a flood or something, he says, and there is no time to negotiate with the ID. Then the farmer organization must be able to solve the problem as best it can quickly by itself, or to contact the engineer right away if unable to handle it.

One cannot know whether these farmers will always act according to such ideals, but they articulate these objectives spontaneously, with both clarity and earnestness. Sena says that by their initiative these farmers are setting an example for this new legal effort. He apologizes that we must be going and will leave them to their constitution making. Waratenne says they will draft a constitution and will invite all farmers to a general meeting to approve it. I suggest they remember that no constitution is perfect. "You should proceed with what you consider best for now. You can make changes later, provided you have included some procedure for making amendments."

They say they will do this, and we take our leave, shaking hands all around. As we walk to the vehicle, Sena is almost ecstatic about what he has heard. Before we came, he had expressed displeasure that Waratenne was president of this organization, not trusting the man based on our past experience. I admit that I shared his fears. But Waratenne's statements and leadership just now have been exemplary. More important, we have seen other leadership talent emerging here that we did not know about before.

Back at Gonagolla: With the Pioneers

It is just 5:00 P.M. when we arrive at the school where we visited the training program in the morning. Sixteen farmer-reps are seated in a circle, and Narangoda is sitting behind a table as chairman of the meeting. We are pleased to meet the first woman farmer-rep, Menike Rajapakse. She is fairly young and a recent graduate from a diploma course at the Technical Institute in Ampare. Narangoda makes a short introductory speech:

> The Sri Lanka economy, as we know, is based on agriculture and especially on farmers. We appreciate very much that you are taking steps to improve the condition of farmers. I want to explain the achievements of the farmer organizations and farmer-representatives from 1981 until now.
>
> I am really very sorry when I think that farmers did not have this system before. In Ampare district, farmers depend on irrigation works. In 1952 the Gal Oya scheme started fairly successfully, but over the years the system deteriorated owing to improper maintenance. The coming of the IO program was thus great good luck for the farmers because the system was so run down. As far as I know, there was no system to manage water at the field channel level before the IO program was introduced and no proper coordination among farmers.
>
> Since establishing farmer organizations, farmers are using the field channels properly and maintaining them willingly. Also, farmers themselves have been able to solve many problems based on water. The requirements of the district as a whole are decided on at the DAC meeting. Farmers are very happy to have members from among themselves elected to the DAC. Today we have the possibility of taking the problems of farmers to the DAC from the field channel level through our representatives.
>
> Farmers have started using credit productively, having worked out arrangements with the Agrarian Services Centre and the cooperative. We expect to build up better relations between farmers and officers. Many important services have become possible for us since launching farmer organizations. So we express our appreciation with full hearts.

This, it seems, is the "good news." He adds what might be considered the "bad news." "Some field officers of the government still refuse to attend Area Council meetings. Perhaps they stay away because they do not have a proper understanding of the farmer organization program. We request you to provide some training for these officers. They will give more services if they understand the program. In conclusion, we wish you long life and success." Narangoda sits down and invites me to speak. I say I am glad to be able to discuss their experience with them and am very pleased with their progress. Great strides have been made, and the next step is to form and maintain strong D-channel organizations. I discuss these groups and say they will be all the more important because of the new schedule planned for distributing water in the coming season. I turn to Hammond, who will describe the changes anticipated.

Hammond asks whether they have heard of the system of "continuous flow." They answer no. We exchange apprehensive glances. The new season is only 6 to 8 weeks away, and this should have been explained to farmers by now. Hammond describes a system that will give the whole area a constant but smaller supply of water rather than issuing water for five days and then stopping the issue for five days. He asks what they think about it. One farmer says: "We are generally pleased with it, but some things are not clear. How will it be possible to give to our fields all the water that is needed during land preparation at the same time if the volume of water is reduced? Otherwise we can't reduce staggering of planting as the government requests." This is a very good question. Hammond tries to explain how water could be distributed for land preparation with this new system.

It has gotten dark, and we need to conclude. For the vote of thanks, Narangoda calls on one of the younger farmer-reps, who stands up and proceeds with unusual self-confidence. He says he knows that the rehabilitation project has been expensive, and "on behalf of the Gal Oya farmers, I want to express thanks to the United States for contributing money for the rehabilitation of this system." He could hardly say this more graciously if giving a toast at a diplomatic reception. I wish the USAID mission director and United States ambassador could be here to hear this.

As I am thinking this, I hear the farmer-rep refer further, in Sinhala, to Cornell University. Sena translates with a smile, "He is thanking Cornell University for its initiative in this pioneering effort." Sena adds that the man used the word "pioneering" in a literal sense. We have thought of the program in such terms ourselves, but it is interesting that farmers, who have *been* the pioneers, see it this way too.

It is now quite dark. I have brought some copies of photos taken during my June visit. We have to move outside where there is some moonlight to be able to make out which photos go to whom. One farmer-rep approaches Hammond to ask if he could come and give a one-day training course on the planned water delivery schedule. Hammond replies that this is planned for all the farmers in the area.

Punchibanda, the DAC member, comes up to Sena and me to tell us about the meeting where the controversy over "the phantom cashew plantation" came up. He says he did not intend to talk against anybody or to make an issue of it, but he had to tell the truth. We question him to learn more about the matter and are relieved that the influential local political figure involved, who might have decided to make difficulties for the program, has instead kept a low profile. This rich person, who is a timber merchant as well as a money lender, got the land allocated to a women's society to curry favor before last year's election with the understanding that the society would start a cashew plantation on it. He then cut down all the timber and sold it privately, without a license, realizing an illegal profit for himself. Only a few coconut trees were planted there. The ADA district manager did not know this, however. Because "politics" is involved, nobody said anything.

As we close the subject, Punchibanda says he is not afraid of any officials and will not back down where the truth is concerned. It is good to see him prepared to stand his ground. We just hope he and the organizations won't get hurt in the process.

I ask the farmers about Sinhalese-Tamil relations in the area now that conflicts have broken out elsewhere. During previous visits, they insisted that Sinhalese and Tamil farmers could live and work together. Narangoda says their views have not changed. They still think cooperation is possible and desirable, and that a single farmer organization can encompass both groups. Narangoda adds that he regards the Tamil TA in their area "as a brother. I will protect him if anybody tries to harm him." Another farmer-rep says, "It is just a few troublemakers on both sides who are causing the problems, the 'Tiger' terrorists, on one side, and some Sinhalese hotheads on the other." I hope these farmers are right.

They ask if we can give Ratnayake a ride back to his home. He is less vigorous now than when we first met him two and a half years ago, when he asked the director of irrigation in 1982 to extend the IO program to other schemes because it had been so valuable to farmers already in Gal Oya. He climbs into the van with us, and we let him off in front of his house. Even in the half light we can see that his step is getting feeble and walking has become difficult. But he tells us he wanted very much to come to this meeting. He was one of the first and strongest supporters of farmer organizations here, really an inspiration. Time and again he has said that these organizations are the best hope for the Gal Oya farmers who come after him.

Returning to Colombo

We leave the circuit bungalow at 6:30 the next morning for a long day's drive back to Colombo. In Ampare we pass an ID bus that is taking twelve farmer-representatives to visit Kimbulwana Oya on a farmer exchange and training program. We pick up Munasinghe at his house so he can travel back with us. He reports that yesterday afternoon the District Minister, while dedicating two bridges at a ceremony Mune attended, praised the farmer organizations. The DM said that the government is spending money not only on physical structures to improve the channels but also on organizations that can change the attitudes of farmers.

This is a rather limited view of what the organizations should accomplish, but right now anything the DM says in support of the program is welcome, since it may come under political attack at any time. The DM also said that the distribution of water is being put in the hands of the people. This is a good way to describe the program. Mune says that when he was asked to speak at the ceremony, he emphasized protecting and maintaining the system now that it has been rehabilitated.

Munasinghe reports that at the Ministry of Education's big training program last summer for its huge new batch of teachers, "our IOs were in the vanguard among the recruits," leading discussions and making suggestions. As we talk on the return trip, we still see various problems ahead, but I am immensely gratified by what I have seen and heard on this visit. The new batch of IOs is living up to all expectations, and some of the "old" IOs have greatly exceeded them. My big concern now is whether we will be able to recruit, train, and field a full-fledged cadre of Tamil IOs.

The farmer-representatives continue to demonstrate a level of initiative and responsibility that, though still not perfect or universal, has begun to change the social and economic landscape of Gal Oya. More than after previous visits, it is clear we should call this a farmer organization program rather than an IO program.

It is marvelous to see how two entirely new roles—institutional organizer and farmer-representative—have become widely understood and accepted. Three years ago nobody in Gal Oya knew what an IO or an FR was. Indeed, we ourselves did not know the latter role. Today tens of thousands of people recognize, interact with, and support these roles because they have contributed to economic and social productivity in Gal Oya. Hundreds of people have moved into (and out of) these roles, performing excellently and fairly interchangeably, and yet each uniquely. Social scientists rarely get a chance to see roles develop and take root before their eyes like this.

The program has developed techniques, roles, and a philosophy that can be effective under field conditions such as Gal Oya. We are now in the stage of having to make the program more efficient (or cost-effective). At the same time we are being challenged to expand it into more schemes on a less intensive basis.[2]

This is like shifting gears in a car. We started in first gear and are now in second. The intended expansion, possibly using farmers or their educated sons and daughters as organizers, would represent third gear. This will present new problems, but we can be grateful that we continue to face new problems rather than being plagued by old familiar and intractable ones.

Like previous visits, this one has been instructive and encouraging. It remains to be seen how well ARTI, the ID, IOs, and farmers will handle the continuing and emerging challenges of the program, particularly as they try to *institutionalize* it without *bureaucratizing* it.

In concluding the trip, I mull over a thought that has become both a caution and a consolation. In development we know there are no permanent victories, but at the same time, we proceed on the assumption there need be no permanent defeats. Any advance, we now understand, can be undercut by budgetary, bureaucratic, political, ethnic, or other problems. Yet setbacks and obstructions need not be accepted. Adversity has been an unwitting ally in Gal Oya. It has brought forth effort, invention, and idealism that nobody, ourselves included, knew was here.

June 1984 Visit: Finding Progress with the Tamil Program but Other Difficulties

Over the next few months, with some effort ARTI got approval to recruit, train, and deploy thirty Tamil IOs in tail-end areas. There was some further attrition of the cadre, but apart from continued worsening in the conflict with separatist terrorists, no major problems were evident. When I was able to return to Gal Oya in June 1984, the most encouraging thing was the good start made in Tamil areas. This was not easy. As we traveled to Bakiella, Nalini and Ranasinghe Perera told

2. This assessment follows David Korten's "learning process" analysis (1980), which extrapolated these three stages from the experience of five reasonably successful rural development programs in Asia. A program cannot be efficient without first being effective, something often forgotten when planners and implementers seek least-cost solutions before they know what will work in the field.

me how three months earlier, while settling Tamil IOs into new locations, they had been stopped on this road by police with drawn guns.
The police offered apologies when Nalini and Rane explained they
were on government business, but after this incident Nalini went to all
the police stations in the area to inform them about the program.
Young Tamils moving from house to house could easily be suspected
of being "terrorists" trying to organize the population. This incident
made it clear how tense the situation had become.[3] Still, the four
groups of Tamil IOs I met in June, who were covering almost 10,000
acres, reported the same kinds of progress we had seen in Sinhalese
areas.

While we were staying at the circuit bungalow, it happened that the
Minister of National Security moved in for a day while inspecting a
nearby army camp. We tried not to get in his way, but a change in his
schedule brought him back to the bungalow while we were meeting
with sixty IOs, both Sinhalese and Tamil. (His security people were
not amused.) He took the initiative to chat with the group and to ask
Rane and me a number of questions about the program. A memorable
half-hour occurred that evening when four wild elephants came grazing just outside the bungalow compound fence, and the minister came
down to watch them with the rest of us. His driver brought a jeep
around so we could see them in its headlights. I had a chance to talk
with him some more about the program while we watched the huge
beasts amble about, disdaining our attention.

At a meeting with all the IOs at the end of my June 1984 visit, several coordinators were quite frank in saying that they thought the IO
program was "at a standstill" and "could break down," given that it
was still not permanent and considering the continuing prospect of
losing IOs to outside jobs. (We were fortunate the government was not
planning to make any more teaching appointments until the end of
1984.) The IOs, while acknowledging their own shortcomings,
pointed to many faults in ARTI's supervision and support of the program. Even if excuses or rebuttals could be made, their criticisms were
essentially valid.

It was gratifying that most of the speeches ended with pledges
of continuing commitment. Ratnasiri closed his remarks by saying:
"I can promise right here, on behalf of the group, that though the IO

---

3. A few months later there was a break-in and theft of weapons at the Central Camp
police station across from where six of our Tamil IOs were living. Their host, the colonization officer, was taken into police custody and interrogated "on suspicion" but was later
released. The IOs thought it best to move to new quarters.

program still has not been made permanent, we will keep doing our duties even as temporary employees. While there are problems from the IOs' side, we can and will correct our own shortcomings. The problems we are most worried about come from others [meaning ARTI]."

A meeting with Gonagolla farmers was the tensest I could recall. They were angry partly because flooding from excessive rains was ruining their crop. They had paid crop insurance premiums to the government but could not get indemnified because the inspection system was slow or rigged. Only four persons in the area who reported crop losses had gotten compensation, and they were all teachers or traders. (It was implied, though not said, that only these persons could afford the necessary bribes.) Narangoda said that farmer organizations were "collapsing," mostly because there were so few results from their meetings with officials. Even farmer-representatives participating in District Agriculture Committee meetings could not get government action.

After hearing many complaints from them, I asked whether they thought it would be better if the farmer organization program had not been started in Gonagolla. One old farmer said emphatically yes. But Narangoda disagreed: "The farmer organizations are fine. But if officials won't respond, they won't last. They will operate 100 percent or not at all." The old man allowed they were better than zero, but he added with a twinkle in his eye: "By the time they reach 100 percent, I will be dead and gone." After the meeting when we went outside to take pictures, Ratnayake, the most senior of the farmer-reps, came up and said in a confidential tone: "We didn't say anything today about the benefits from this program, but spoke only about the problems, because we want to see it strengthened. I hope you will understand this."

In Weeragoda farmers were more optimistic, and the organizations were engaged in a variety of efforts to promote savings, crop protection, fertilizer use, and so forth. A meeting we observed between ID engineers and sixty farmer-reps was quite constructive, with farmers being self-critical as well as critical of the program. A meeting with IO coordinators showed that they were taking their responsibilities seriously despite the uncertainty of their futures.

After the field trip, we met with the director of the Irrigation Management Division in Colombo, K. D. P. Perera, who had been helpful in resolving an impasse with the deputy director of irrigation in Ampare three years before. He was now drafting a cabinet paper to approve appointing twenty-seven IOs in his new IMD to oversee the introduction of farmer organizations under a proposed World Bank rehabilitation project in six more irrigation systems in the north. We persuaded him that a core cadre should be included for Gal Oya.

Though the cabinet paper did not get approved at that time, when I met with the Secretary of Lands in July he expressed appreciation for the IOs' work. USAID was also supportive and included appointment of a core cadre of IOs as one of its conditions for additional funding of the Gal Oya project and in its plan for a follow-up project improving irrigation systems management in Polonnaruwa district.

Unfortunately Jeff Brewer's tour finished in July, so Cornell was not represented at ARTI at a time when it started losing staff and coherence for work on water management. Vida left without any notice to become a United Nations volunteer in the Sudan, and Rane was going to be leaving for graduate study in the United States soon. We lost Raj to a scholarship for study in Thailand.

Mahinda, an economist who accompanied me on the June trip, was put in charge of ARTI's Gal Oya work. He has limited interest in the IO program, so operational responsibility for it fell to Nanda. With only the status of a research assistant, he was expected to keep the program going, with Sena's and Nalini's support and with some help from Mahinda and Bhawani. The climate within ARTI was becoming less and less congenial, perhaps reflecting the "atmospherics" of the country at large.[4]

Farmer Convention Success and Setbacks

While ARTI's capacity and the IOs' continuity were tenuous, the farmer organizations moved ahead on their own. In October 1984 they held a huge convention in Ampare to which they invited the Minister of Lands, the Minister of Agriculture, and the District Minister as guests of honor. They organized this meeting themselves, with some IO advice, through a planning committee that raised 13,000 rupees from farmers to cover expenses. To keep the event

4. A senior member of the ARTI staff wrote to me in October 1984: "Life at the ARTI is busy and frustrating (as usual). Its social environment has further decayed and it provides an ideal example of Hobbesian anarchist community in microcosm—all are at war with each other—sometimes for personal gains and quite often, simply out of jealousy and suspicion. Anyway, very little work has been completed by the ARTI staff as no one is accountable for anything." This may have been too gloomy an account, but it reflected how the situation appeared to some.

nonpartisan, they decorated the stage and approaches with white bunting, avoiding the colors of any party.

Reports on how many farmers attended ranged from 2,000 to 5,000, with most guesses about 3,000. Because the school auditorium they rented could hold only 900 people and most who came milled around outside, listening over loudspeakers, there was much coming and going, and nobody thought to count the participants. Certainly the convention was a major event in Ampare's history. The "success" of this event caused some euphoria at ARTI and in the organizations, though when I visited three months later, grounds for disgruntlement had appeared.

About the same time as the farmer convention, word filtered back to ARTI that somebody, presumably a "terrorist," had advised Tamil farmers not to have anything to do with "government people," which included IOs. Whether the program could continue to work in tail-end areas became a moot question in November, when most Tamil-speaking IOs were given permanent teaching positions by the Ministry of Education.

Sena wrote on December 3 to say that ARTI had decided to suspend the farmer organization program in Tamil areas and not to recruit new Tamil IOs, due to "intense terrorist activity" in the lower reaches of the Gal Oya system. He also informed me that there had been another heavy dropout of Sinhalese IOs. Arrangements were being made, however, to recruit another batch of twenty-five by the end of the year. They would be trained in January, when I planned to visit Sri Lanka again. Sena felt obliged to tell me: "There is a lot of tension because of terrorist activity. Terrorists now attack farmers in settlement projects [in the north]. We have curfew off and on. We are really having a tough time." Still, with characteristic optimism, Sena concluded his letter: "With all the problems, we move forward undaunted. Do not worry about our Gal Oya program. We shall succeed. With kind personal regards and compliments of the season. Yours sincerely . . . "

7

Growing Tensions and Evident Progress (1985)

This January I am making my eighth field visit to Gal Oya to observe the IO program in operation. This has become practically a semi-annual pilgrimage to monitor and learn from the experiment that by now should be called the FO (farmer organization) program. I bring with me on this visit the usual apprehensions about the many ways the program's efforts may have gotten bogged down or been thwarted. With me are Sena and Bhawani; Nanda, the research assistant overseeing the program; and Tilakaratna, an economics lecturer at the university who will be doing some economic analysis for our program.

The bad news is that once again we have lost most of our organizers; all but sixteen have left for teaching jobs or other more permanent positions, reducing the cadre to one-quarter of its level on my June visit. Most of those departing, like IOs before them, I am told, said they would rather remain with the program if they had some prospect of making this work a career. But the government has not yet established permanent posts for IOs to continue spreading and supporting farmer organizations. The good news is the farmers' convention held in October, perhaps more properly referred to as a rally, organized to mark the third anniversary of "their" program. I am curious to find out more about the convention, since I have heard there is some dissatisfaction with the way it was conducted.

We have a problem of leadership for the program, since Ranasinghe Perera leaves soon for graduate study in the United States. Nanda is filling in for Rane, but he cannot exercise any authority. Sena, though only a consultant, has been given authority to sign vouchers, approve travel, and such, because ARTI is so shorthanded. Since Bhawani no longer has a Tamil IO program to supervise, she helps out with the

173

Sinhalese program, but it is difficult for her to travel and move about by herself, given the ethnic tensions. We are pleased that she has married Manoharan, the engineer in Ampare who was so helpful to us the past two years. But he has been transferred to Kirindi Oya, some fifty miles west of Gal Oya, as resident project manager, and he would understandably like her to join him there, so we may lose her soon.

Civil Defense Duties

We have been driving nine hours by the time we reach Gonagolla, still twelve miles from our destination, Ampare. We stop to see Jinadasa, the area coordinator, to start arranging our program. This trip was made without any advance notice, so we will have to improvise our schedule, as usual. Jina is not in the farmhouse where he has rented a room for several years now. His host says he is at the Village Council library half a mile away, reading the day's newspapers. There is not much else to do in the evenings in Gonagolla.

When we find him at the library, Jina says the farmer-reps in this area have a meeting planned for tomorrow morning. They will be glad to have us attend, he says. If we come to his house by 9:00, he can tell us the time and place. What is the current situation with organizations? Jina tells us that farmers are now taking turns providing patrols to help guard the main reservoir and its bund, sluice, and powerhouse against possible terrorist attack. There have been many incidents in recent months, including bombings of public property by the separatist guerrillas.

It is the Gonagolla farmers' turn tomorrow to send a truckload of volunteers for all-night patrol duty. The three Area Councils—Gonagolla, Uhana, and Weeragoda—take turns sending thirty volunteers each afternoon about 4:00. They take their supper with them and return about 6:00 the next morning. Transportation is provided by local merchants who lend their trucks. All logistical arrangements are handled by the farmers-reps, with some IO assistance. We did not anticipate civil defense duty as one of the functions of farmer organizations, but then many things have evolved beyond our original plan.

Closer to Ampare, we stop at Munasinghe's house to arrange our schedule. Since we have come with no advance notice, I say it would be unfair and disruptive to set up formal meetings. For the meeting tomorrow morning in Uhana, let the IOs just invite a few farmer-reps with whom we can talk informally to get ideas. But Mune says, "It will be better if the IOs contact all D-channel organizations and invite them each to send one representative." Most D-channels have organizations now. I appreciate that farmers should themselves decide who speaks for them and think to myself that this will be an unplanned test of the organizations: How well can they make decisions and communicate on short notice? I ask whether we can meet all the IOs on Wednesday at the circuit bungalow, as usual. Mune says this will be easy, "because we are now a very small group."

We head into Ampare in the darkness. Every couple of miles we have to stop for vehicle and personal identification. The barbed-wire barriers are "manned" by youngsters in their teens. The atmosphere is relaxed, but the concern with security is obvious. A few months ago, near where we met Tamil IOs in June, eight soldiers were killed when their jeep drove over a radio-controlled mine. That the terrorists consider themselves at war only with the military and the police does not give much comfort.

Postmortem on the Farmer Convention

After we pick up Jinadasa the next morning, we drive to the house of Punchibanda on G1 distributary. This is not the Punchibanda who represented farmers so boldly on the District Agriculture Committee. I am sorry to learn that he is in disgrace among his peers. When the Minister of Lands made an impromptu visit to Gal Oya in early October before the convention, to get a sense of the farmer organizations for himself, he met with a dozen FRs—under a tree, Sena says. "Tall Punchibanda," who had often been their spokesman, got carried away with his opportunity and ended his presentation by telling the minister about some personal problems, like having to raise money for his daughter's dowry. This greatly embarrassed the farmer-representatives. They have established a rule that their role should never be used to present purely personal interests. Afterward, Punchibanda was reprimanded by the FRs and was given no role in the farmer convention. As the farmers assemble, I notice that tall Punchibanda is not here. I hope he can be forgiven and taken back into the program, to which he has given so much leadership in the past.

"Short Punchibanda," our host, has also been active in the program from the start and has emerged as a current leader. He points out the room off his porch that served as a headquarters for the steering committee of farmer-reps who planned the convention. He shows us a copy of the souvenir program the farmers prepared in Sinhala and distributed at the convention. With us are also Narangoda and Ratnayake, two stalwarts of the program, and four younger FRs, including the woman FR I met last time.

Narangoda leads off the discussion, saying that the farmer convention was the main thing since my last visit. It was arranged by the farmers themselves but unfortunately did not come off as well as expected. There has been considerable disappointment among farmers, since they did not get the concession they asked for on payment of operation and maintenance fees.

Narangoda hands us a copy of the resolutions prepared by the farmers, and Nanda translates the last one for me. It says: "Since Gal Oya farmers are making, and will continue to make, contributions of labor to the operation and maintenance of the system—not just on field channels which are their responsibility but also on D-channels which are the ID's responsibility—we propose that the O&M charge should be reduced in Gal Oya Left Bank from 100 rupees to 75." One farmer-rep says, "We thought such a reduction would be fair, since farmers here are being asked to pay the same amount as farmers in other irrigation systems who make no contribution of labor to O&M." (The proposed reduction amounts to one dollar an acre, about half a day's wages.) "We are ready to pay some amount. But farmers are being forced to pay the full amount under threat of legal action by the government, even if the ID has not yet made any improvements in O&M here." He reminds us that the rehabilitation program has not been completed in Gonagolla. "Farmers are discouraged and losing faith in the organizations."

Ratnayake, the oldest and most supportive of the farmer-reps, doesn't like the negative tone of the discussion and intervenes: "The convention was called to evaluate the farmer organization program and to show everyone the progress we have made since 1981 and the benefits achieved. Farmers were pleased to be able to arrange the convention and to express to the ministers our gratitude for the program, and for the facilities given by the government. However, the O&M tax has discouraged farmers. There has been a 40 percent decline in attendance at the organization meetings." I say that I have heard the farmers' proposals were not presented correctly to the ministers. Yes, says Narangoda, this happened. "The representative from Uhana who has chosen to read the resolutions only paraphrased them, and when it came to our proposal about reducing the O&M charge, he said farmers accepted it, without stating any qualifications. We organized and financed this event to put our views forward, and then this was not done."

Why was there no follow-up effort to present the farmers' proposal correctly? Narangoda says they discussed sending a delegation to meet the minister and correct the impression. "This was agreed, but then the decision was reversed." Why? Some embarrassing political considerations were pointed out. The farmer-rep who presented the resolutions was formerly prominent in the opposition party. (Sena whispers that it was Kalumahattea, the influential farmer we now know very well.) Since he has spoken publicly in support of the O&M charge, it is awkward for farmer-reps who are government supporters to appear to be challenging government policy.

"We do not want any political divisions to come into our organizations," says another farmer-rep, taking over from Narangoda. "We want to distribute water equally to all, Tamil and Sinhalese, SLFP and UNP. This means we must avoid party problems, even if it means backing down on this issue now."

If farmer organizations are able to undertake some of the maintenance work on contracts from the ID, I suggest, they can earn back part of what they pay in O&M charges. Punchibanda says they are already doing work on some D-channels, which are the ID's responsibility: "It cleans the channels at most once a year and seldom on time. If farmers want to get water to the tail, they have to clean the D-channels themselves. We would be glad to assume responsibility for D-channel maintenance in return for some reduction in the O&M charge."

I ask whether they have organizations at the D-channel level yet. No, because there are very few IOs now, "only Jinadasa." Also, there is disillusionment among farmers after the convention. But one young farmer-rep says these are not good reasons, only excuses. "We can do the organizational work ourselves. After all, we put on the convention ourselves, didn't we?" They had planned to start several D-channel organizations, someone says, grouping channels LB29 to LB32 into one organization, G1 to G4 in another, and then G5 on its own. But then came the convention and the loss of IOs, and the effort lapsed. "It is time to start again," says the enthusiastic FR.

I understand Narangoda to say that holding the convention was a mistake, so I ask the farmers if this is their opinion. "No," is the reply. Narangoda tries to clarify: "The convention raised farmers' hopes that the O&M charge would be reduced. When this did not happen, it created disappointment." The women FR speaks up to say that, even so, half of the Gonagolla farmers have already paid their tax. Ratnayake leans over to me as the others are talking, and Nanda translates his whispered remarks: "The government has invested a lot of money to restore this system, and if it is not well maintained, our area will become a desert again. I'm ready to pay the O&M charge because if we let this system run down again, there will be nothing to give to future generations. That's why I try to persuade other farmers to pay. This system is being put in our charge, and we need to take good care of it."

What problems are the farmer organizations facing? Punchibanda says the field channel groups, being informal, have no power. "Being informal is okay, but the D-channel organizations will need to be formal and to get recognition from the government so they can have legal rights." He says their efforts have gotten stalled because of the dropout of IOs. Sena, worried by this, asks why they are so dependent on the IOs. Punchibanda insists they are not dependent. "We have been making progress, as seen from the convention. But the combination of the O&M tax and the IOs' leaving has set us back." Ratnayake joins in, saying: "The fault is on our side. The deputy director of irrigation came and encouraged us to get registered so that he could give us contracts to do work with the ID. But we have not followed this up. There is a balance of about 1,200 rupees from the money raised for the convention. This could be divided up among farmer groups so each could open a bank account." Punchibanda agrees but says it would be good if some IOs could help with coordination. It was always our intention, I say, to have a few IOs stay on with the farmer organizations in each area. Narangoda says with a smile, "If we have Jinadasa with us, that is enough."

The subject of patrols at the main reservoir comes up. How did these got started? One farmer says that after the convention the Minister of Lands told the President about the organizations. The President got the idea of having them guard the facility and called the Government Agent in Ampare to suggest this. The GA then met with farmer-reps from different areas, and they agreed to take on the responsibility. (The GA later verifies this. How this farmer came to know the details of a presidential initiative, I don't know.)

What other changes have there been? The Government Agent now has a lot of confidence in the farmer organizations, someone says. Yes, says someone else ironically, "the GA now calls on us for everything, even to contribute our lives." (Sena explains this is a reference to the fact that the families of servicemen killed on duty get large payments from the government, but no such promise has been made to farmers if anything happens to them while on patrol at the reservoir.) The GA has been impressed that there are now no conflicts, no complaints, and no murders over water, when before there were so many problems, someone says.

We pause to eat some of the mango and guava slices that Punchibanda is cutting and passing around. The conversation returns to the convention. Some of the farmers are still upset that their proposals were not properly presented by Kalumahattea. Was this because he got flustered? Because time was short, he was told to paraphrase the proposals, not to read them word for word. One farmer suggests that a man so experienced in public speaking would not make such a mistake. Someone uses the word "traitor," but others are more charitable, saying that Kalumahattea was a good representative on the DAC, very energetic. He does a lot of work in Uhana now, taking responsibility for organizing their participation in the patrols. In 1982, when the Irrigation Department tried to limit planted area to 5,000 acres, it was Kalumahattea who led the fight to get more acreage authorized, someone says. Harsh words have been spoken, but Narangoda says, "We don't want to hold a grudge." I hope this will apply to tall Punchibanda too.

It is nearing noon. Nanda thanks the farmers on our behalf, and Narangoda from the farmer side expresses their thanks for our visit and for the training we have given them. I comment that we are learning more from them than the other way around, and I thank them for having set a good example for the whole country in combating ethnic divisions. Narangoda says this is the right thing for everyone to do. After taking snapshots, we pile into the ARTI van and head to Uhana.

Assessments in Uhana

After lunch, we drive to the junior high school where I have met so often with IOs and farmer-reps. We are early, since it is only a little after 2:00 and the meeting was announced for 2:30. We set up chairs in a circle outside under a tree, where we will get more breeze. P. B., one of the first batch of IOs, arrives, and I ask him about the situation on UB9, where farmers were setting up a D-channel organization last time. They have officers but still no constitution. The chairman, Waratenne, is ill, and the secretary, a dynamic young fellow, has left to take a job in Moneragala. Gunaratne, who impressed us at that earlier

meeting, now heads the organization very effectively. As we are talking, indeed, he arrives on his bicycle. Several other farmers arrive
(from UB1, 3, and 12) and also D. S. on his motorcycle.

We sit down to start talking. What about the O&M charge? Farmers are not happy with it, but in Uhana the charge has been paid for all
4,000 acres. (This is where the rehabilitation has been completed, and
water supply is good; still, 100 percent payment of any tax is noteworthy in Sri Lanka.)

What about the scheme for sending farmers to guard the reservoir
facilities? Gunaratne says, "It is our tank, our bund." But, he adds,
"we do not have weapons and fear for our lives if an attack ever
comes." Should farmers have been asked to do this? "Why not? It is
our property, and we should protect it," he says. (What a change in
attitude from when the program started and farmers expressed no
sense of responsibility for the system, which was seen as "government
property.")

Several more farmers-reps arrive, so we start the meeting. What criteria should be used to judge a good D-channel organization? I ask.
They suggest: "Meet frequently, once a week if necessary; ensure that
water distribution is equitable; look after the field channels and make
sure they operate properly; make decisions collectively, preferably by
consensus." One farmer-rep says, "There are two main things: maintain unity and cooperation, and evolve methods of equitable water distribution." Another says there is now a good sense of unity and also
safeguarding of structures. "Earlier there was breakage, but now no
more." I remind him of the breakage in March 1983. "Oh yes, but
nothing since. That was during rehabilitation, when some mistakes
were made from the ID side." One farmer says the new organizations
should be given ID contracts for construction and maintenance work,
to ensure that good-quality work is done. The FR from UB11 says his
organization did the work on their D-channel and three field channels,
with a contract worth 43,000 rupees.

Have others done such work? No other examples are given. Why
not? "Other organizations were not strong enough at the time," is the
answer. Then someone says there were lots of delays in getting approval of the UB11 contract. The representative says he had to make
more than a dozen trips for this. "The TAs like to give out contracts to
their own people," volunteers one farmer. Also, the payments were un-
duly delayed, someone adds. "It took months to get reimbursed for the
work." This would surely be a deterrent to other farmer organizations
considering seeking ID contracts.

We had hoped such work would lead both to better construction and to stronger organizations, but these goals have failed to materialize. I say we can talk to ID officers about this problem. One farmer says the deputy director is already aware of it and has transferred one TA out of the district for misconduct in such matters.

How satisfied are farmers with the rehabilitation work overall? The FR from UB11 says "60 percent." He says that one of the drainage causeways on Uhana branch canal, between UB10 and UB11, has washed out several times. Farmers finally decided to repair it themselves, though he adds that their repair cannot be considered permanent. Did the TA approve of their work? No, because they took away an opportunity for him to give out another contract. "It had been fixed once already by a private contractor, but that repair did not last."

How satisfied are they that their own repair will last? "It should hold for a while," a farmer says. The previous work was done by packing dirt around the drainage pipes and then putting stones on top. The farmers made the structure narrower and deeper, digging down four feet and putting stones around the pipes. Will this embarrass the engineers? "That is their problem," is the reply. Farmers needed the structure repaired, because otherwise tractors could not drive across it to get to certain fields. How much labor did the repair require? Two days of work, with twenty farmers the first day and sixteen the second. One farmer says he heard later that a contractor brought some trailerloads of stone, even though the farmers had reused stones already at the site, and charged the ID for the work farmers had done.

More farmer-reps have joined our meeting, and I count fourteen now, twice the number I expected on short notice. I ask about rehabilitation work on other channels and get figures of 60 percent and 40 percent satisfaction. Two structures are reported at the tail of UB9 with openings that are too small. When it rained, the water washed around them and made them useless. The ID has promised to replace them. Another farmer says that whenever there is heavy rain, somebody should close the D-channel gates off the main canal. "A farmer-rep could do this if given the key," says Neeles, who has arrived representing UB16. A young farmer-rep, looking rather "mod" in a brightly colored shirt casually open to the navel, interjects that this could create a greater problem, of overflowing and breaking the main channel bunds. He speaks like an engineer. I had not thought of this myself right away. (The number of FRs is now up to eighteen.) A farmer-representative reports 100 percent satisfaction with most of the rehabilitation work on UB12 channels. Neeles, however, says that the work on UB16 was not so good and the ID has agreed to redo it: "Farmers complained while the work was being done that the bunds were not being compacted properly. They are so soft one can sink into them. We kept telling the overseer to compact better, after every half foot of fill." He proudly says that the farmers on LB16 did a lot of shramadana work to clear the drainage channel. I cannot verify whether the various complaints are correct and can only note that farmers are critical of the technical standard of work. In their view, if they had had a larger say in the design and construction, various problems could have been avoided.

I ask whether farmers are trying to reduce their use of water. "Yes, it is in our interest." Why? The following reasons are given: "If there is too much water in the fields, the bunds of the fields break and have to be repaired; letting water flow without control wastes water; too much water affects plant growth adversely; if the water is flowing rapidly, it washes away valuable soil, and fertilizer nutrients are lost; the plant roots do better if they have some periods of dryness and more sunlight reaches them." The last reason is not quite right (oxygen rather than sunlight needs to reach the roots), but I am impressed that farmers so quickly give many reasons for reducing water use. When we first came to Gal Oya, farmers hardly ever attempted to economize on water use. The IOs, together with Irrigation and Agriculture Department personnel, have gotten the message about conserving water across to head-enders in Uhana.

One farmer says that the greatest waste is still in the channels, not in the fields. "Only if farmers have control over the gates and can close them when the flow is not needed will real water savings be made. At present whenever we see need to change the settings of the gates, we must send a messenger to the jala palaka, who has the key. If we cannot find the JP, there is water waste. The keys should be handed over to the farmer organization." Gunaratne asks, somewhat rhetorically: "How much will it help if individual farmers save water? We need a project-level committee to plan and coordinate water deliveries better between farmers and the Irrigation Department. Also, we need D-channel organizations, since the critical management point is the D-channel gate. Then we can make adjustments when necessary."

How are relations with the jala palakas now? There is near unanimity that working relations are good. This is remarkable, because these officials used to be very unpopular with farmers. On my first visit to Gal Oya in 1980, some farmers we talked with referred to their JP as the *jala balla,* meaning "water dog" rather than "water manager."

I ask about the convention. The farmers are very pleased with the event and with having planned it themselves, but they express the same grievances as at Gonagolla about the way it was conducted. "It was our idea. We organized it. But the expression of views was controlled by the government and bureaucrats," says Gunaratne. Another farmer says, "Up to the time of the convention, everything was done systematically, by farmers. But the event itself became a convention of officers, not farmers."

It is after 5:00, and we do not want the farmers to have to bicycle home after sundown in these unsettled times. I thank them and say that their examples of initiative and responsibility have encouraged the government to plan to extend this kind of farmer organization to other schemes. We will need their help in this. There are murmurs of agreement. A young farmer-rep gives the vote of thanks, saying they have appreciated being able to talk about the convention and their difficulties of the past six months, and getting guidance on some of their problems. They will talk about these with the Government Agent and DDI. After some picture taking, we head back to Ampare.

Progress of the Weeragoda Farmers' Association

Next morning we stop by the Irrigation Department to see the deputy director and the new irrigation engineer in charge of operations, Godaliyadde, who has taken Manoharan's place. Then we drive to the middle of the system, to the head of M5 canal, where we meet Sarath Wijesiri, our IO coordinator, in front of a small tea stall, and from there we go to the home of Weeragoda Farmers' Association chairman, also named Ratnayake. To reach his neatly kept house, we walk through a carefully swept yard, past two dogs, big and friendly but very old, like their master. The chairman and three other farmer-reps are there, with two IOs.

Organizing work started here a year and a half after Uhana and Gonagolla, but it has moved more quickly. M5 in particular is very complex hydrologically, encompassing 2,500 acres under one distributary channel, branching into field channels, subfield channels, and sub-subfield channels. Because there are so many farmer-representatives on M5, thirty-five in all, a separate organization has been established for the longest channel, M5.2. Its fifteen FRs meet separately and send a representative to meet with the rest of the farmer-reps for M5. The M5.2 FR speaks for an area of about 1,000 acres, and he communicates decisions back to these farmers. These are made by consensus rather than by majority vote. How many farmer-reps are there in the Weeragoda area now? 105. How many come to Area Council meetings? 85–90. The same ones every time? "No, but there are some who do not come very often. If someone misses many meetings, we take action to get a new representative appointed." I comment that in Uhana, about 10 percent of the representatives were replaced in the first year and a half. Ratnayake says that some have been replaced here too. On M5.2.6, for example, he went and talked with an inactive farmer-rep and asked him, "Why not get a new man?" This was the replacement, he says, pointing with evident satisfaction to the M5 secretary, Jayasinghe, who smiles broadly.

How about having regular rotation of farmer-representatives? It was decided initially to have fixed one-year terms, says Ratnayake, but this has not been implemented. "We want to keep any good farmer-representative as long as possible." I tell him about the traditional system of village organization in Japan. Every six to ten households were grouped into a small unit (*kumi*) with the chairmanship rotating annually among all the household heads. This helped identify leadership talent within the community, and the most effective persons were then elected to higher offices. An incompetent or irresponsible household head could not do too much harm in one year at the kumi level.

Ratnayake says that this method is okay, but still they like to keep their energetic farmer-representatives. I suggest that such talent can be used at higher levels of organization, with rotation expected only at the field channel level: "Without change, you would have missed Jayasinghe's talents. Also, we know that giving people responsibility often brings out talents not otherwise seen. There has been a great deal of human talent in Gal Oya all these years, but it took the establishment of the farmer-representative role to bring this talent to the surface and to put it to work for the benefit of the whole community."

Ratnayake says that even among farmers who don't participate actively in the organizations, there have been changes: "They now close their pipes and follow the advice of the farmer-representatives. One characteristic of the M5 area is that people here have for so long had so little water that they do not complain so much about shortages. They are glad for any water as long as it is shared equally. They even tell us when they are getting too much water." This represents a dramatic change in behavior since introduction of farmer organizations.

I respond that this is especially good to hear because when we first visited M5, it was reportedly the worst area for irrigation. "Now it is the best," says Ratnayake proudly. The difficulties arose partly because the command area overlapped four different settlement units, he says, which made getting cooperation difficult. But also there were many problems with the Irrigation Department: "There were so many fights with the ID. Sometimes the engineers had to run for their lives when farmers got angry and came with their machetes. And there used to be lots of fights among farmers over water, even murders. You can check the records of the police if you don't believe me. Now there are no more." What about the organizations' previous plans to collect money from members for various group purposes? Ratnayake says they wanted to collect 100 rupees per member for the whole Weeragoda area, but this didn't work, because they tried to start it after the yala season when there had been large crop failures due to flooding. One of the farmer-reps says the M1 organization has started collecting 200 rupees per farmer. Ratnayake says they want to encourage this for all D-channel organizations.

What would the money be used for? "(1) For a revolving fund to be drawn on when there is a pest attack. Now when there is an attack, not all farmers have ready cash to pay for spraying, and if the counterattack is incomplete, we cannot eliminate the pest. If money is on hand to pay for spraying the whole area, it will be more successful. Members can pay back the loan later when they have cash in hand, with no interest required. When the money is not in use, it can be in the bank drawing interest. (2) For making emergency loans to members. These would have to be repaid with interest, but at a lower rate than banks or moneyleaders charge." (Banks charge 22 percent per annum; moneylenders get as much as 25 percent per month.) One FR points out that the loans could be made quickly, without paperwork. Another says there are many instances where farmers who had to take emergency loans from merchants ended up losing their land and becoming laborers in their own fields. This seems a very good idea, setting up savings and loan services to help protect farmers against the loss of their land. Sarath says we need to be getting along to our next visit, since we are already half an hour behind schedule. We take pictures and say good-bye.

An Archaeological Find

When we arrive at M5.5.4, about twenty farmers join us as we walk along this subfield channel. Sarath turns sharply to the left, and we proceed along a sub-subfield channel that carries no water. Sarath explains that unfortunately when this channel was put in thirty years ago, it went uphill and could carry no water. Farmers are now trying by shramadana to dig it down low enough to bring water to about twenty acres currently cultivated without irrigation. As we walk through the barren scrub brush, what is here called "jungle," I can't imagine this being turned into productive, irrigated land. I begin to wonder whether the plan is technically feasible and get a little angry thinking that farmers may be wasting their labor in a fruitless effort. If they do all this work and then can't get water to their fields, their enthusiasm for organization will surely evaporate. Should IOs be promoting or even supporting such initiatives? Perhaps we have erred in not giving them more technical training so they will know better than to let farmers embark on technically inadvisable tasks. I ask Sarath, somewhat testily, whether there is any chance this barren area can get irrigation water from the sub-subfield channel. He says the farmers are confident they can dig the channel down far enough to make it work. I hope he and they are right.

The families here surely need the water. With us is a young man carrying a child as emaciated as any I have seen. The footpath we are following makes a turn, and we come to a couple of farmhouses. Sarath points out an irrigation channel in front of us, the tail of M5.4.4, which he says did not function at all until this year, when it was deepened and cleaned by farmers. Now twelve households who previously got no irrigation water have it for their fields. The farmers walking with us confirm this with broad smiles. I can feel revived enthusiasm for the program "flowing through my veins," replacing the doubts that were accumulating as we trekked through the wasteland.

About fifty yards upstream, we reach a point where a sub-subfield channel has recently been dug. The farmers point to a concrete turnout structure and explain something excitedly in Sinhala. Sena tells me this is an "archaeological discovery." When farmers started to dig a new channel here, they hit this structure with their shovels and excavated it. It had been put in thirty years ago, but because water seldom flowed in the channel, nobody bothered to remove the mud and silt that accumulated around it during floods. Eventually it became completely covered, and people forgot it existed.

The newly dug channel below the turnout has brought fourteen acres into irrigated production. The original design for this area planned for it to be irrigated, but when water did not come, it was abandoned and the allottees assigned to it were given land elsewhere in the colony. In the meantime, of course, the population has grown, and there is great land hunger. Fourteen young farmers, children of original allottees, got together and developed this area jointly by their labor, assisted by others in the community, a welcome demonstration of solidarity. I take a picture of the broad sweep of newly established fields with the young farmers standing proudly on the glistening, weed-free bunds.

Back in the van, I comment to Sarath that the IOs do not seem as demoralized by their loss of numbers as I had expected. The work seems to be going on as before, perhaps less intensively but in the same manner. Sarath says that appearances can be deceptive. The IOs remaining are disappointed that the program is still not institutionalized. Most of the IOs would not have left if it had been permanent, he adds.

He says that he himself left the program last fall to take a teaching position, but he stayed for only one day. Once he started the new job, he realized how much he preferred to work as an IO, so he resigned and returned to our program. He was willing to gamble that the program would offer him a future. Sena tells me later that Sarath's decision to return was a real boost for the program, because he is one of the most effective IOs, as can be seen from the accomplishments here on M5.4.4.

Reducing Bribery and Thuggery

We stop in front of a house on M5.2.1 to pick up Piyadasa, a farmer-rep I have not met before. As we drive along this channel, I can see from the condition of the houses that this is a somewhat more prosperous area, better served with water than M5.4.4, which we have just left. We drive up to a house where several farmer-reps are gathered. The IO here is Keerthi, one of the few women remaining in the program. I recall meeting her shortly after she first came into the field— she seemed so shy that I wondered how she could work in the field as an IO. But her performance in the training program had been outstanding, and it was a better indicator of her potential than my first impression. She now appears quite confident and outgoing.

The farmers explain about their D-channel organization for M5.2, which Ratnayake told us about this morning. The secretary is a young farmer from M5.2.8. Our host is the farmer-rep for M5.2.3A. There are FRs also for 5.2.1, 5.2.4, 5.2.5, and 5.2.6. The numbers reflect the complexity of the channel network.

What does the D-channel organization do? "Equitable distribution of water is our primary concern," Sumitipala says. "Beyond this we can expand into activities like credit, but the main activity involves water." (Glasses of soda are brought around to us.) "Before the farmer organizations, we couldn't get adequate water. There was a lot of taking water by force, in illicit ways, with a lot of thuggery. Now connections with the ID and other officers have improved." How did you get farmers who had previously fought each other to cooperate? "We were able to convince them. We met and discussed everyone's problems. It took a lot of persuasion. There is not yet 100 percent cooperation." How many do not yet cooperate? Piyadasa says, "Less than 10 percent." Who are those who don't cooperate? "Those with more land, usually. Persons with private landholdings or encroachments who need and take more water than they are entitled to, who must get water somehow." Through bribery? "Not now," is the emphatic answer. "Before there were corrupt practices, but now that is reduced."

One farmer-rep tells us how on M5.2.6 there were a lot of problems with the head-enders on his channel, who did not allow much water to reach the tail. "There was lots of thuggery to intimidate the tail-enders. Then the IOs came." He nods his head toward Keerthi. She smiles demurely. I ask how she managed to overcome the "thuggery." The farmer rep says the IO was able to persuade the farmers to deal with the problem of water shortages as a group. They developed a rotation system within the field channel, where the tail-end farmers get all the water for five days, and then those at the head get it for five days, and back and forth. Whose idea was this? The FR says it was his own, since he is a tail-ender.

The farmers now describe the rotation system developed for the M5 area as a whole. It corresponds to what the IOs have told me. By closing all head-end channels for two days to give all of the water first to the tail and then giving all to the head, almost the whole M5 area can now get water. About seven hundred acres previously left uncultivated in yala season have been brought into production. (The value added from this increase in production amounts to 900,000 rupees, roughly the total cost of the IO program per year.) A farmer says: "This system was discussed with the ID, but it was first decided on by farmers and it got water to the tail of M5 for the first time in years. The irrigation engineer said we may be able to manage with perhaps 25 percent less water than was previously issued to M5 if we follow a rotation. He personally came and thanked us for our cooperation." Who closes the channel gates to make this system work? The jala palakas. "But since they still work as government officers and do not come on Sundays or holidays, on those days they give the farmer-representative the key to the gates. There is a close association now with the JPs, and we work cordially."

How did this change, which Uhana farmers also reported, come about? "Previously the JPs didn't listen to farmers, and farmers made duplicate keys to open the gates whenever they liked. Only after the farmer organizations were formed did relations improve. Before, ID officers gave the JPs specific instructions to do this or that. No deviations were permitted. That is why JPs wouldn't listen to us. Now that farmer-representatives speak on behalf of an organization, they are listened to. Before we spoke only as individuals." This is a process described by many farmers, how the creation of the organizations affected the behavior and also the values of officials while it changed farmers' actions and attitudes.

After some discussion about the rehabilitation work, with complaints about the quality of the work being done here—"head areas always get the best of everything"—Piyadasa asks about the future of the farmer organizations: "The rehabilitation project will end in December 1985. What will happen to the farmer organizations after that?" I ask whether he attended the farmer convention in October. Yes. Didn't he hear the Minister of Lands say that the organizations will get recognition? "Yes. But there are only four IOs here now, and we are afraid that the organizations will get washed away if there are no IOs to help." I explain that ARTI has already recruited twenty-eight new IOs who begin two weeks of training next Tuesday. The new IOs will be coming to Gal Oya for further training in the field, and we would like farmer-representatives to help train them. There seems to be some puzzlement on their faces. I say they can be helpful by telling new IOs how to work most effectively with farmers.

One objects, "Not all farmer-representatives have good attitudes. Some behave like officials themselves." I say these are not the ones who should be called upon to work with new IOs. "Others are very devoted to the program," Piyadasa adds. I tell them, "You are in the best position to judge which FRs should be asked to assist in training IOs." Did you heard the Minister of Lands say this program would be extended to other parts of Sri Lanka? I ask. They answer yes. We would like their help in this. Piyadasa says they would be glad to help with organizing efforts. "We are willing to go to other colonization schemes to tell other farmers of our experiences, but we still have some weaknesses in our own organizations to overcome."

I tell them we appreciate their willingness to help extend the program, as well as their self-critical attitude. In the United States, farmer organizations were established by farmers without government aid. I comment that when my own mother was Keerthi's age, having had a few years of higher education, she was employed by the North Dakota Farmers Union as their education director, and she worked for several years helping to organize and educate farmers at their expense. Until this moment, I have not thought of how my own family history relates to this program in Sri Lanka, but it has popped into our conversation and it interests the farmer-reps and Keerthi.

It is time to leave, since we have to attend the meeting between the Weeragoda farmer-reps and the irrigation engineer at 2:00. We thank them for their time, and Piyadasa thanks us for listening to their experience and giving them advice. We drive to where Sarath stays. He has arranged a nice rice and curry lunch for us, which we must eat quickly, however, because it is almost 2:00.

An Amicable Reduction in Water Supply

When we reach the TA's office, the meeting is already under way. Godaliyadde, wearing blue jeans, stands behind a rickety table while half a dozen farmer-reps gather around it. Ratnayake is representing M5 rather than the Weeragoda Farmers' Organization. It is socially significant that they all stand around together rather than having the engineer seated "at the head" with farmers on benches like pupils as is usual. They are talking about water distribution. Godaliyadde asks whether water is reaching all fields. Yes, is the answer. The discussion seems relaxed. The farmer-rep for M16 says, "We should try to save more water for the yala season." Ratnayake endorses this, as does the FR for M12, who says that they have adequate water and that farmers are satisfied.

The purpose of the meeting is to get agreement on cutting back the issues of water, usually a very touchy subject. This time it is quickly agreed to. Dissanayake from M16 restates his opinion that there is adequate water: "It is time to reduce deliveries. It is no use sending water down the drainage channels." This should be music to any irrigation engineer's ears.

Dissanayake and Ratnayake step aside to talk between themselves. I wonder if there is any dissatisfaction. It is not polite to eavesdrop, but I feel we should know if there is any unspoken unhappiness over Godaliyadde's proposal. Maybe they are just agreeing to the reduction to please the ID and will be resentful or uncooperative later.

Setting scruples aside, I suggest that Nanda find out if they are really against the reduction in water issues. He comes back quickly; no, they are discussing how to get transport for the farmers from Weeragoda who are supposed to guard the facilities at the reservoir this evening. They are agreed on a cut in water supply, so these leaders have already turned their attention to other matters. The IOs are all in one corner talking with the farmer-rep from M6. They finish and rejoin an increasingly informal meeting. Godaliyadde asks in a loud voice: "Do you want water continuously or not?" Dissanayake answers for the group, "We don't need continuous water, only adequate water."

It appears that the meeting has adjourned, since people are drifting away. We ask Dissanayake his impressions of the meeting. "It is good to be able to speak with the engineers in an informal way. Earlier, this kind of meeting with an engineer about changing water deliveries would have taken hours. But now, as you saw, we could take care of everything in twenty minutes." Are there no problems now? "Still a lot of rehabilitation work remains to be done," is Ratnayake's reply. Dissanayake mounts his bicycle and heads off. We give Ratnayake a ride in the van back to the main road. He says that Godaliyadde is a wonderful person, adding that the TA too has changed, becoming more friendly. At first he did not want to deal with farmer-reps. "He even said that if we wanted to come to him, we should come as individuals only, not as farmer-reps. But now he sees the value in our representative role." How does Ratnayake like his responsibilities? He says he has meetings almost every day. Isn't this a burden? Shouldn't the job be rotated? He laughs and says meetings are a kind of "entertainment" for him. He does not have much work to do these days, since his sons do most of the farming, so it is no inconvenience.

At a junction the driver stops so Sarath and Ratnayake can get out. Sarath explains that they have not yet been able to arrange a truck to take Weeragoda farmer-volunteers to the reservoir later this afternoon, so they need to visit some local truck owners. He and Ratnayake, separated by almost forty years in age and about a dozen years of education, walk together into a small shop as we drive away.

An Almost Regular D-Channel Organization Meeting

We pull up to a community center building just as the monthly meeting of the M1 D-channel organization is beginning. We are pleased to be able to attend a meeting not set up for our visit, though it will surely be affected by our presence. The building is a low structure with half walls and a thatched roof. Wooden chairs are arranged in a big circle. We are given seats at the head of the circle with the chairman, who calls the meeting to order. I count fifteen persons, but some are local officials, not farmer-representatives. The chairman makes a short welcoming speech, then calls on another officer to explain the meeting's purpose: "This is our regular monthly meeting, the first of the new year. We have seen a lot of good results from our work so far, and we expect as in 1984 to have very successful programs in 1985 with the cooperation of ARTI, the ID and the IOs. We expect they will help and guide us in our D-channel organizations. If there are any obstacles, we expect they will help us overcome them also." This is rather too deferential a speech, but it may be just a manner of speaking. The secretary stands and reads the minutes of the last meeting, which it turns out was held October 3. So these are not monthly meetings, at least not yet. The chairman asks if there are any mistakes in the minutes. Someone says that the date of the farmer convention was the fourteenth, not the twentieth. This correction is agreed and made. (I am impressed that the exact date is remembered three months later.) One FR rises to propose approval of the minutes as true and correct. There is agreement.

The chairman says that the farmer organizations were set up to use water in a proper way, to distribute it fairly to all, and to save water whenever possible. According to information given him by the ID, farmers are still using more water than necessary. So their cooperation is requested to reduce water use. He turns to me and invites me to give them some "guidance." I say that I am better at asking questions that at giving answers. "Farmers should be able to come up with good answers. What are your ideas for saving water?" Someone says that if they use less water now they will have more for the next season. What methods would he suggest for saving water? "Farmers should take just the amount they need and then close their pipes." Should they close them completely or just reduce the flow? "Completely close it. After a few days you can start again."

By how much could water use be reduced if they manage the water more carefully? I ask. Someone says, "30 percent." Another says this might be difficult, because there is a lot of seepage in their canals. "The soils are very sandy in this area, and it can't be helped that a lot of water flows into the drains." I say that I am not suggesting any particular amount, only that they experiment to reduce water use. "What if there is drought?" someone asks. "In that case you will need more water. But if you have been willing to save water when possible, your requests to the ID for more water in time of need will be more credible."

One of the IOs says that farmers should work out a plan for water and then inform the ID of their decisions. After all, the extension agent has told them they can save water without disturbing their plants' growth. There is a lot of talk among the FRs at this, and I am not sure if they are agreeing or disagreeing. The meeting gets somewhat disorderly. The chairman seems unsure of himself and asks Sena to speak. Sena says we are not the best persons to address water management questions, since our area of work is farmer organizations. One of the farmer-reps asks me to tell them about experience with farmer organizations in other countries and how this compares with what I have observed in Gal Oya. This is a fair request, and I try to oblige by talking about farmer water management efforts in the Philippines and Nepal. I stress the idea of self-reliance and tell them, "Next time I visit, I hope you will not ask us for solutions but will instead tell us about your own solutions."

The chairman thanks me and says it is very good that by coincidence some people from ARTI were visiting on this day and could attend their meeting, to help them better understand the work to be done at field channel and D-channel levels. "It has been some time since we were able to hold a meeting," he acknowledges. "But we got very good encouragement from your visit." He says they have been working "shoulder to shoulder" with the IOs and will continue to do so. He calls on one of the farmer-representatives to give the vote of thanks, which echoes most of those we have heard before.

As we drive back along the road beside the Mandur branch canal, we see and give a lift to Wimalasena, a former IO now teaching near Weeragoda, and his wife, formerly an ARTI statistical assistant who helped with farm record keeping in Gal Oya. They have their small child with them. Wimalasena was one of our outstanding IOs in the difficult M5 area. I ask whether he would be willing to rejoin the IO program if there were a permanent position for him. Most of our conversation has been through Nanda's translation, but he understands my question and replies in clear English, with a beaming smile: "That is my dream, sir." His wife says this is true.

Meeting with the Remaining IOs

As we are finishing breakfast the next morning, the two Tamil-speaking IOs remaining with the program arrive. Abu Bakr worked previously with ARTI as a field investigator collecting data in Gal Oya, and he does not want to leave the work. The other, Sugatadasa (Sugi), was offered a teaching job but passed it up in order to stay with us. (He majored in philosophy and was one of Nalini's husband's top students at the university.)

When the other Tamil-speaking IOs left the program, Sugi and Abu Bakr were told by letter to join the remaining Sinhalese IOs, assuming they could not continue to work in the areas they were covering below M18. But they do not want to drop the work they have started with farmers on M22 and M23. Since we still hope to revive the program in Tamil-speaking areas and expect to recruit a new batch of Tamil IOs, we will propose to ARTI that they continue working in their present area. Bhawani can review their advance program and work reports (in Tamil) while Mune handles administrative matters.

Three Sinhalese IOs arrive. Ketsiri greets Sugi and Abu Bakr heartily, while Ranjani and Keerthi talk with Bhawani. Three motorcycles arrive in close succession, bringing Sarath and Priyantha from Weeragoda, and D. S. and P. B., Anula and Jayalath from Uhana. Jayalath is now acting training coordinator, since Ratnasiri was demoted after an unexplained accident with his motorcycle. (It appears he had been drinking too much, an occasional weakness in an otherwise fine IO.) D. S. greets us and also Sugi and Abu Bakr. I tell them we still hope to be able to resume the Tamil IO program, and D. S. says in English, "We are very keen to continue the Tamil program."

By 9:20 all the IOs are assembled. D. S. begins with a pleasingly simple introduction. It will be an informal meeting, he says. He welcomes me and says they want to know my observations "after six months away," so they can know better what to focus on and what to improve. I tell them how pleased I am to be back with them. "I know it has been a very difficult six months and am happy that you are carrying on despite the difficulties, which I need not enumerate. I am glad also that farmers seem to be making such good progress and are so laudatory of your efforts." I go over my suggestions for agenda items and ask if they have any to add. Someone suggests we talk about training activities for farmers and officials, and another asks about extending the program to new areas (units 1 and 2 at the head).

We start with rehabilitation progress, since Munasinghe has not yet arrived. He is supposed to put in some time at the district headquarters this morning. Field channel work is progressing, but slowly—"farmers are not 100 percent happy." The ID's standard of D-channel work in Weeragoda and Gonagolla is not up to that done in Uhana. I ask whether farmer organizations have been given contracts for minor rehabilitation work as previously agreed. "No, the ID agrees at the top, but things do not happen that way at lower levels. Sometimes the TA is not cooperative," one IO says, citing the UB11 situation we heard about on Monday. "But one victory for the program was that we chased away one uncooperative TA."

It is clear that the system for getting contracts awarded to farmer organizations is not working properly, and we as a program need to work to improve it. The bottleneck appears to be not so much getting farmer groups registered to do the work—the ID will do that on Munasinghe's recommendation—but setting up bank accounts into which funds can be paid. It would be good for the IOs to form a committee to investigate the procedures for opening accounts and then to disseminate information on this to all groups. Before recommending anything, however, they should have tested it with farmer groups to be sure the steps are clear and practical. We need to keep an eye on the development of the *program*, not just on progress in their respective geographic areas. Along with some definite area responsibilities, each IO should have also some program tasks, such as assisting in the formation and functioning of a Project-Level Committee.

One IO asks whether the program will be continuing. I say that both the World Bank and the USAID irrigation rehabilitation projects expected to start this year and next plan to have farmer organization with IO guidance. We may experiment with IOs who are not college graduates, working under the direction of experienced graduate IOs. The IOs say that graduates are better able to make farmers understand the importance of cooperation among themselves and to get cooperation from officials. I say they are probably right about this, but the government wants to experiment with lower-cost methods of organization.

D. S. now brings up the farmer convention. He asks, having learned of the concerns I expressed in a letter beforehand to Ranasinghe Perera, "How could you predict in advance, from the United States, that the convention would get politicized?" I tell them one doesn't have to be a political scientist to expect that anything like this in Sri Lanka will attract political interests and pressures. He says that although there are good reasons for some of farmers' unhappiness, they learned a lot from it. The ministers who came referred to the event several times in the newspapers and television, which made farmers proud.

It is after 1:00, and we break for lunch. Afterward we review our organizing strategy and discuss how experienced farmer-reps could help the program by introducing new IOs to farmers we want to get organized. Everyone agrees this is a good idea. We talk about transportation logistics, whether bicycles are still practical now that IOs have to cover areas of five hundred to one thousand acres. We originally rejected motorcycles because of the image of IOs this could establish in farmers' minds. Perhaps now that farmers have such a high regard for IOs it will not be harmful for the program if IOs use motorcycles.

The IOs raise a number of "housekeeping" problems with my ARTI colleagues, such as the bicycle repair allowance, which has not been resolved since my last visit. The IOs have not yet documented their costs of operation, which could give ARTI a basis for changing its present allowance of 25 rupees ($1) a month. We tell them they must do this before ARTI can act.

It is now after 4:00, and we thank the IOs for coming to spend the day with us. They should all feel much satisfaction from the progress made since last summer, despite the loss of many IOs and the continuing communal tensions and other difficulties in the country. There have been some significant changes in the way farmers approach their irrigation responsibilities and how engineers deal with farmers. Ten years from now, Sri Lankans may look back on this experiment and wonder how such a small group as the IOs contributed to such a large and important change.

The vote of thanks is offered by one IO to close the meeting: "We have had successful discussions all day. We get new spirit from these discussions, from the new ideas that are raised, and from the proposed strategies. It is up to us to try to introduce them in the field. We have some golden opportunities. We wish you good health and a safe journey, and look forward to your next visit." After shaking hands and taking pictures, now a ritual, they head back to their temporary homes. As he is leaving, Ratnasiri tells me he is sorry he let the program down (referring to his accident and demotion). But he says in

English that he is "not demoralized" and will keep making a good contribution as "a regular IO." Like all of us, he has his weaknesses, but we benefit from his strengths.

That evening, Godaliyadde joins us for dinner. One of the first things he says is that he is "a changed man": "When I was the operating engineer at Mahakandarawa, I had a reputation for being unapproachable. I used to shout at farmers. But I have learned to be more cooperative with farmers, more empirical." We tell him we were impressed by the way he worked with farmers at the meeting yesterday in the Weeragoda area. He says Uhana farmers have also agreed to a 30 percent reduction in the water issued to them. Since this was the first maha season using a system of continuous flow, and since there was a good supply of water in the reservoir, he started the season with a generous distribution to show that water could reach all parts of the system. Now that farmer confidence has been built up, he wants to cut back deliveries, and so far he has had cooperation. There have been no attempts by farmers to reopen closed gates. This represents a real accomplishment of our program, but it also reflects the climate of confidence the present ID leadership in Gal Oya has helped create.

As usual, the trip back to Colombo takes most of the next day, but we have good discussions en route. Responsibility for the IO program has now fallen to Nanda and Bhawani, the youngest members of the ARTI group. I have watched them carefully over the past few days and feel confidence in their commitment and skills. Of the original team, only Sena and I are left, and we work in advisory roles. Each time there has been a change in leadership at ARTI, there has been some anxiety about how the program would fare. But a succession of good persons have picked up the baton in turn. The same applies to the IOs. We have repeatedly wondered how the program would manage when it lost so many of its experienced members. Yet there has been a continual infusion of talent from the new IOs, and those who remained have displayed talents previously unrecognized. This applies also to the farmers, since new faces and new energy are seen on each visit, reinforcing stalwarts like the Ratnayakes, the Punchibandas, and Narangoda.

Rice Plants and Weeds

Back in Colombo I have three days to meet various persons concerned with promoting participatory water management. The

Secretary of the Ministry of Lands and his director of water resource development, the head of the new Irrigation Management Division, and the ID's deputy director for water management all express support for continuing and expanding the farmer organization program with an ongoing role for IOs. Unfortunately, they can make no firm commitment to create a permanent cadre of IOs and an administrative home for the program. Since the farmer convention, there seems to be more appreciation of the program; the farmers' initiative in this respect has been very helpful. The program itself is no longer viewed as an experiment. It has become an accepted part of the institutional infrastructure for improving irrigation management.

My most interesting conversation is with the Additional Secretary of Agriculture, Dixon Nilaweera, who was being visited by a former director of agriculture, Chris Panabokke, when we met. Dixon was Government Agent in Batticaloa when the IO program started, and he has been supportive, though always from a distance. Chris is a highly respected agronomist who is about to join the professional staff of the International Irrigation Management Institute. When he served previously on ARTI's board of directors, he was a friendly critic of our work in Gal Oya, being skeptical about "action research" as an approach ("not real research") and sometimes objecting to views Wijay and Lucky presented on irrigation problems.

Chris had just visited Gal Oya the previous week with an international group sent by IIMI. He tells me this was his first visit there since the rehabilitation project started, and he is very impressed with what he saw and heard. He first went to Gal Oya twenty years ago to conduct a soil survey for the Irrigation Department. In the following years, he says he saw the physical system fall into disrepair and the social and administrative systems deteriorate. On this visit he could see great changes among farmers—their care for the system and their now cooperative attitudes—and among engineers—their willingness to work with farmers and their attention to agricultural matters. He tries to come up with the right word to describe the changes. He first calls them "a revolution" and then decides the best word is "transformation." Without seeing them, he says he would not have believed the changes that have come about in just a few years. Because he is so knowledgeable about agriculture in Sri Lanka and was previously a skeptic and critic, his assessment is especially appreciated.

Dixon recounts how favorably farmers spoke of the help they got from IOs right from the start. He would like to see something like the IO program started in smaller irrigation systems. The Ministry of

Agriculture is currently rehabilitating small tanks and handing them over to farmers, but without any preparation. He asks my thoughts on this. I respond: "The crucial factor in Gal Oya has been the quality of the leadership that has been brought forth from within the farming community. I am sure that persons of similar capability, concern, and integrity can be found in most rural communities, not just in Sri Lanka but around the world. But they are not likely to come forward unless the conditions are right." Then an analogy comes to mind that characterizes certain realities in terms especially appropriate for Sri Lanka:

Conscientious local leaders, willing to work for the common interest and advancement of all farmers, are like rice plants. They are very productive but also very fragile and vulnerable. In contrast, self-aggrandizing leaders are like weeds—aggressive, greedy, crowding out other plants. If there are more than a few weeds, they will take over, and the rice plants will not grow.

Now if enough rice plants are established around them, a few weeds can be kept in check. But if there are more than a few weeds, they will take over the whole field, because rice plants cannot or will not fight back. Unselfish leaders will not fight to advance or entrench themselves because they are being productive for others rather than for themselves.

Rice plants need a well-prepared seedbed to grow in. IOs are like hardworking farmers who create the right conditions for the plants to germinate and take root. They keep back the weeds until the good plants have established themselves. Giving leaders training is like providing fertilizer for plants. It helps them to grow faster and taller and to give more yield.

The analogy keeps growing:

Most government programs that have attempted to establish farmer organizations have proceeded like farmers who broadcast seed on dry, unplowed soil. Meetings were called and elections were held without any preparation. We should not be surprised if mostly weedlike leaders emerge. Leaders who care more for others than for themselves will not come forward without encouragement, or if they do, they are likely to be crowded out. Once discouraged from taking a leading role, they will not come forth because they have neither the temperament nor the incentive to fight for themselves.

Dixon likes this analogy, which throws light on the social processes we have observed in Gal Oya. It helps to explain both the success of our program and the failure of so many other efforts. It also gives reasons why a government should invest in getting these processes started properly. If it does not understand this and isn't willing to deploy IOs as catalysts, it should be prepared for a lot of weeds and for poor harvests.

Farmer-leaders who are ambitious and domineering would take exception to being thought of as weeds. (Actually, several of the weeds in Gal Oya—Kalumahattea and Waratenne, for two—were domesticated and turned into productive leaders for the program, though never fully with us or fully trusted.) IOs, however, would surely be proud to be compared to farmers practicing organizational husbandry, and conscientious farmer-representatives who have contributed the most to this program would surely find few characterizations more pleasing than to be likened to rice plants.

Carrying On

Before leaving in January 1985, I spent two days with the new batch of IO trainees and was impressed with their energy and enthusiasm. Nanda wrote in the middle of February to tell me that at the final session of the training program, after field training had been completed, the District Minister and senior ID engineers gave very supportive speeches. The training program itself he characterized as "very successful": "We could bring about a lot of changes in the trainees gradually and improve their leadership qualities and hidden abilities." One of the trainees, he wrote, was offered a job as a project officer with the Freedom from Hunger Campaign but left the post after a short while and rejoined our program. Nanda closed by saying that his own position at ARTI had not yet been upgraded from research assistant to research and training officer, but his contract had been extended until July. "Anyway, I will do whatever possible for the progress of FO program."

With the new batch of IOs, the Sinhala-speaking area covered came to about 25,000 acres. There was still no progress in getting IOs permanently established, however. When the government of Sri Lanka and USAID agreed in 1984 to launch a new project that would assist in improving additional irrigation schemes in Polonnaruwa district, building on the experience gained in Gal Oya, one condition for proceeding with its approval and funding was establishing a permanent IO cadre. We were not happy that a somewhat coercive element was being introduced into establishing a home for the IOs, but the Sri

Lanka government was being asked to do something already approved of by several ministers, secretaries, and directors. The source of resistance everyone pointed to was the Treasury, which was blocking the creation of new government positions as a way of controlling the government's ever-widening budget gap. If further foreign assistance hung in the balance, maybe this would make it willing to approve the new cadre.

In early June, Sena wrote with news. Some was tragic and much was bad, but there was some good news too. First, one of the new IOs had been killed in a bus accident near Ampare, and two more had been injured but had now recovered. The news of accidents did not end there. D. S. Ranasinghe, our project coordinator, had a near-fatal motorcycle accident in April and would be many months recuperating. Nanda also had a motorcycle accident near his home village in Anuradhapura, but he was not seriously injured and only had to take some weeks of leave. Both Raj and Mahinda had left the Water Management Group to begin studies in Bangkok, and Bhawani resigned in May to join her husband in Kirindi Oya. She had tried to continue working at ARTI, but she had difficulty working out satisfactory living accommodations in Colombo, where there were ongoing ethnic tensions.

About these, Sena wrote:

> The situation in the country is horrible. You must have read in the papers of the massacre in Anuradhapura [in February 150 Sinhalese were gunned down by terrorists near Buddhism's most sacred shrine]. Every day there is some incident of death and destruction. There have been a lot of problems in the Tamil and Muslim areas of Ampare district. There was a riot [in March] between Tamils and Muslims in Ampare and Batticaloa districts [with maybe 30,000 temporary refugees]. The situation is near normal now. We have had no problems so far in the Sinhala-speaking areas of Ampare. There were several incidents of armed gangs' stopping and robbing vehicles between Maha Oya and Gonagolla, so travel to Ampare [by the northern route] is very risky. . . . Actually, there is really no problem in traveling to Gal Oya through Ratnapura. . . . We usually take this [southern] route now.

The good news in Sena's letter was that it was finally agreed to form a Gal Oya Project Committee with representatives of the Area Councils and all the top officials, and that the ARTI board had agreed, at Sena's and Jayantha's urging, to extend the IO program for another three years. This would be at a reduced level, but after five years of effort this was appropriate. The Irrigation Department, under a new director of irrigation, was willing to pay for eight IOs plus six field investigators through 1988 from its own funds, though it could commit only one year's budget at a time.

The ID's willingness to fund the program until it was finally given some permanent administrative home decoupled the IOs from the USAID-funded Gal Oya project, which was ending in December. Attrition within the IO cadre continued, however. Whereas we had forty-three IOs at the middle of February, by the end of May that number was again down to thirty-one. With D. S. and Nanda out of action, with Bhawani and Raj gone, with Wijay and Rane doing graduate studies in the United States, the life of the program rested more than ever with the IOs themselves. We still had Munasinghe as supervisor, with Sena and Nalini overseeing the program. These, plus the remaining IOs, their numbers dwindling all the time, were strong but few reeds to lean on. No wonder my next trip made me as apprehensive as any, at a time when the program was being increasingly acclaimed.

July 1985 Visit: "Snake and Mongoose" Work Together

My ninth trip to review the situation in Gal Oya was made with Nanda and Sena, as well as a newcomer on the ARTI staff, Karu, who was to edit a water management newsletter. No ARTI or Cornell staff had been to the field during the preceding two months. Terrorist activity had become frequent and violent enough that field travel was suspended for a while. Now a cease-fire had been agreed to between the government and separatist groups, with peace talks starting in Bhutan the day we arrived in Gal Oya. We did not find so many barriers and checkpoints as during the January visit, when they were a nuisance and continual reminder of tensions.

We found only twenty-four IOs still in the cadre. With D. S. recuperating at home from his motorcycle accident, Sarath Wijesiri was the acting coordinator. He was being considered by the district minister for appointment as district manager of the Agricultural Development Authority, an important post vacated by the transfer of the official Punchibanda had clashed with over "the phantom cashew plantation." Losing Sarath's services would be a great loss to the program but a gain for the district and its farmers.

The program was now covering 25,000 acres. A newly formed Project-Level Committee (PLC) consisted of ten top officials and ten farmer-representatives. An official had proposed that a farmer-rep serve as chairman, but the farmers, fearing this would diminish the new body's authority (it had no legal standing), proposed the Government Agent instead. He accepted, and the deputy director of irrigation became its secretary, with a farmer-rep chosen as joint secretary.

To have a good system of organization below the PLC, farmers decided to have all Area Councils meet before the fifteenth of every month to consider ideas to be taken up at higher levels, with D-channel organizations meeting before the Area Councils. Of special interest was the farmers' idea of having the FRs on the Project-Level Committee form a special committee to meet separately and regularly, since the PLC would meet only every three months. Its planned meeting was set for the next Thursday, so we could attend it.

The IOs reported that D-channel committees of farmer-reps were being formed to review and approve the annual maintenance plans and budgets of the Irrigation Department. Farmers were expected to pay the disputed "service charge," but in turn they would have a voice in how the money was used. They could suggest what maintenance work needed to be done. IOs cited some impressive examples of farmers' initiative in preventive maintenance already.

One loss was the reassignment of the Government Agent, Ariyaratne, back to Colombo after five years in Ampare. He had become a real friend of the farmer organizations, and they organized a big farewell party for him. Several hundred farmers attended and gave him a going-away present. Sena said the GA had tears in his eyes, a rare sign of emotion, when he said in his farewell that "farmers are golden people." This figure of speech, especially meaningful in Sinhala, was commented on several times by the farmers we met during the visit. Sena reminded me of the disparaging remarks the GA had made to the IOs about Gal Oya farmers four years earlier at the inauguration of our program—that they were a rough and incorrigible lot.

We visited a meeting of the UB9 D-channel organization, one of whose founding meetings we had attended in 1984. Gunaratne was the new chairman because Waratenne, sometimes friend, sometimes foe of the program, was ill. The main topic of discussion was setting up a savings scheme to reduce farmers' dependence on moneylenders, who charge exorbitant interest. The joint fund could also be used to buy inputs like fertilizer in bulk and thus more cheaply and to get equipment like sprayers for crop protection. Due care was being taken to safeguard the fund and to be sure it had the support of all members.

The chairman of the D-channel organization for UB1 took us on a drive along a mile-long stretch of road that had been impassable since it was washed out in a flood twenty years before. Once farmer organizations got established on LB15 and LB16, they repaired it through a massive shramadana involving 120 people for five days, assisted by ID heavy machinery for three of those days. The chairman commented that relations between farmers and officials were now quite good, though when the program started in 1981, "We were like snake and mongoose." I refrained from asking who was which.

As we drove along the newly rebuilt road, one farmer informed us, "Before, tractor drivers used to charge us 175 rupees to haul out a load of rice to the market, when the path was so terrible. But now that it is renovated, the charge has fallen to 100 rupees." Collective action had increased the value of every load of rice produced by farmers in this area by 75 percent. No wonder farmers were pleased with their organizations.

It was particularly pleasing to find the Gonagolla farmers, so disheartened in January, now more optimistic. Narangoda opened our meeting by saying: "The organization here when started in 1981 was like a child, walking, falling, getting up, falling again, getting up. Now we are at the stage where we are more steady on our feet." The cause for this confidence was their D-channel organization's getting and executing a contract worth 44,000 rupees to do rehabilitation work on LB29. True, it took at least a dozen trips all the way to Ampare to get approval. They repaired the channel bund and rebuilt the road along it wider than specified (ten feet instead of seven) and stronger (filled and packed twelve to eighteen inches deep instead of nine). They also improved the paths along their field channels, not figured in the cost estimates for the contract. They paid themselves the standard wage that the ID paid its laborers, 40 rupees a day, and still had almost 20,000 rupees remaining as profit. This reflected both how much harder and better the farmers worked on their own behalf than did wage laborers and how much profit margin was built into the contracts being awarded by TAs in collusion with private interests. After 15 percent was put into the organization's treasury to cover various expenses, the rest would be used to buy fertilizer at bulk rates for the next season, giving all members a big bonus.

Gonagolla farmers now felt they could do without IOs, but they asked us to make a special request of "high-level people" in Colombo, to make the program permanent so as not to lose all the IOs. "If Jinadasa goes," Narangoda said, "we are nowhere." Didn't this contradict what they said earlier about self-sufficiency? Old Ratnayake explained: "We are suggesting this because our plan is not limited to the Left Bank or just Gal Oya. We want to see this spread to all of Sri Lanka. We want to avoid the need for rehabilitation in the Mahaweli and Minipe schemes. Don't let them face in the future the same problems we have faced in Gal Oya without farmer organization." They still felt a need for someone like Jinadasa to help coordinate activities. Farmers, after all, have to look after their agriculture. One solution they had thought of was to find a nice local girl for Jinadasa to marry, "with a dowry of five acres of irrigated land and two acres of highland." Everyone laughed as Jina blushed. Our host, "short Punchibanda" recalled: "Last year when we met at the Village Council office, you may remember that I said this organizational effort was a failure. But today I would like to tell you, with the same mouth, that this organization is a success. You told us not to get discouraged, and we did not."

Organizing work had been begun in units 1 and 2 at the head of the system. The IOs encountered unexpected difficulties because another division within ARTI had started what amounted to a competing small farmer organization program there, with FAO funding. Fortunately the IOs were able to establish good working relationships with the local officials, and an Area Council had been established with farmer-representatives from both programs working together. Even at the head end of the system there were enough failings in irrigation and agriculture that farmers had reason to cooperate once appropriate organizational channels were opened up to them. One distributary, LB5, was so silted up that tail-end farmers had to divert water and run it along a *paved road* for some distance to service their fields. Once IOs began working with farmers, two days of shramadana remedied this situation.

A Reunion, Two Losses, and Words of Encouragement

I brought back from Gal Oya numerous positive observations. We were approaching the end of the project as far as USAID was concerned. The government had not yet created a permanent IO cadre, but its promises seemed more credible now that good results were so evident in the field. A follow-up project would utilize IOs to improve irrigation management with farmer participation in half a dozen more irrigation schemes.

The first week of August, we held an in-service training course in Colombo for the remaining IOs—an act of faith. Further training of IOs had been repeatedly delayed as a likely waste of resources, since the departure of many if not all of them was imminent. But this negative view undermined the strength and enthusiasm of the cadre. So we vowed in July to proceed with additional training even if only two IOs remained. The training was mostly self-assessment of the program, augmented by lectures on technical aspects of agriculture and irrigation, about which IOs always wanted and needed to know more.

This week was capped off by a reunion to which we invited all the IOs who had left the program. We couldn't make contact with everyone, but several dozen returned for the festive occasion, with the District Minister attending as the guest of honor. He as well as the director of ARTI and the director of water resource development in the ministry made strong statements of support for a continued and expanded program, so the event was both a morale booster and politically useful.

Shortly thereafter the program suffered two losses, one actually advantageous, the other inestimable but unavoidable. After much political jockeying, the District Minister went ahead with Sarath's appointment as ADA district manager. UNP partisans in Ampare had tried to block it because Sarath's father had been active in the opposition party. This meant we lost an excellent IO but gained a strategically located and sympathetic official whose commitment to farmer interests had been repeatedly demonstrated. This was entirely unplanned, but it was welcome. Now even if the IO program collapsed, as long as Sarath was in Ampare, farmers would have a friend and advocate in high places who could oversee and guard their organizations.

Sena Ganewatte, after five years with the program, giving it more time and heart than we could ever purchase, regretfully took a position as a consultant with another USAID-funded project. Since our project was ending in December, Cornell could not offer him employment beyond then. He needed to find another position, and this was a similar one, introducing a participatory approach into the Sri Lankan government's water supply program. He maintained an active interest in the Gal Oya program and contributed time when needed and when he could.

Nalini, with no Tamil program to assist, moved into the vacuum created by Sena's departure. Training consultants could be assigned almost any responsibility, but unfounded rumors against her—that she was sympathetic to the separatist "Tigers"—impeded her ability to work for the program. Ethnic feelings continued to eat away at the program within ARTI even as Gal Oya farmers persisted in their efforts for unity.

Before leaving Sri Lanka at the end of July, I was able to spend an hour with the Minister of Lands, Gamini Dissanayake, one of the leading members of the cabinet. I thanked him for having allowed the program to remain nonpartisan, something many ministers would not have done. He could have spoiled it by politicizing the IOs' selection or their work in the field or by trying to influence the leadership and performance of the farmer organizations. He had visited Gal Oya several times and was very impressed with the farmer-representatives he met. I emphasized the need for careful nurturing of service-oriented leadership from the start, sharing with him the "rice plants and weeds" analogy. This appeared to make a strong impression, as it did when I shared it with the Secretary of the Ministry of Lands when we discussed the program's future.

Both the Minister and the Secretary expressed appreciation for what the program had accomplished and could contribute. They expressed no doubts about the strategy or philosophy underpinning our program. Yet no firm commitments could be made because the cabinet had approved their new INMAS program to improve the irrigation sector with the condition that no new positions would be created. This boxed them in, though they thought they could find some way to create a permanent core cadre of IOs. These would manage a larger cadre of short-term IOs working on contract before gaining permanent employment elsewhere. (The government had formally agreed to this when it approved the new USAID project.) Having a large number of Sri Lankans with this kind of grass-roots experience would itself be an asset for the country. The head of the Irrigation Management Division suggested that new junior engineers be trained as IOs to work with farmers before assuming their duties as technical assistants.

Yet when all was said and done, despite the evident progress we could point to, no institutional base for the program existed—in ARTI or the Irrigation Department or the Irrigation Management Division. The encouraging words from the highest levels of government had yet to be translated into policy. And the overall political and economic situation was worsening for Sri Lanka. What would be possible and sustainable beyond the project's end in December?

8

The Project Formally Ends
(1986)

This trip is my tenth semiannual field visit to Ampare. It should seem like a momentous occasion since the tenth of anything usually represents a milestone and, moreover, the Gal Oya project of USAID ended on December 31, so our introduction of farmer organizations has been officially completed. But I have learned a number of disquieting things about the program since I arrived in Colombo on January 17. With me are Nanda, finally appointed by ARTI as a full-fledged research and training officer, and M. G. M. Razaak, a new R&TO who will be helping with the farmer organization program. He is trained in sociology and speaks both Tamil and Sinhala, as well as good English.

The Irrigation Department agreed last spring to continue ARTI's work with its own funds for another three years at a somewhat reduced level, but the new director of irrigation has not sent ARTI any written confirmation of his predecessor's commitment. So the remaining IOs have not gotten their letters of appointment for 1986, and this must heighten their long-standing concern about their future. I share this concern but am also worried about the present. The number of IOs has fallen from twenty-four last summer to only eleven now, and three of these may leave for more permanent jobs within a few weeks. The government is moving closer to creating twenty permanent IO cadre positions, which could reenlist some of those who have left, but this has been promised for so many years that it may be a mirage.

As we drive to Ampare, I make some notes on the back of an envelope listing my various apprehensions. The first item on the envelope is "*Project-Level Committee—no meeting since June.*" A meeting scheduled for November got postponed while Munasinghe was in the Philippines for a training course. The previous deputy director of

208

irrigation, who supported farmer organizations, left last September for graduate engineering studies in Italy. His successor as the committee's secretary should have rescheduled the meeting, but he has not shown much interest in farmer participation, so the PLC is now stalled.

The next note reads, "*LB 29—contract work redone, no profit, TA negative.*" Last summer I was pleased by the initiative one Gonagolla D-channel organization showed. Through persistent effort, it got a contract to rehabilitate the LB29 channel bund and road. Its members did more work than specified in the contract and still ended up with a profit of almost 20,000 rupees out of the 44,000 rupees it was given for the job. But later the technical assistant told farmers he had not asked them to do enough work for the money, so they must resurface the road again, even though he previously told them the work was satisfactory and authorized payment. This suggests some Irrigation Department staff, now that Senthinathan has left and foreign involvement has been withdrawn, are reverting to earlier modes of behavior, and that farmers are acquiescing.

Next comes "*Charges of farmers' breaking structures.*" ARTI received a copy of a memo written by a senior ID engineer charging that farmers have done a lot of damage to irrigation structures. Copies went to the District Minister, the Government Agent, the director of irrigation, and other officials. The memo concluded that there did not seem much point in having farmer organizations if they could not prevent such damage. Munasinghe told me many of the locations listed are outside the area of farmer organization, and farmer organizations should not be reproached for the disappearance of bridge railings within the town of Uhana. We suspect the memo was written to deflect attention from the many allegations of faulty construction and financial misdealings against this engineer and others. But even untrue charges can hurt the organizations' reputation.

Then, "*Namal Oya experiment canceled.*" Last summer, IOs agreed to try to establish farmer organizations in this small, water-short irrigation scheme not far from Gal Oya, at the urging of the ID's work supervisor. This looked like a good opportunity for collaboration among IOs, farmers, and officials, experimenting with less intensive methods of organization. But now ID officials have squelched this, saying they will do the organizing work themselves. As we approach Ampare, I ask myself how, if my fears are confirmed, I can determine whether on previous trips to Gal Oya I have been more deceived or self-deceiving. Nanda and I go over the list of remaining IOs. Some of the veterans are still with us, as well as some of the more capable and energetic IOs from recent batches. That is reassuring.

Outside Ampare, our car must stop for inspection at an army checkpoint. As we get out to walk through a long barbed-wire passage, I realize I have left off the list an additional adversity facing the program. Ampare is on the edge of the escalating military-ethnic conflict gripping Sri Lanka. I have found little support for the secessionists among my Tamil friends and acquaintances, but a few hundred well-armed and desperate youths can inflict great costs, human and financial. We were advised to travel to Ampare by passenger car rather than by jeep because the latter is a popular target for hijacking. Even the Government Agent had his jeep stolen by terrorists last month, Nanda tells me. Fortunately, they did not recognize whom they had in their grasp, and he and his driver were able to walk away unharmed.

We drive to Munasinghe's house near Uhana. It is Sunday afternoon, and we hope to find him home. Instead, to our pleasant surprise, we meet Chuti, one of the stalwarts from our fist batch of IOs. She left the cadre in 1984 to become a teacher in Ampare district, which meant she could still interact with IOs. Her husband, also a teacher, was recently transferred to Mannar district in the north, so she is renting the house next to Mune's while teaching in Uhana town. She carries her young son on her hip. Mune is at a meeting at the junior school, Chuti tells us. She says she doesn't know what the meeting is about. I wonder if it is concerning "security" matters, since we have to alight from our vehicle for inspection once more when passing the Uhana police station on our way to the school.

Revival of the Project-Level Committee

As we approach, we see from the many bicycles parked outside the school that IOs and farmers are there. Nanda told me Sarath would be out of Ampare this week for training in his new duties as district manager for the Agricultural Development Authority, but he recognizes Sarath's jeep with the bicycles. As we enter the school, I see farmer-representatives from the four Area Councils gathered together. Mune and the others are surprised to see me. He learned of our plan to visit Gal Oya only this morning by phone, through a neighbor, and he whispers to us that he didn't expect us to arrive until evening, "given the roads."

Narangoda is chairing the meeting, and Tilakasiri is sitting next to him as secretary. Dissanayake, whom we got to know last summer on M16, is speaking about the problems farmers have in selling their paddy crop. I recognize most of the other FRs, but I have to look twice to be sure about Neeles, one of my longest acquaintances in Gal Oya (see chap. 4). He has had a haircut, and his laundered white shirt and sarong make him look very respectable. (Mune tells me that one of his daughters is now attending university and has had a good influence on him, including getting him to stop drinking.)

Sarath is attending as ADA district manager and is sitting with three other officials, the divisional officer for Agrarian Services at Namaltalawa, who has been very cooperative; Tilakaratne, the newly appointed project manager for Gal Oya under the ministry's INMAS program;[1] and the Assistant Director for Agriculture for this district, a lively fellow Tilak is staying with until he gets permanent quarters. The farmers do not seem uncomfortable to be airing their problems and planning their strategy for the next Project-Level Committee meeting scheduled for January 24, in front of these officials. Sarath, of course, they have good reason to trust, but the others also seem sympathetic.

Dissanayake says the Paddy Marketing Board (PMB) must agree to buy seed paddy from farmers or the private dealers will force the price down, because farmers cannot store their paddy until a better price is offered. Also, farmers need to be given at least fifty rifles for self-defense, to deal with wild animals (boars, I guess) and terrorists. There has been a lot of damage from floods (Ampare has just had a week of heavy rains) and some canals need repair. Another water issue is needed because the crop is in the booting stage, when the grain is forming. "If we don't get water soon, there will be less harvest." With this appeal, Dissanayake sits down.

Neeles stands to say that farmers in his area are concerned about getting loans to produce subsidiary crops, which need less water, in the coming dry season. Narangoda says this will be noted and halts the discussion to welcome us as new arrivals. Gunaratne, the chairman for UB9, then picks up Neeles's point. His neighbors also want to grow subsidiary crops, but because they have no certificates proving they were allotted the land, they cannot get loans. He asks that the

---

1. This Integrated Management System for Major Irrigation Schemes (INMAS) was set up under the Irrigation Management Division. It operates without external assistance and will establish farmer organizations in other major irrigation systems along the lines developed in Gal Oya as well as at Minipe.

promised certificates be issued. Next a complaint is made about the rehabilitation work. Channel levels in his area have not been measured properly, the farmer-rep from Rajagalatenne says. The TA there did this work himself—not openly, of course, because such arrangements are not officially permitted. Although some desilting was done, the water does not flow as it should. Now the TA is telling farmers that it is their duty to put things right. They have asked him in return why farmers were not given the contract to do the work properly in the first place.

Narangoda starts to speak in harsh, stern tones that surprise me. I think he is starting to sound like an "official" in the way he speaks to farmers—until Nanda translates for me. Narangoda is explaining, with anger in his voice, that they intend to introduce a resolution at the next Project-Level Committee to say that until all rehabilitation work has been completed and judged satisfactory, no TAs or engineers should be allowed to transfer out of the area. "Any persons responsible for work that is found unsatisfactory should be called back for an inquiry. They shouldn't be able to run away," he says, implying that some have left already with illicit gains.

The secretary stands to summarize the discussion so far. "Anything else?" One farmer says everyone needs to repair the bunds of any fields that were damaged in the recent floods so they can cultivate next season. Neeles adds: "A lot of houses have collapsed in the rain and need to be rebuilt. But there are no trees available for cutting timber. We need access to wood. One of the reasons houses collapsed was because their materials were of poor quality. We can't cut wood without official permission, but some influential people clear whole areas without permission. Farmers aren't allowed to cut trees even for necessary purposes."

Heads nod in agreement. One farmer says the best solution is not to build houses in low-lying areas, because flooding can happen again. He turns to Munasinghe and says in a jocular way, "You can give us permission to build houses in the highland, can't you?" Mune shrugs, and there is good-natured laughter. Neeles does not back off his point: "Big merchants in Ampare can cut down trees with impunity. We should fight for our rights in this matter." This will be taken up with higher authorities, Narangoda says.

He next invites D. S., our IO coordinator, to make some comments. It is good to see D. S. back in Ampare, though walking with a slight limp from his motorcycle accident. D. S. says he hopes they will cooperate with the program outlined by the new project manager and also that the PM will listen to the suggestions of farmers. Since the farmers in Gal Oya are better organized than in other projects in Sri Lanka, this creates opportunities for more progress. Earlier, D. S. adds, there was some doubt about this program: "It was thought that once the organizing work was done and the IOs left, farmers would be left helpless. But we can see that farmers are self-reliant now and can handle responsibility. We IOs can leave with a peaceful mind to work elsewhere." This is the most optimistic statement I have heard so far from IOs. D. S. concludes by repeating his request for farmers to work with the project manager and sits down.

I am invited to say something, so I ask their opinion about the success of the Gal Oya project. One farmer says: "In terms of farmer organization, we can say about 90 percent successful. However, the rehabilitation work is not successful." I ask, In what way? He says it is "only about 75 percent successful," with 25 percent of the work still to be done or not done satisfactorily. One farmer exclaims: "The old structures built thirty years ago by the Americans were so solid they had to be blasted away with dynamite, whereas the new structures built under this project can be damaged just by running water." There are ripples of laughter.

I ask whether they are satisfied with the distribution of water now. One farmer says "now 100 percent," and others nod in agreement. I ask about the availability of agricultural inputs, saying that an evaluation report just completed for USAID said there was no problem with credit, fertilizer supply, or marketing. "Did the report writers talk to any farmers?" asks one farmer-rep. A few, I reply. "They should have talked to more of us," he responds. Farmers are not satisfied with progress in this area. It is judged "maybe 50 percent successful," with complaints focusing on lack of credit at a reasonable interest rate. Private moneylenders still charge as much as 25 percent per month.

I explain that with the end of the project, the role of Cornell and the IOs is diminishing and so the role of the farmer-representatives must expand. I say that I hope to keep visiting the area from time to time, not in any official role but to see how they take advantage of the opportunities they now have to make suggestions and take responsibility for their own development. "I wish I could say that the Gal Oya project has solved all your problems, but it has not."

Narangoda says he is pleased that farmers are now willing to work to strengthen their organizations. "Farmers realize that they must protect their own organizations." Nodding toward Tilak, he says the project manager has worked with Gonagolla farmers previously as representative of the Crop Insurance Board. With a gracious smile he says: "We farmers in Gonagolla know him well." This sounds like a compliment and is accepted as such, but the board has been notoriously poor in making payments for crop damage (chap. 7), so it may be a veiled criticism. Narangoda adds that farmers in the Gonagolla area have great attachment to their farmer organizations and that the project manager knows this.

Narangoda then says, probably for my benefit, that the Project-Level Committee has not met since last summer because of "communication difficulties." The farmers' secretary, Tilakasiri, did not know how to convene a meeting when the deputy director of irrigation did not do so. The project manager should in the future facilitate PLC meetings, Narangoda adds. He closes his remarks by qualifying my comment that they have been successful: "We will consider our organization in Gal Oya a 'success' only if this system of farmer organization is spread throughout the island." This bold assertion is immensely gratifying for us. The secretary gives the customary vote of thanks to all who attended, and the FRs and IOs break up into small groups to discuss plans and problems. It is good to see the easy camaraderie between the two groups. There is more visible friendship and interaction among the farmer-reps than before. Mune says this may be due to the study tours ARTI arranged where farmers from Gal Oya visited other irrigation schemes in Sri Lanka.

Nanda and Mune start scheduling my time for tomorrow. There is going to be a regularly scheduled Area Council meeting at the head end of the Left Bank in the afternoon, so I can attend that, visiting Weeragoda and Gonagolla farmer-reps earlier in the day. After some informal discussion, we get into the ARTI car to travel back to Ampare. Not far from the school, Narangoda flags us down by the side of the road. He wants us to visit a new medical dispensary just opened for service. It is operated by a doctor who is a brother of one of the farmer-reps, which is why Narangoda is involved.

Munasinghe says as we walk in that he was the "guest of honor" at its dedication this morning, cutting the ribbon. There are still balloons and crepe paper decorating the austere white walls. Bananas, biscuits, and soda pop are given to us. We talk a bit about medical facilities in the area, but talk quickly turns to current problems caused by the ethnic conflict.

Farmers are unhappy that the Paddy Marketing Board is no longer buying seed paddy. Terrorists burned down one PMB warehouse, and now the Marketing Board no longer wants to have any stocks on hand that could be destroyed (and for which managers could be held personally liable). This presents a serious problem. To whom can farmers sell the seed paddy they have grown? And for next season, will there be stocks of seed available when needed for planting?

We discuss whether farmers could be given space in empty PMB warehouses to store seed at their own risk, perhaps mounting their own guards. I express the hope that the terrorists would not destroy "people's property." But there is doubt about this. The war is getting closer. Killings are now commonplace only twenty miles to the east, toward Batticaloa. Suspected terrorists are shot by army and police, and suspected informers are killed by competing militant groups. Occasionally bystanders get caught in the crossfire. It is a very unpleasant situation, but life is going on in the Left Bank area almost as before.

Irrigation Department Visit

The next morning we get to the Irrigation Department offices by 8:00. Unfortunately, Godaliyadde, the engineer in charge, has not returned from a trip to Kandy, so we cannot talk with him. A lot of the credit for improvements in water distribution that farmers spoke approvingly about yesterday goes to him and his staff. Farmers' complaints about the ID are directed mostly to its construction wing, not those handling operation and maintenance. The TA on duty pulls out figures on water availability and issues.

At the end of December, the level in the main reservoir was only 234,000 acre-feet—30 percent full, not very good. But six days of heavy rains since then (thirteen inches in one location) have produced 60,000 acre-feet of inflow to the reservoir. The water issues for this season have come to only 1.3 acre-feet per acre so far, and the total issue for the season is not expected to exceed 2. This will be a great accomplishment, since the national norm is 3 acre-feet per acre and at least 5 used to be issued here before the project started. That farmers' fields are getting sufficient water with reduced issues means it is being well managed by them at field and distributary channel levels.

The TA we are talking with is now responsible for water distribution in the Right Bank area, but he previously worked in the Uhana area and knows the farmer organization program. He says that cooperation has become very good there and that farmers now have confidence in the ID. For example, water distribution was stopped during the recent rains so that channels would not overflow and be damaged. But the rice is now at a stage where it needs water again, and deliveries will begin tomorrow morning at 6:00. Farmers have been informed of this through their representatives and were told to get their fields ready for water, repairing and bunds that were damaged by the flooding so there will be no waste of water when flows resume. The TA is confident that farmers will have done this by now and that the water delivery will go smoothly.

"We've Got It on Our List"

We drive to the small store and house where two IOs, Dissa and Priyantha, rent a room together. Ratnayake, the elderly chairman of the Weeragoda Area Council, is there waiting for us along with a young farmer-rep who is the council secretary. Eventually there are half a dozen representatives who have come on short notice. Ratnayake has recently attended a farmer seminar organized by ARTI, and he says that farmers from other areas were amazed by the progress Gal Oya representatives could report.

They are obviously feeling good about their organization. Ratnayake says that with only two IOs left for the Weeragoda area (about 7,000 acres), it is necessary to change their mode of operation. It is impossible for IOs to set up all the meetings, for example, so farmers now have a regular schedule of meetings for the whole area. The Area Council meets on the twentieth of every month. Actually, he amends the remark, it is the Area *Committee* that meets on this date, not the Area Council, which includes all farmer-reps for the whole area—over one hundred. The Area Committee functions as an executive committee comprising the chairman, secretary, and treasurer from each of the nine D-channel organizations, plus a chairman and secretary elected from and by the whole Area Council. Each D-channel organization meets on a regular date in advance of the Area Council meeting.

Ratnayake says there have been 311 farmer meetings in the Weeragoda area over the past four years, and he has attended all but four of them, giving reasons for each of these absences (wedding, illness, etc.). Who presides over Area Council meetings in his absence? The vice-chairman, he says, pointing to a farmer-rep who has just arrived. To shift the conversation from Ratnayake, I ask the vice-chairman, "How strong are the D-channel organizations?" He thinks a bit and says, six very strong, three just strong. Ratnayake interjects that one of the latter, M16, should probably be considered weak. It is in an area where terrorists operate.

Why are the three weaker than the others? Because there is a lot of mortgaged land in these areas and participation in the organizations is poor. Why? Because sharecroppers are reluctant to invest time in the organizations. They seek to maximize their benefit from the current season. "Maybe they will be cultivating on a different channel next season." What percentage of farmers are sharecroppers? "As many as 40 percent in some areas."

Are there still water problems? No, is the answer. To play devil's advocate, I ask, Why should people participate in farmer organizations now if water is adequate and timely? Ratnayake says there are still many other problems facing farmers. They need to work on agricultural improvement and on alleviating debt burdens through group savings. "If farmers carefully follow the instructions of the Agricultural Department, they can get very high yields. Before, yields were 40 to 50 bushels per acre. Now it is possible to get even 100 to 125 bushels per acre, double or more. Farmer organizations should promote such improvements." Unfortunately, when there were serious water shortages before, some farmers here who could not get enough water for a decent crop sold or mortgaged their land. "Now that there is good water supply, farmers can get good yields. In unit 16, land used to sell for 5,000 rupees an acre. Now the price is over 20,000 rupees per acre if the water supply is reliable."

Is the supply of agricultural inputs adequate? "Merchants have chemicals and fertilizer in their stores, but the cost of private credit is prohibitive. If our organizations are recognized by the government, we could build up capital through savings and loan activities." Would the merchants give the organizations trouble if they tried this? "No, we farmers are a large number and they are few." Have they given the organizations any trouble so far? No, some are even members. This gives me pause, and I express concern that they might gain undue influence. But Ratnayake says they want to draw in people who have ten to fifteen acres, far above the legal limit of four. "We want to get them even as farmer-representatives and then to try to convince them to return at least some of their holdings to rightful owners through close association with those who have lost their land." I say this is a risky strategy, and they laugh knowingly. To myself I think, such a conversation with the leaders of farmer organizations in most other South Asian settings would have me face to face with the large landowners and merchants who are doing the exploiting. Here we have mobilized a cadre of middle-sized and small farmers into leadership roles.

Ratnayake says they have already gotten two acres returned this way: "Everyone knows that the transfer of land is illegal in Gal Oya. But even when the government gives land back, farmers have sometimes returned it to their creditors because they feel obliged to render some benefit to those who have assisted them in time of need. Changes will only come through farmer solidarity." I express approval for such efforts, saying that the American funds provided for rehabilitation of Gal Oya were intended to get benefits to the majority of small farmers, not to improve water supply for rich farmers who captured land when the water supply was inadequate.

Ratnayake takes off on this, saying he knows that millions of rupees were spent for rehabilitation: "But much money was wasted because work was done under contracts that some Irrigation Department officials gave to themselves improperly. Much of the work is incomplete or of poor quality. Farmers are trying to get this remedied." He tells us about the resolution that Narangoda said yesterday they intend to introduce at the next Project-Level Committee meeting. Obviously farmers have been conferring on this.

I suggest that they should make a detailed list of the various deficiencies, carefully checking out each allegation. They should not make any minor or frivolous accusations, such as those the engineer Hassan made against farmers in his memo that charged them with breaking a lot of structures. The farmer-reps become agitated, all speaking at once. They say they know about Hassan's memo. Ratnayake says flatly: "It was sent to cover up his own mistakes. We have in fact made a separate list of the faulty work for which he was himself responsible. Unfortunately, we weren't able to give it to the District Minister before Hassan got himself transferred out of Gal Oya to another posting." When presenting the allegations, I say, they should be careful not to criticize the whole Irrigation Department for the misdeeds of a few. Ratnayake nods: "We know this. We have already thanked the department, especially engineer Godaliyadde, for improvements in water distribution."

I pursue the matter of documenting rehabilitation shortcomings. Ratnayake says they have received fifty letters from farmer-reps in the Weeragoda area about substandard work: "These we have screened down to twenty-five complaints judged serious enough to take to the Project-Level Committee." They are already a step ahead of me, I am glad to see.

Ratnayake says that some of the TAs who left the project area have spread the story that they completed all the work satisfactorily and handed over responsibility for the channels and structures to the farmer organizations—so any shortcomings in the system now are the farmers' affair: "We have not received anything from anybody. We can maintain the system if it has been correctly rehabilitated. If given this responsibility publicly and if we are satisfied that the structures are in working order, we will accept the responsibility. But only if it is handed over properly by the GA or DDI, not by a TA." I ask for an example of the kind of rehabilitation faults they have documented. On M5.4.3, the stock tank bund was rebuilt by a TA who took the contract and used sand instead of earth to repair the bund. Also, he did not dig the footings for the gate deep enough: "The recent rains washed away the bund, and two nearby fields were spoiled by the sand that now covers them. The gate is buried under the sand. It is there for anybody to see." This sounds like a valid complaint of poor work.

How, I ask, can you keep out merchants or politically ambitious persons who would use the organizations for personal advancement? Ratnayake answers: "All farmers know that politics can destroy the farmer organizations. They will keep out any persons who threaten their organization." How can you be so sure? I ask. "Remember the so-called Independent Farmers' Organization started here in Weeragoda some time ago?" he replies (chap. 6). I nod yes. He makes his point with a wonderful figure of speech. "Well, it vanished like the fizz in a soda bottle."

Ratnayake says that though he himself was involved in party politics earlier, since being elected a farmer organization officer he has remained aloof from politics. "I vote of course, but I do not take sides. Otherwise others would too. Both the UNP and SLFP organizers in the area are members of the farmer organization. They are very friendly within the organization. No backbiting."

Will the organizations here need IOs much longer? Ratnayake says they assume IOs will be gone after this year. "But we will not require any. We are getting ready to manage completely by ourselves." I explain that we hope to have a few IOs continuing to work in Gal Oya to give assistance when needed, but I am pleased they expect to be self-reliant. A farmer-rep says it will be good to have some IOs, but they know they should not depend on such help anymore. We have to leave for our next meeting so we say good-bye. We stop in Uhana for a quick lunch at a place that serves notoriously hot curries.

A Farmer Success Spoiled

We arrive at Narangoda's house and find a number of farmer-reps waiting there. He is not at home, having forgotten yesterday when we agreed on the meeting that he had to attend a court proceeding this afternoon. But Punchibanda and others usher us inside. Jina, our IO coordinator, is here, as are Ratnayake, the most senior farmer-rep, and Menika Rajapakse, the only woman FR. Eventually fifteen FRs join the discussion, including Narangoda, who unnecessarily apologizes for not being home to welcome us.

I tell the farmers that I have already heard about their setback after their victory last summer doing rehabilitation work on LB29 (chap. 7). I ask short Punchibanda, who spearheaded the effort, to tell me more about the unhappy experience. He shakes his head and says the TA instructed them to redo the work, saying he had given them the wrong specifications. They were told to build up the roadbed twenty-four inches instead of the nine previously specified. (At their own expense, they had already raised it twelve to eighteen inches to make it sturdier, and the TA approved payment on this basis.) This sounds quite unjustified. They spent more than 12,000 rupees on this added work. This and other work leaves them now with a profit of only 5,000 rupees.

Punchibanda says that Hassan, the construction engineer, told him after I left last summer, "It was a mistake to have told the American professor about the profit." I see that if I had not reported their success to others, they would have been spared the extra work and could have kept their profits. I feel bad about this and say the mistake was mine, not theirs. My apology is shrugged off. "We learned a lot from this."

How are relations with the TA now? "Before, they were very good. He even praised our work after we first did it. But now we know he is a great cheater. He was very unhappy that the extent of profits possible under the contracting system was revealed. He took his profits and ran off. Many TAs were in a hurry to leave the project without completing their work."

What is the situation now? Punchibanda says that the organizations in Gonagolla are "now going downhill." But another farmer says "the trend is going up, especially since the study tour." Punchibanda defends his statement, saying there is much disillusionment because of the way the rehabilitation work turned out. He tells about a problem on LB31, a D-channel off the main canal that is 1.5 miles long.

Officially it serves twenty-three farmers and ninety-six acres, but an additional one hundred acres at the tail are cultivated without authorization.

The Irrigation Department classified this as a field channel and did not rehabilitate it, saying this was the farmers' responsibility. But an adjacent channel, LB30, with only thirteen farmers and thirty-nine acres, was fully rehabilitated by the ID as a D-channel, and LB32 got the same treatment. When LB31 farmers complained, ID staff conceded that at least some rehabilitation work should be done but said their funds had run out. So if LB31 farmers wanted their channel to function properly, they would have to do most of the work themselves. When farmers protested that this was unfair, they were brushed off with an analogy: "The LB30 farmers were lucky, just like some people who win in the lottery. You on LB31 happen to be among the unlucky ones who lost out." This explanation did not satisfy them, since they feel entitled to the same treatment as farmers on the channels above and below them. The ID subsequently offered to rehabilitate half the channel using its maintenance budget if LB31 farmers would first clean "their half" of the channel. But, these farmers are now so angry about the unfairness of the situation that they have been unwilling to do the work requested.

Could they get the "lucky" farmers from neighboring channels to join in shramadana? Ratnayake says it would take five or six days of work even with a tractor because of the length involved, and it is hard to get anyone to give up use of a tractor for that long. "Besides," growls one farmer, "what happened to the money allocated for LB31?"

Hard feelings are coloring their thinking. I ask whether they want to pursue their grudge with the ID or get water flowing. Someone says the priority should be getting water to farmers on LB31, and others agree. They will see if they can get farmers from several channels to help LB31 farmers clean half the length so they can pressure the ID to come through with its part of the bargain, making the best of an unfortunate situation.

How is water distribution now compared with five years ago? They say it is five to six times better. Tall Punchibanda, who has been silent so far, perhaps because he was ostracized for some time, says "There is now 100 percent success in water distribution." That he has not hesitated to criticize official shortcomings in the past makes his statement of satisfaction all the more credible. Ratnayake says: "Five years back, the gap between farmers and officials was so wide. [He gestures with his hands.] Now there is no significant gap. [He draws them together.] Relations are very amicable." His assessment also deserves to be taken seriously.

What about paddy yields in Gonagolla? In 1981–82, they were thirty-five to forty bushels per acre, farmers report. Now they are fifty to sixty bushels. These yields are not as good as reported this morning in Weeragoda, but soil and water conditions are not as favorable here. The increase is still an impressive 50 percent. The USAID evaluation team, after reviewing government and ARTI data, said there was no evidence of any increase in yields due to the project. But these farmers have no reason to exaggerate gains.

What has contributed to the increase in yields? Better water deliveries, is the first response; better rice varieties; more agricultural extension advice. "Farmers had no scientific knowledge before," says one farmer. They say, however, that the number of extension agents has now declined. We discuss whether the farmer organizations could help to spread agricultural information more efficiently. Narangoda says: "Our farmer organizations have neglected to work systematically on agricultural improvement so far, because we have been preoccupied with rehabilitation problems and with water management. But now we will turn our efforts to agricultural work." What kind of assistance will the farmer organizations need in the future? Narangoda says they are worried about coordination between the D-channel organizations and field channel groups if the IOs are all removed. For example, it is difficult to travel between Gonagolla and Kotmale colony, he says. I respond that I hope the IOs are not needed as messengers, and the point is accepted. I describe the Weeragoda plan to schedule all meetings regularly so the need for ad hoc communication is reduced.

We need to excuse ourselves to go to the Namaltalawa Area Council meeting, but Narangoda says first there must be a vote of thanks. He rises and says that this farmer organization program started as an experiment, and now it is to become a national program. He hopes the ministry will do everything to extend it adding "It could become a national program because of the farmers' strength." They understand this now, which is good. Previously they credited ARTI and Cornell too much for the program's success. Narangoda concludes: "In 1982, we in Gonagolla were proud to be working at a higher level than other areas. We recognize that at present this is no longer true. We have fallen below the level of other areas. But I promise that with the help of my colleagues here, next time you visit us we will have regained our leading position." The farmer-representatives clap their hands in endorsement, even sadder but wiser short Punchibanda. I am glad to see that our friend, tall Punchibanda, who fell from grace in October 1984, now appears to be accepted back into the group.

"Farmers Always Lose"

The meeting of the Area Council for the head of the Left Bank system is already in progress when we arrive at the school where thirty-five farmer-reps are sitting with half a dozen officials. A manager from the Paddy Marketing Board is standing, responding to farmers' comments and questions. He looks embattled but has not lost his composure. A farmer is saying that the cooperatives won't buy paddy from farmers because they lack the capital, and the PMB won't buy because it hasn't enough storage space. (The implication is that each has what the other lacks.) So farmers have to sell their paddy to private traders at whatever price is offered. He asks whether farmers can be given one of the empty PMB buildings to use as a warehouse.

The PMB manager says farmers should not expect his agency to buy all their produce: "There is a problem only if you can't sell your paddy for at least the guaranteed floor price. Once we tried to buy paddy through the farmers' cooperative, even providing sacks. But we got no paddy because the farmers were selling their crop to private dealers who offered more than the guaranteed price." He concedes that whenever the PMB opens a buying operation, private dealers raise their price until the PMB closes down its operation for lack of business. "Then they lower their price again, paying less than the guaranteed price. But what can PMB do? We can't have buying operations everywhere. We do not have enough warehouses."

A farmer-rep stands and says: "If the cooperative were buying regularly at a fair price, private dealers wouldn't come into the picture. But the co-op lacks funds and staff, so private dealers can exploit farmers. Why can't the PMB buy directly from farmers?" The manager responds that farmers' paddy is "too wet." I have heard this complaint before. Disagreements are common over whether the moisture content of the paddy being sold is too high. A farmer counters: "How is it that the private dealers to whom we sell our rice at less than the guaranteed price can turn right around and sell the same rice to the PMB for the guaranteed price? How is this possible if the rice is too wet?"

The PMB official is flustered, and farmers nudge each other, smiling. Fortunately, no tempers flare. A farmer says that private merchants come right to the threshing floor to buy paddy: "They make no complaints about the quality of our produce, about its moisture or about sand. But when we take the very same paddy to the cooperative or the PMB we get complaints from its buyers who want to pay us less." The PMB manager says: "If you find any shortcomings in our operations at the storehouses, come and inform me. There is no point in making complaints in public meetings." He seems uncomfortable to have us as outsiders hearing this give and take. The chairman tries to initiate some constructive action: "We can undertake not to sell our production to anyone else in the coming season, but to sell it all directly to the PMB. We farmers will agree to give you only quality paddy. If we do this as a group, we can save on transport costs, which are nine cents a kilo."

One of the farmer-reps who is a leader in the Small Farmer Development Program (SFDP) sponsored by FAO and being introduced separately in this area by ARTI[2] adds: "If the PMB is going to discharge its responsibilities to the farmers, as farmers we will do the same toward the PMB. But we want you to keep your buyers from forcing us to give them an extra sack of paddy for each load purchased. The person who tests moisture content always says it is too high. There should be fair testing of paddy." If paddy is indeed "too wet," any subsequent weight loss during drying must be made up personally by the buyer, so he can reasonably request some extra amount of undried paddy at time of purchase. But as the farmer has just suggested, this requirement can become a routine extraction, not an adjustment for excessive moisture.

The representative continues with graphic imagery: "Rules are set in such a way that farmers always lose. Everything is said to be our fault. We are like the man who got stabbed but whose death was judged by the coroner to have been caused by his having fallen down." The PMB

2. ARTI introduced a Small Farmer Development Program funded by FAO to the head end of Gal Oya late in 1983, without consulting Wijay or Lucky and for reasons never made clear. What need was there to start this new program when ours was already underway and making good progress, and when so many other areas needed farmer organization? We surmised that some persons within ARTI who were jealous of the attention we were getting wanted to upstage our program. The SFDP program was first started at Galgamuwa in the north, following principles of organization we judged unpromising. Its base groups were very large, sixty to seventy instead of fifteen to twenty, and it used the not very popular yaya palakas as organizers instead of independent persons like our IOs. The incentive given farmers to organize was access to subsidized credit. Once ARTI withdrew from Galgamuwa and foreign funding ceased, so did the organizations.

manager can't help smiling at this, but says: "Sometimes farmers are at fault. Often you bring low-quality paddy to the PMB, having sold your good stuff to private traders for a premium price. You expect us to pay standard price for substandard grain." In his own defense he says that the one kilo per bag that farmers say is taken by buyers doesn't get passed on to the PMB managers, thereby conceding that such extractions do occur. He repeats his complaint that farmers sell their poor paddy to the PMB and that this hurts his agency. The SFDP farmer-representative promises they will provide only "quality paddy" to the PMB.

The chairman suggests that they experiment with this in the coming season. "Farmers should let their crop lie several days in the field after cutting and before threshing it so as to reduce the moisture content." The secretary says that some farmers won't want to sell their whole crop at one time, and also some farmers are used to selling their pro-duce to a private dealer at a satisfactory price. He is right when he says, "We can't commit all farmers in this area to sell all their crop to the PMB."

Lilaratne, one of the enthusiastic young SFDP farmer-reps, stands and says earnestly: "Organizing farmers is not like collecting hens or dogs and putting them into a cage. It should be based on knowledge. We should try to understand people, to educate and convince them, not force them. Earlier it was said that getting farmers together is very difficult. But now we farmers have so many meetings that officials and IOs can't attend them all."

The chairman says it is time to select officers for the coming year. The Area Council covers all channels from the head of the system down to LB12. He asks them to resolve that "all farmers here will live as members of the same family" and to select such members as their officers.

Lilaratne stands and says that unfortunately this Area Council has not been as effective as others: "In selecting new officers, there should not be any prejudice or favoritism. We should select the best person who will work sincerely with full devotion to his colleagues." Is he making a campaign speech? Against the current officers? For himself? He implies that current officers show less than full devotion. But he sounds a conciliatory note, saying they must all try to be more coop-erative and more unified in the future.

I wonder how the politics of officer selection will work. The secre-tary proposes that the chairman, Peter, be reelected. Neither of them is from the SFDP "faction" that has become evident during the meeting.

To my pleasant surprise, this nomination is seconded by Sirisena, the most articulate of the SFDP representatives. But Peter, to everyone's surprise, says he prefers not to be renominated. "I recognize that there have been shortcomings in the work of this Area Council, including a lack of unity. Maybe you can find someone who is more qualified than I." Sarath, perhaps improperly, cuts in, saying that the chairman should not withdraw himself but should let others decide who is most qualified. Lilaratne, who made the implied criticisms of the chairman, says he should not withdraw. "We will help you do a better job." Peter bows his head and says, "All right."

There are no other nominations, so Peter is declared reelected, and he proceeds to nominate Sirisena for treasurer. Another SFDP leader is proposed as secretary, and then Lilaratne as assistant secretary. Gunasekere is put forward as vice-chairman. There are no other nominees. Except for the chairman, all the other officers are from the SFDP group. We will see whether this reflects greater energy and interest on their part or a kind of coup.

Nominations are invited for the executive committee, one representative from each D-channel. It is getting dark as the chairman goes "down the system," getting an acceptable nominee from each command area. When this is done, he consults with Kularatna Perera, the IO helping here, standing with his hand on Kule's shoulder in a friendly manner. There appears to be no other business pending. They will meet again on the twentieth of next month and will invite the cultivation officers and extension agents to talk about agricultural problems.

The treasurer says he wants to go over the budget from their *pirith* religious ceremony last month, to be sure there are no misunderstandings about the accounts. But it is so dark that he can hardly read the figures. The secretary, who reminds everyone that he has been serving only as the acting secretary, invites the newly elected secretary to give the vote of thanks, which is done very nicely. The officials who have put in a long afternoon with the farmers show no sign of resenting the lateness of the hour. Everyone appreciates their having come.

Farmers break up into small groups and slowly disperse on their bicycles or on foot. We excuse ourselves and head back to the circuit bungalow, where we have dinner with Sarath and Munasinghe and also the new project manager and his friend, the district head of the Agriculture Department. It is a relaxed evening, and it is evident that the farmer organizations and IOs are taken for granted now as part of the institutional landscape of Gal Oya.

Meeting with IOs: Learning from an Evaluation

The next morning, IOs gather at the circuit bungalow for a group meeting with Nanda, Razaak, and me. Sarath is able to join us as well. We begin by discussing a new self-evaluation system for farmer organizations that we have developed (Uphoff 1988a). Clearly, ARTI did not start this out in the correct way, proceeding in a "top-down" manner in contradiction to our philosophy. The IOs think they can make the process work in a more participatory way now that all understand better what is required.

The IOs ask about the conclusions of the USAID evaluation team (ISTI 1985). They fear that the farmer organization program may have been misunderstood and unfairly criticized because the team did not talk with many farmers who know much about its accomplishments. Indeed, the evaluators declined to meet Gonagolla farmers because they were known to be better than average rather than typical. This caused some unhappiness among these farmers, who thought they should have been given a chance to tell the team how the organizations had improved their lives. Anula says that one group of farmers the team stopped to talk with along the road at random were recent settlers. "They know little about our organizations because they get their water from drainage runoff." She says that while not all the farmers in her area know about the organizations, most know a lot about them, and all the farmer-reps are very knowledgeable.

I say that the IOs' apprehension is understandable, given the very short time the evaluators spent in Gal Oya, but even so their report was quite positive about the organizations' contribution to better water management and about the IOs' role in creating the organizations. I read from some sections of the report, and they are satisfied. Anula asks whether the team made any suggestions for improving the IO program. Our collection and evaluation of data about program operations and effectiveness were faulted—correctly—as being unsystematic, I say. They agree that we need to improve on this.

We next talk about two statistics the evaluation report cited as indicators of a decline in the effectiveness of farmer organizations. We learn how measures can be misinterpreted when their context is not understood and they are viewed statically. According to the IOs' own figures, the percentage of farmer groups cleaning their field channels by shramadana has declined over the past two years, from about 80 percent to about half. Does this mean that the field channel organizations are now less effective? Munasinghe says this is not a good indicator of effectiveness: "At the start of this maha season there was the best channel cleaning ever. Almost every one was desilted and repaired before the first water issue. Half the time farmers decided it was simpler for them each to clean an assigned length of channel rather than to gather and do the work together on a designated day or several days." Such an assignment of individual responsibility was the nominal system for field channel cleaning before our program started. But few did the assigned work, especially not head-enders, who got water whether the channel was cleaned or not, and who might even get more water if the work wasn't done. It made no sense for farmers downstream to clean their part of the channel if those upstream did not, so the channels deteriorated.

Now that the organizations are in place, the IOs say, there is an expectation that every farmer will cooperate. Often farmers find it preferable to do their portions of the work at their convenience. When the IO program started, we used shramadana as a device for organizing farmers, to get the process of group formation and cooperation started. That strategy is no longer needed. Also, channels were then in such run-down condition, not having been cleaned for five, ten, even twenty years, that individual efforts appeared futile. Now that channels are in reasonably good condition from previous group action, individual efforts can suffice. So the amount of maintenance work done by shramadana is no longer a good test of the organizations' real strength.

The evaluation acknowledged that field channel maintenance appeared to have been much better this past season then in previous years, but it attributed this to the Government Agent's having issued a stern warning threatening legal action against any farmers who did not do their share of maintenance work. This improvement was cited as evidence that strict administrative enforcement could be successful, but Munasinghe says this interpretation is wrong: "The farmer organizations themselves requested the Government Agent, through me, to issue that statement. They still lack standing to bring legal action against any farmers who do not comply with their decisions, so they wanted to strengthen their hand against uncooperative persons. IOs put up posters of the GA's proclamation all around, and our farmer-representatives distributed thousands of copies of the statement printed up in a leaflet. So the GA's 'threat' was publicized through our program. Getting any stubborn farmers to cooperate was mostly up to the FRs anyway."

I ask how much maintenance would probably have been done without the GA's pronouncement. Maybe 80 percent. And how much if there had been only the GA's pronouncement and no farmer organization? May 40 or 50 percent. (Before 1981, similar statements produced perhaps one-third compliance.) This appears to be a good demonstration of the value of complementary approaches. A combination of bureaucratic and participatory means produced "99 percent compliance," according to Munasinghe's estimate. Even if only 90 percent, this would be a real accomplishment.

The other statistic concerns less regular meetings of field channel organizations. The evaluation team cited data to this effect from a survey that ARTI and Cornell are doing for their own evaluation. Whereas about 80 percent of field channel groups met at least monthly at the outset, that figure is now about 50 percent. The IOs acknowledge that this is true, but D-channel organizations and Area Councils are now meeting regularly and often; they did not exist when the program started. Most of the problems that can be solved at field channel level have been taken care of, so efforts have now shifted to problem solving at higher levels.

The IOs think the whole structure of organization should be evaluated together, not inferring too much from any one level. Also, since the farmers who get water from a single field channel see each other frequently, they think they don't need so many formal meetings now. Once a season is often enough.[3]

D. S. reviews the plan of work for 1986 that the remaining IOs worked out last week. I write it down as fast as I can, impressed by its thoroughness and insightfulness.[4] The IOs have not been demoralized

3. When I discussed this with the social scientist member of the evaluation team, she pointed out that she had been using measures of organizational effectiveness that we ourselves had used in the first years of the program: the proportion of channel cleaning done by shramadana and the frequency of field channel meetings. We were being "hoist with our own petard," so to speak. This shows how as an institutional development effort proceeds, the criteria by which it is evaluated should be reconsidered and changed to reflect new circumstances and needs.

4. The plan was as follows: "(1) We will give more priority to working with D-channel organizations, trying to strengthen and consolidate them, assisting field channel organization through the D-channel organizations and officers. (2) As water problems become less, we will give priority to working on agricultural problems in the coming year. [This anticipates what we discussed the day before in Gonagolla.] (3) We plan to get all the farmer organizations registered as legal entities. (4) We will extend our training activities to cover more farmers, farmer-representatives, and officials, provided we get budget for this from ARTI. (5) To raise funds for operation of farmer organizations, we will try to persuade the ID to give them contracts for channel maintenance. (6) We will see that every D-channel organization has its own constitution. (7) We will try to keep better records on

or deflected from their task by their shrunken numbers and the continuing uncertainty about their future. I ask Munasinghe, Who put this plan together? "We all did," he says. "Priyantha took the lead, and of course I helped them. But it was a group product."

Could it be that our training and supervision were really so good? Probably not, though we must have been doing something right. The experience of working in Gal Oya itself seems to have focused fine but latent intellectual energies at the same time as it mobilized the physical effort required to make an impact on this 25,000 acre area and beyond. Mune deserves immense credit for all this, but of course it is the IO cadre that has come through five years with so much tenacity and conceptual growth.

D. S. notes several things needed from ARTI: support for maintaining the IO unit office at ID headquarters in Ampare; increased traveling allowances to cover larger areas; providing motorcycles for all IOs; redesignating IOs as "area coordinators," since they are no longer really "organizers"; budget for training programs; and funding for farmer study tours to well-managed irrigation systems elsewhere in Sri Lanka. Such tours, they say, have been quite beneficial to the program, as we ourselves have heard from farmers.

Mune describes the impact such a tour had in Uhana, where farmer activity had been "somewhat lagging." The farmer-rep from the tail of UB2.3, who impressed me at the Project-Level Committee meeting on Sunday, has emerged as a strong leader there partly as a result of the trip, which he took responsibility for organizing. Sumitipala got a farmer-rep who had gone on a previous study tour to come and brief the Uhana farmers who had been chosen to go. Among other things, they were cautioned against drinking on the bus and spending money on lottery tickets along the way. So Sumitipala issued an edict: no drinking until after the last stop on the trip, and no gambling! "Everyone came back very enthusiastic about what was seen and heard in other systems and pleased with our relative progress in Gal Oya.

---

all organizations. (8) We will schedule meetings to meet farmers in each area on fixed dates, with set places and times. [This had already been decided for Weeragoda.] (9) To achieve these goals: (a) IOs should spend about four days a week meeting FRs and one day a week in report writing and program business; (b) we should work as much as possible through scheduled meetings; (c) more emphasis should be given to planning and implementation meetings; (d) instead of meeting farmers individually, we will work with FO officers to focus attention on organizational and training matters; and (e) rather than get involved in the operational details of FOs, we should be standing back more to observe their functioning and make them more efficient. (10) IOs in each area should meet fortnightly to review the situation in their area. Process documentation reports for the whole area should be prepared monthly, with separate sections for each area."

Sumitipala made them all report on their trip to their field channel organizations." A stronger sense of fellowship and of purpose among Uhana farmers has now emerged. Forging personal links among the farmer-reps was possibly more important than what they learned from the visits.

The IOs have been thinking about what can be done in Gal Oya if we lose still more IOs before the program is made permanent. Sarath expects to remain in the area for some time to come and to be able to oversee the program in any case. The options they have discussed are (1) Give responsibility to some of the most active farmer-reps and give them some allowance for their work. (2) Co-opt some of the officials in the district who have shown keen interest in the program and have them assist it, getting some extra pay for their time. (3) Get some of the former IOs who are working as teachers in the area to devote some time to overseeing the work. Any of these might be effective, but I hope the government will soon make the IO program permanent and assign at least several IOs to remain in Gal Oya to ensure good operation and maintenance (O&M) of the organizations built up over the past five years.

It is time for lunch. A splendid rice and curry meal is laid out for us on a long table. Because our numbers are so reduced, we can make this a sit-down meal together, and the cooks have prepared some extra dishes to make this a more festive occasion. I am impressed that the IOs have made fewer requests for themselves and have expressed less apprehension about their future than previously. Maybe it is because they have given up hope and are all expecting to abandon the program? The quality of their 1986 work plan does not suggest this. Those who remain are fewer than the number of permanent positions likely to be created. Maybe this makes them more relaxed about their futures.

After lunch, we discuss how to strengthen the program's links to officials of various departments. Not all come to farmers' meetings when asked to do so. Sarath can help from his position as ADA manager, but it would be good to get letters from district heads of departments telling their staff that they should attend meetings when invited. Also, the Area Councils can send letters to all the officials they work with announcing their regular meeting date, saying that officials are always welcome to attend but that they will receive a special invitation when there is an item on the agenda that concerns their department.

It is time to conclude our meeting if we are to get back to Colombo by midnight. We have been advised to complete at least the first part of our trip while it is still daylight because of continuing security problems around Gal Oya. There is the usual vote of thanks by D. S., and I express my satisfaction that they are carrying on the program so capably under conditions that continue to be difficult and uncertain. There have been many disappointments, and the program has not met all of our hopes or others' hopes for it. It has had many shortcomings, which we know better than anyone else. Nevertheless, we can all take real satisfaction in the progress so far.

The program, I suggest, is like an infant that just won't give up on life. Even when abandoned by others, it manages to survive. As I say this, I realize that this figure of speech is probably too graphic, possibly impugning motives in the ID, ARTI, or Ministry of Lands. Yet people outside Sri Lanka have found it difficult to imagine how a program can persist and progress with such a huge turnover in its cadre (now 95 percent), with six different supervisors within three years.

We go outside to take pictures of the IOs. How much their numbers have shrunk is evident from the ease with which I get everyone into my camera's viewfinder now. Parting is, I am glad to say, less emotional than after most previous meetings. Everyone seems very matter-of-fact, with signs of camaraderie all around. There is little evident apprehension, despite the larger uncertainties that surround Ampare district and the country as a whole. The program itself seems reasonably established. I ask myself, What will it look like on the next visit?

The Trip Back: Accounting for Momentum

On the way back to Colombo, with Munasinghe traveling with us, we do a lot of reflecting on the program. I comment that we were fortunate to have had such good support from the District Minister and from the previous Government Agent and deputy director of irrigation, who are now gone from Ampare. The farmer organizations would have been stillborn had the minister not let them remain free of partisan politics. The GA several times stated publicly that any official who was not prepared to work with farmer organizations could leave the district, and the energetic and sympathetic assistance that Senthinathan gave the groups has been documented often in my previous trip reports.

Still, it appears that a complicated process of social, institutional, and attitudinal change has been involved. Munasinghe says he doesn't want to take anything away from these three gentlemen, but they were not all so favorably disposed at the start, though they became strong supporters. The initial successes of the farmer organizations seemed to get the ball rolling. A favorable image of the program justified and attracted the support of district officials, which was of course critical for further progress. How personally inclined the officials were to support farmer organizations at the outset is hard to know and is a moot point anyway. What counted was the *interaction* between farmers' and IOs' initiative and performance, on the one hand, and the endorsement and assistance they received from higher levels, on the other. Each reinforced the other and contributed to a momentum that helped the program through its many difficulties.[5]

It will take more time and more reflection to construct an adequate explanation of how and why the process has unfolded as it has. But since it is apparent that this program has accomplished more than anybody (myself included) expected, such an analysis should be undertaken. It is tempting to try to explain this all idiosyncratically, as due to certain personalities or events. Such factors can be easily overestimated because they are so visible. For changes as widespread and persistent as those we now see here, there must be other explanations, in addition if not instead, and ones possibly quite subtle though systematic. In the meantime, I hope the government will follow through on its earlier agreement that if this experiment proved successful it would become the basis for a national effort.

A Subsequent Trip and Then a Long Wait

Because 1985–86 was a sabbatical year for me at Cornell, I was able to return three months later, this time with official sponsorship from USAID and the government of Sri Lanka.[6] By April 1986 the security

---

5. When he was participating at a seminar at Cornell University in May 1987, I asked Senthinathan, without advance notice, to tell us how engineers came to support farmer organizations. He said that when he returned to Gal Oya in 1981 as DDI, he was not particularly well disposed toward farmer participation. Chuti and another IO came to meet him in his office and explained the program to him. "I was not prepared to accept it fully, but I put my hand out partway. As I got better acquainted with the program and saw good results, I stretched my hand out a little more, and a little more. Eventually I was fully persuaded." This retrospectively describes the process we had been observing.

6. My first trip in 1986 had been supported from my Cornell research account, because once the USAID-funded project ended, it was more difficult to get the necessary approvals from both the United States and Sri Lankan governments to make an official visit. I was able to cover most of the cost of travel by linking it to the farmer organization work we were starting to plan for irrigation improvement in Nepal.

situation had gotten worse around the Gal Oya project area, though I traveled to Ampare without trouble with Razaak and Nalini, both Tamil-speaking, via the northern route. Just to be safe, we drove through the most insecure stretch at midday. A few days before our visit, terrorists had attacked the police station at Vellaveli, ten miles east of Gonagolla, at the tail of a branch canal we were organizing along two years before. And the day before our visit, police raided a terrorist hideout and seized weapons taken from their station, killing five and capturing fifteen in the attack and losing a sergeant in the process. So the conflict was more intense than in January.

Yet the program had moved ahead. On the day before our visit a new Project Committee met, composed of twenty-five farmer-representatives and almost as many officials, including the District Minister and Government Agent. Farmers were unhappy, however, because a new subcommittee of the District Agriculture Committee had been created to oversee irrigation matters with only official members. There were no farmer members on it, as on the DAC itself. The project manager had offered to have one farmer-representative on the subcommittee, but only one, because he thought having one from each of the four Area Councils would make the body too big. When farmers refused his offer, he decided that *he* would represent farmers. This, they felt, undermined their access to official decision making and reestablished top-down relationships. The new bodies and roles undercut the four-tier structure that was in place in January.

The future role of Institutional Organizers, on the other hand, seemed clearer, since there was cabinet approval for twenty Assistant Project Managers for institutional development in major irrigation schemes, including Gal Oya. Former IOs could be appointed to these positions. The need for catalytic roles within a conventional system of management was recognized. Area Council attendance was down in Gal Oya since farmer-representatives were not being reminded of meetings by IOs, whose number had shrunk to eight. To compensate for this, farmer organizations had set up regular monthly meeting times so that reminders should not be so necessary.

As it happened, the District Agriculture Committee was scheduled to meet the second morning of our visit, and we were invited to observe. The easy interaction between farmer-representatives and officials there was gratifying to see and now appeared institutionalized. Members of Parliament supported the idea of farmer organizations' being given contracts to do maintenance work up to 150,000 rupees in value, particularly because of general dissatisfaction with the work done by private contractors.

Farmers were still unhappy with some of the work done by Irrigation Department staff, but no action had been taken on their complaints since January. The new engineer in charge of operation and maintenance, however, said his department accepted responsibility for any substandard work and would correct it as soon as the government approved its budget for 1986. The promises of this new engineer were accepted by farmers as well intended. One continuing rough spot was implementation of a government policy that farmer-representatives should review and approve the annual maintenance plan for each area, since it would be their O&M fees being spent on the work. Engineers were finding (or saying) that it was difficult to prepare estimates in time to meet the procedural requirements for release of funds. There was also still some foot dragging on giving maintenance work contracts to farmer organizations.

The overall performance of the system was markedly improved. The total water duty for the preceding yala season had been 5.2 acre-feet per acre, down from 8–9 previously, and it was expected to be under 5 acre-feet in 1986. In the preceding maha season the duty had been only 2.1 acre-feet per acre, two-thirds of the national norm. Farmers reported substantial improvements in crop yields. The vice-chairman for the Weeragoda Area Council, echoing what I was told in January, estimated that yields were up from thirty-five to fifty bushels an acre before the project to seventy-five or even one hundred bushels now, doubling production. No wonder farmers seemed satisfied with the outcome of the project.

Unfortunately, the IO program was in a precarious situation. ARTI had handed over responsibility for it to the project manager assigned by the Irrigation Management Division to oversee activities in Gal Oya. He was not accustomed to the mode or philosophy of operation we had evolved. For instance, when the IOs suggested having farmer training sessions on Sundays, because the schools were not in use then, they were told this was impossible, because "Sunday is a holiday for officials."

"We are dealing with two different approaches to farmer organization," said one IO. "It is difficult to merge them." To make matters worse, ARTI had withdrawn the motorcycles it had gotten from USAID for the project and had auctioned off the IOs' remaining bicycles. This loss of mobility was a great setback, since the handful of remaining IOs had to deal with over 10,000 farmers in an area of about two hundred square miles. Still they pledged to continue doing their best when Razaak, Nalini, and I left them at the end of our visit.

Before I left Sri Lanka, I was able to take part in a national workshop on "Participatory Management of Irrigation Schemes in Sri Lanka," organized by the International Irrigation Management Institute near Kandy. K. D. P. Perera, the new director of irrigation, said candidly and without qualification at the workshop: "Without active involvement of the farmer, I don't think any irrigation system can succeed. . . . It is necessary to motivate and educate farmers, and also the officials. . . . No government officer can distribute water at the field channel level. We must rely on cooperation with the farmers." In conclusion, he said that "the concept of farmer participation is now accepted" and added with appealing candor: "At the beginning, there was certain doubt and resistance, I can say. . . . There was no concept [then] of getting farmers involved as we have today. USAID brought the IO program and farmer participation in design. We were not very convinced. But now we can look back and see that we have been making useful changes. We are learning and continue to learn."

My trip report for April–May 1986 ended with these observations:

There now appears to be fairly broad acceptance of the idea that some form of organized farmer participation is valid and valuable for irrigation. This is as widely assumed now as five years ago it was believed that farmers did not need to be organized to get efficient irrigation results; they only needed to follow the instructions of technicians and administrators. There are new modes and patterns of organization to be worked out as efforts are directed to new irrigation systems. Sri Lanka is also likely to remain enmeshed in costly and agonizing conflict over separatist demands that cannot be met without great losses to most Tamils as well as to Sinhalese and Muslims. So the scope for institutional innovation may be constrained, and even current gains could be undone if no continued attention is given at high levels to the evolution of farmer participation in irrigation management.

Still, although the band of IOs has dwindled to a handful and the country's future is clouded by terror and violence, the farmer organizations have a life and leadership of their own. The new head of the IMD, Joe Alwis, is committed to a substantial role for farmers in irrigation management. Wijayaratna, who headed ARTI's efforts in Gal Oya for the first three years, is back on the scene with his characteristic zeal, and IMD wants to draw on his knowledge and skills. He wants to attract the most energetic and committed of present and former IOs to fill the twenty positions in the permanent IO cadre. If we can reenlist these talents, with Government of Sri Lanka and USAID backing, the prospects for further mobilizing new talent, ideas, and ideals in the years ahead should be good.

In August 1986, as part of a Cornell undertaking to assist in implementing a USAID irrigation management project in Nepal, I started working with colleagues in that country to introduce the role of association organizer (they did not want the program to be an exact copy of the one in Sri Lanka, so they adopted a different designation). Coincidentally, at the same time USAID and the Sri Lankan government formally approved the follow-up Irrigation Systems Management Project intended to extend and extrapolate institutional lessons learned from Gal Oya to half a dozen major schemes elsewhere in Sri Lanka. Cornell was not to be involved in implementing this, however, owing to some USAID decisions back in Washington.[7] The most welcome news was that Senthinathan was once again deputy director of irrigation in Ampare, having returned from his studies in Italy a year early. The program had survived without him, but it was good to know that Senthi was back in charge now that the program faced more difficulties than ever.

At the end of October, Razaak and Nanda wrote long letters. Four of our IOs had been sent to Anuradhapura district to work in major irrigation schemes there that had been physically rehabilitated by the World Bank. We had proposed that the Bank project use IOs from the outset, but the Bank's design team had rejected this suggestion as "too expensive." Our benefit-cost estimates at the time suggested a 50 percent rate of return for investment in IOs (Wijayaratna 1985), three to four times higher than for most Bank projects. But Bank minds had been made up, calling IOs "gold plating." Now our IOs, assisted by thirty new recruits, were expected to improve the operation of systems that might have been better rehabilitated with active farmer organizations from the beginning, before planned improvements were all literally "set in concrete."

Nanda also reported that a permanent Institutional Development Officer cadre was now supposed to be created by the end of the year.

7. USAID decided that this project should be implemented by a "minority" private contractor, which subcontracted with Colorado State University, though I was listed as a short-term consultant for the project at the request of officials in the Sri Lanka government and the USAID mission in Sri Lanka. My inputs to this project were never elicited directly, however. I was able under a different USAID project to keep visiting and to make observations and suggestions periodically.

This was long overdue and still not implemented. He closed by asking: "Don't you have plans to visit Sri Lanka? We are very much eager to meet you and talk." My expectation was to return in January 1987 for the new USAID project, to offer some input on the next stage of institutional development. But the new director of the Irrigation Management Division wanted no input from short-term consultants until his long-term advisers were on top of their tasks. So no invitation to work on the project came, and it would be a year before I would see the program in the field again.

When I next went out in June 1987, my assignment was to assist in institutional development for the irrigation sector at large, now that there was clear government support for participatory approaches.[8] My visit to Gal Oya would be "unofficial"—for observation's sake (really for old times' sake). First I was to meet with the experienced IOs and their teams of younger organizers now working in Anuradhapura district to improve four irrigation schemes that a World Bank project had left nominally but not functionally rehabilitated. Then I would spend some time in Polonnaruwa district, where USAID's new project planned to use IOs to upgrade system performance with farmer participation in another four irrigation schemes. Wijay and I would spend several days in an informal training program at ARTI with two dozen INMAS project managers. They are now responsible for introducing irrigation improvements and water user associations with IO assistance in many parts of the country. But the highlight of my trip to Sri Lanka would, of course, be a visit to Gal Oya.

---

8. On April 30, 1987, a high-level policy meeting on participatory management in Sri Lanka's irrigation schemes was held at ARTI, organized by the International Irrigation Management Institute. One recommendation was that the secretary of the Lands Ministry brief the minister on the key issues discussed and, if he agreed, develop a cabinet paper on appropriate policy and law for implementing farmer roles in irrigation operation, management, and development.

9

The Program Continues
and Expands (1987 to 1990)

This trip in June 1987 is made when the program faces the most difficult circumstances since it started six years ago. The armed conflict over demands made by a militant Tamil minority to create a separate state in the north and east of the country has raged around the project area more intensely this past year and has penetrated parts of it. The American embassy discourages travel to Ampare district "because of the security situation." Two weeks before my trip to Gal Oya, a bus carrying young Buddhist monks on pilgrimage, most of them novices twelve to sixteen years old, was stopped by terrorists about ten miles north of Gonagolla. All thirty-two travelers were hacked or shot to death. The terrorists seek to provoke the majority Sinhalese community into retaliating with atrocities against their Tamil neighbors so that India may be pressured by its own Tamil population to intervene and halt the Sri Lankan government's military campaign to wipe out guerrilla enclaves in the north.[1]

The Water Management Unit of ARTI, once again headed by Wijay, has tried to assist the farmer organizations from afar and with occasional visits. But visits to Ampare by ARTI staff have become fewer

1. This struggle is very poorly comprehended outside Sri Lanka, being portrayed, as the separatists desire, as a conflict between the Sinhalese majority (74 percent) and the Tamil minority (18 percent). (The remainder are classified as Muslims [Moors] or Burgers [Eurasians].) In fact, the demand for a separate state comes mostly from a minority of northern Tamils, a community from India established in Sri Lanka for eight hundred to one thousand years. The separatists use violence against their own people to coerce them into supporting their movement for independence. No statistics have been kept, but even a cursory reading of news reports shows that more Tamils than Sinhalese have been killed by terrorist violence. The rest of the world sees this misleadingly as an ethnic struggle rather than as an ethnically based *political* conflict. The most detailed and balanced analysis I can recommend is Gunaratna (1987).

and briefer. Munasinghe, the IO program supervisor, has been discontinued in this role and is now working full time as a district land officer in the district headquarters. Having invested so much of himself in its progress, however, he continues to look after the program on a voluntary basis.

The number of institutional organizers (IOs) has been reduced to just four. They no longer have motorcycles or even bicycles, so they must get around the project area on foot or by bus. Each is responsible for over six thousand acres and about three thousand farmers. They are working as best they can with the five hundred-plus organizations and farmer-representatives, but their task is extremely difficult logistically, quite apart from the problems of isolation and uncertainty created by the warfare going on.

Fortunately, Senthinathan has returned as deputy director of irrigation in Ampare. Unfortunately, during his absence his replacement did little with farmer organizations. This period of weakened Irrigation Department support coincided with the reduction in IO strength and mobility and with the misunderstandings that arose because the project manager, did not mesh well with the farmer organizations when he arrived. So I wonder how the organizations have survived these buffets. Each time I travel to Gal Oya, I do so with both anticipation and apprehension. Having been away a whole year, I feel more than the usual unease as I return to the field.

With me is Wijay, the only ARTI staff member willing to make the trip in the present circumstances. No cars have been interfered with along the northern route through Maha Oya for six months now, so Wijay figures that is the better way to go if we travel during daylight. If one is rationally calculating the odds of death on the highway, traveling the country's main road between Colombo and Kandy is several times more dangerous than this route per passenger-mile. Still, there is a grim twenty-mile stretch where terrorists have demolished the homes of settlers in half a dozen villages. The army has burned back the tall grass and brush on both sides of the road to reduce the opportunities for ambush, so it looks like a war zone. Our driver, anxious to get through it as fast as he can, pushes the car up to seventy miles per hour. As soon as we come to some settled areas and an army camp checkpoint, we feel more secure.

Reaching the Left Bank main canal near Gonagolla has never felt better. We are in familiar and friendly territory. Since it is only 3:00 we stop at the house where Jinadasa, our most senior IO here, stays. He is taking a bath in the canal, so we wait for him to return. He confirms that some of the Gonagolla farmer-representatives can meet us to-morrow morning at 9:00. How are the farmer organizations function-ing? The Area Council meets on the tenth of every month, and the fifteen D-channel organizations in the area usually each meet before that date.

The Area Council is made up of the president, secretary, and trea-surer from each DCO. Average attendance at Area Council meetings is twenty-five to thirty out of the possible forty-five, not perfect but not bad. Usually at least one person comes from each DCO, except from units 34 and 36 below Rajagalatenna. Though these are Sinhalese ar-eas, it is not thought safe for persons there to travel about. Jina says he organized a meeting there some time back, but it was adjourned when they heard gunshots nearby.

How are relations between farmers and officials? Cooperation with the project manager has improved since a year ago, and generally the officers of different departments are working with the farmer groups. Was anything ever done about farmers' complaints about low-quality rehabilitation work in this area? Talk to the farmers, suggests Jina. We decide to drop in on Narangoda, who raised this matter of substan-dard work during a previous visit.

The Stalwarts' View from Gonagolla

It is late afternoon, and we find Narangoda at home, playing with his infant son. Fairly soon old Ratnayake, the elder statesman of the organizations here, walks into the house to join us, having seen our car pass from a distance. My first question is, How are the farmer orga-nizations functioning, compared with an ideal level of 100 percent? Narangoda says, overall about 35 to 40 percent. How about Go-nagolla? It is estimated to be about 60 percent, which they consider too low. I must look disappointed, because they clarify that this refers to field channel groups, not to the higher levels, which work more satisfactorily.

Happily, they point to good performance of several organizations in areas that IOs entered after 1983, not part of the original effort here. D-channel organizations on G4 and LB22 have taken and executed construction contracts from the Irrigation Department. Does the ID give farmers contracts without the kind of resistance we saw previ-ously? "Yes, but for some good organizations like that on LB29 there is no more work to be done at present." I express appreciation that

LB29 members underwent so much difficulty, expense, and disappointment to establish farmers' right and competence to do such work.

How is maintenance of channels now? Not perfect, but better. There are no disputes about water among members, says Narangoda, though some encroachers still try to open the gates to take more water than authorized. Farmers are helping the ID reclaim reserved areas along canals that are not supposed to be cultivated because this interferes with operation and maintenance of the channels.

I tell them that last week Wijay and I met Kuruppu, the former ADA district manager who had a dispute with farmer organizations over the "phantom" cashew plantation (chap. 6). He is now the project manager for an irrigation scheme in Polonnaruwa district. Though formerly wary of farmer organizations, he is now one of the strongest supporters they will find in Sri Lanka, I say. They smile and suggest that perhaps he learned some good lessons here in Gal Oya.

What is the security situation in this area? About once a week someone gets a threatening message and has to go into hiding, they say. Do they ever go into Tamil areas? Yes, some go. The jala palakas go there sometimes, but not often. "We live in harmony," says Narangoda, repeating the position he and his neighbors have always taken: "We have an understanding with our neighbors in the Tamil area. They will try to prevent the terrorists from coming into our area and will give us warning if possible. We have said that we will try not to let the army move into their area, and to let them know of any raids." Despite all the killing and animosity around them, it is gratifying to know that farmers here retain their idealism and sense of solidarity.

How are working relations with officials? With the Irrigation Department, no problem. How about the divisional officer of Agrarian Services, for example? (He has resisted the farmer organizations for most of the past five years.) "Now he is fine. It is good we retained him here and corrected him instead of getting him transferred. Otherwise he would have been a burden for other people." This might seem an arrogant comment, but it is said matter-of-factly and is probably correct. Why did he give them so much trouble? They are not sure. "Perhaps he thought the organizations would not last. Perhaps he suspected we would act against him."

How about the operation and maintenance fee the government introduced in 1984, which farmers tried to stop or reduce? Are farmers paying this? Some pay it voluntarily, others because they must in order to be able to transfer their permit to someone else. Is the money being used well? "It is now being applied in priority areas. Now the ID is being very careful in its expenditure on maintenance. We discuss the maintenance plan in the subproject committees and in the District Agriculture Committee." This is a big change. Last year the ID would not give farmer-representatives estimates of maintenance work for them to review and approve. Field channel organizations are now asked each year to prepare a list of repairs and work they think should be done on their channels. The suggestions from all the field channels are discussed by the D-channel organization, and a list of priority work is drawn up. This is taken by the DCO president to the Area Council, where all work is assigned a priority and submitted to the ID for estimates.

Are they happy with this? The size of the contracts given to farmer organizations is still very low, only 5,000 rupees. Wijay tells them that the limit has now been raised to 25,000 rupees, and the ID can divide larger projects into 25,000 rupee blocks, so they can do these too. So far, mostly very small contracts have been given, about 2,000 rupees (under $100). "That much can be spent by a technical assistant on fuel," Ratnayake grumbles. "The contracts being given private persons are much larger."

I ask what happened to their request to investigate and recover funds where faulty construction work was done under the rehabilitation project. "We suggested three alternatives: appoint a commission; bring back the TAs and engineers for an on-the-spot inquiry—some of the structures washed out before they had even left; or get the irrigation engineer to take responsibility for everything. We sent a letter to Colombo, and the senior deputy director for water management started an inquiry into this, but when he left the country for an overseas assignment nothing more happened." This is unfortunate. They report this matter-of-factly. Either they are resigned to no resolution of the problem, or it is not viewed with as much vehemence as a year ago.

What is the leadership situation like now? Are new people coming up, or is it the same group? Ratnayake, who has been sitting slumped back in his chair, now sits forward and speaks vigorously despite (or because of) his advanced age.

I am happy you asked that question. For example, I have been trying to resign as farmer-representative for my channel for some time, but the group has not met for months, and nobody else will take responsibility for the organization. Since the IOs were reduced and stopped working at field channel level, there have been few meetings. We know that leadership should change. I have not been well. I have heart trouble and want to get rid of the burden. I want to train someone else—in fact, several persons—in this work before I die, to impart the skills I have learned. But there is no chance of this. The field channel organizations are dead.

This is alarming. I ask whether this applies to all field channel organizations. Narangoda says that the D-channel organizations, the Area Councils, and the Project Committee are functioning fine, but the weak point is at field channel level. "Practically all of these groups are weak, but other levels are functioning." He expresses doubts about the future of the organizations because IOs were taken out so abruptly. Wijay asks if someone from the community could perform the IO role if given means of transport and some financial assistance. "It's worth trying," says Narangoda. "It could be done," adds Ratnayake.

I ask whether the dispute a year ago over the number of farmer-representatives to sit on the DAC's subcommittee for irrigation management has been resolved. The project manager had insisted that only one farmer could be appointed, since the group was already so large because of the number of officials appointed; farmers wanted at least four FRs, one from each Area Council. Four farmers sit on the subcommittee now, with Narangoda representing the Gonagolla area.

Ratnayake does want us to pass over his concern for the weakness of the field channel organizations: "The lowest level of organization is gone now. It needs to be reactivated. We need IOs again, or we need to have farmers with some support doing this work. All they need is a bicycle or some facility. No payment is needed." To drive his point home, he uses a marvelous and mighty simile: "The farmer organization system is like a tree The root system [the field channel groups] has been damaged. One day the tree will fall over if the health of the roots is not restored." Ratnayake knows that the program is now being extended to other irrigation schemes in Sri Lanka and asks: "What good is it spreading the organizations elsewhere if they are becoming weaker here?"

He notes sadly that the links between ARTI and the farmer organizations here have been weakened: "ARTI people don't come much anymore." Wijay explains the difficulties they have had in getting to and from Ampare. The staff have not been allowed to travel. I tell them that I had expected to come last January but the trip was canceled because I did not get the necessary approvals from the government in time. I add that I am sorry to hear the field channel organizations have weakened but am glad they have developed institutionalized linkages with officials at higher levels, right up to the district level. It is good to see them again, and I look forward to meeting them again tomorrow morning.

After taking pictures, we drive to Munasinghe's house in Uhana to confirm tomorrow's schedule. Then we go on the Ampare and get settled in the bungalow behind the Irrigation Department office. Our usual resting place, the ID circuit bungalow at Kondawattawan has been booked by some high officials. Senthinathan has arranged supper for us, and a good one it is.

Next Morning

We have breakfast early so we can leave by 7:00 for Inginiyagala, below the dam creating the reservoir for the Gal Oya scheme. Last night Munasinghe told us that the father of Sarath Wijesiri, our most active IO and now the ADA district manager, has died. We cannot come to the funeral in the afternoon because of meetings previously scheduled with farmer-representatives, so we go to pay our respects early in the morning. After going through the rituals of a funeral visit at Sarath's home and talking with him and his wife, Keerthi, who was one of our best IOs and is now a community development worker here, we drive toward Gonagolla. While passing through Uhana after picking up Mune, we happen to see Dissanayake, our remaining IO for Weeragoda, about to board a crowded bus near the market. We stop to confirm the afternoon's planned meeting, and he squeezes onto the bus just as it pulls away.

In Gonagolla, several farmers are already at the home of short Punchibanda, together with Jinadasa. Two of the farmer-reps, for LB30 and G10, are new to me, so maybe Ratnayake was exaggerating yesterday when he said there is no new leadership coming up. Wijayatunga from G10 is particularly confident and articulate. When I learn that he is also a yaya palaka, we discuss the possible conflicts between the YP and FR roles. He suggests the YP role be abolished and its functions handed over to the farmer organizations. The salary due the YP under existing law (one-quarter bushel of rice per acre per season) should be given to the farmer organization, he says. "If it wants to give part of this to the representative as compensation for his time, that should be up to the organization." Someone says that most farmers have already stopped paying anything to YPs.

How are relations between Sinhalese and Tamil farmers in the area? There is no problem, Wijayatunga says. "The situation is one of armed neutrality," someone else suggests. "We live very peacefully," says another. "There is no problem over water, and there is exchange of labor between the communities." "Certain outsiders are making the problems. Parents here don't allow their young fellows to create problems for others. If there are any problems arising, we have agreed to inform each other in advance." This supports what I was told yesterday. Narangoda arrives by bicycle.

What about the opening of the project manager's office last fall in Ampare? I have heard there was some controversy about this. The Secretary of the Ministry of Lands was expected for the occasion. The project manager had a poster printed saying that "the farmers of Gal Oya" invited people to come to the opening, but since farmer organizations had not been consulted on this in advance, some objected to this invitation. In Weeragoda they insisted that the reference to "the farmers of Gal Oya" be cut off the poster—an unusual assertion of farmer independence—and it was. When the meeting was held in Ampare, relatively few people came even though free transport was provided.

The problem, farmers say, was that the project manager did not discuss the program with the farmer organizations in advance. It is unfortunate that he relied mostly on posters and public address systems instead of working with the FOs as channels of communication. I am told that working relations with the project manager have improved somewhat. "We have known him for a long time," one farmer says, alluding to the fact that the PM was the crop insurance representative in Gonagolla for some years before his promotion. That program is notorious for irregularities, though I have heard no allegations against the project manager himself. It is said that farmers must pay 1,000 rupees (about $30) to get an official to come and estimate crop damages, certifying them for compensation.

An interesting discussion of the crop insurance program ensues. There is some laughter as a recent incident is recalled. The Government Agent and other top officials inspected a field where crop damage had been minimized through nonchemical means under an FAO project for "integrated pest management." Yield was measured to be 92.5 percent of the norm. Then the next day a crop insurance inspector came and recommended compensation for that field of 3,500 rupees. How could such a loss have occurred in one day? a farmer asks. Payment was stopped after a farmer organization objected, but farmers are unhappy because the reason for recommending payment was never explained.

Munasinghe says the Government Agent is trying to improve the insurance program with test checks of claims. It was proposed at a District Agriculture Committee meeting that farmer-representatives accompany the inspectors to guard against irregularities. But some FRs are reluctant to have anything to do with the program because of its bad reputation. Another suggestion is to publish a list of all persons receiving payments so everyone knows who is claiming crop damages, properly or not. One proposal already made to the DAC is to put up a red flag on every field where compensation has been asked for, so others can see for themselves what damage there is.

Narangoda points out that if an epidemic occurs, farmers cannot all wait for an inspector to come before harvesting their fields. Wijay suggests that field channel organizations certify claims for compensation; even if they cannot determine exact amounts, they can classify fields in terms of degree of loss. But he and I do not want to propose any specific plan for dealing with the problem. Rather, the farmer organizations should come up with a proposal that they can introduce to the Project Committee before the start of the next season. Now that water management problems have been reduced, if the organizations are to remain vital, they need to start dealing with new and urgent problems.

Old Ratnayake voices his concern that field channel groups are declining in strength: "At the beginning, there was a system where we started discussing everything at the field channel level and then took it up to higher levels. This is not happening now. Fewer farmer-representatives go to meetings now. There is no flow of information up and down. The system is growing weaker." Wijayatunga says that the problems stem from two basic factors: there is no legal status for the organizations, and officers of some departments still do not accept the organizations. I ask if this is true of the Irrigation Department. "No, the ID gives 100 percent cooperation," says tall Punchibanda.

Do farmers maintain a system of water rotation in the field channels? "In difficult channels, yes." This is a reasonable answer; the effort required to manage a rotation should not be expected unless there is some need owing to water shortage. How about cleaning and maintenance of field channels before this yala season? "Where farmers were responsible, it was about 85 percent. The ID did not do all of the work it was supposed to do." Where was work not done? "On G10. Elsewhere it was OK."

How much work was done by shramadana? "We use that system only about 50 percent of the time now." Seeing a disappointed look on my face, Punchibanda explains: "Most of the time we don't need to work by shramadana. Not everybody can conveniently participate at any one time. So everyone does his assigned length of channel cleaning. We needed to work together in 1981, when the channels were in such terrible condition. But there is no such need now. We did, however, build up a damaged channel on LB31 by shramadana this past year, with everyone coming out for two days." I ask about the 15 percent of maintenance work not done. "There are always a few black sheep," says someone. "Most of those who don't clear their length are tenants, not owners." The lapses in channel cleaning are scattered, so the effect on flow is limited.

What are the biggest problems facing the farmer organizations? Politics? No, politics are not allowed in water distribution, someone says. But someone says, "There will be elections coming, and we can't know for sure what will happen." Wijayatunga says, "The problems are now less. I can't see any problems if we implement what we have discussed." Punchibanda says, "Lacking legal recognition of the organizations is a problem. We could do a lot more if we get recognition." He adds that they have requested this through the Project Committee. Narangoda, who has been silent so far, letting others talk, says: "We brought a proposal to the last Project Committee meeting to send a delegation to Colombo, to discuss this matter with the ministry at the national level. But the project manager would not give a green light, saying we need to discuss this first at the district level. We went to the DAC subcommittee on the fifteenth [five days ago] prepared to discuss this, but we didn't get a chance."

Wijay and I say that ARTI can help get legal recognition for the organizations, but it is up to them to revive the field channel groups and also to diversify their efforts into areas beyond water management, such as crop protection. I tell them I hope to see a variety of activities when I visit next time. Tall Punchibanda says with a laugh, "Maybe you won't see us after six months," referring to the threat of the terrorists, very much on their minds.

Wijay tells them about plans to have four farmer-representatives travel to Colombo on Sunday to meet a visiting delegation of engineers from Egypt, sent by USAID to learn about farmer participation in irrigation management in Sri Lanka. The visitors were supposed to come to Gal Oya, but given the security situation it was decided to ask some farmers to come to ARTI instead, at USAID expense. From Gonagolla, Appuhamy has already been chosen, and he is looking forward to the trip. I say we will see him tomorrow in Colombo. We take pictures and say good-bye.

We are late getting back to Ampare by noon to meet Senthinathan, who has invited me to come home with him for lunch in Kalmunai. He participated in an international workshop held at Cornell in May and stayed several days at my home, and now he wants me to visit his home. I am pleased to be invited, though Sinhalese regard Kalmunai, on the coast, as "behind enemy lines." But everyone I checked with agrees that traveling with Senthi anywhere in this area should be safe enough. Everything appears normal on the way to and within Kalmunai, though "the Tigers" operate freely here. Senthi points out as we drive where several dozen terrorists were recently killed by a rival faction in pitched battle. Bhawani, who has been living in Batticaloa since she got married and left ARTI, joins us with her infant son for lunch, as does Ravindran, one of the junior engineers with whom we started working in Gal Oya in 1981. He is now a senior engineer in Batticaloa. It is a very nice lunch, but we get back late for the afternoon meeting.

Uhana and Weeragoda: Rise and Decline?

I reach the familiar Uhana school a little after 3:00, where Wijay has already started the meeting with fifteen farmer-reps plus our two IOs remaining in this part of Gal Oya, Sarath Silva and Dissanayake. Ratnakaye, the Weeragoda Area Council chairman, is speaking. They are talking about crop insurance, raising the same problems we discussed this morning. The Area Councils have already made some suggestions for improving the system. Farmer organizations themselves should make the initial assessment of crop losses and submit claims from each of their areas. Then farmer-representatives would observe the official inspections. All payments should be posted publicly, so that there are no secretive dealings. Farmer-representatives will present such a proposal to the Project Committee for improving the system before the start of the maha season.

What happens if claims from Tamil areas are made for payment? someone asks. "No officer can go there to inspect." I inquire about relations between the two ethnic communities in this area. Wijay tells them about the plan for mutual protection we heard about in Gonagolla this morning. The same arrangement exists in unit 16, they say, where Sinhalese and Tamils "live in harmony."

I ask to have introductions all the way around the circle of farmer-representatives, since I do not know half of them, though there are also several familiar and appreciated faces in the circle. I ask whether there are still water problems, and the answer is none. Farmers here have been quite willing to voice criticisms in the past, so that is a good sign. The driver we sent to bring some soft drinks returns, and Sarath and Dissanayake open the bottles and pass them around with small glasses.

An IO says that the field channel organizations "need to be revived. Matters should be dealt with by the organizations, not just by the farmer-representative." One FR says that ARTI trained the IOs very well, but not the farmers. "Maybe this is one reason for the weakening." Another reason, adds a second FR, is the 1984 farmer convention. "That caused big damage to the farmer organizations." I knew there was unhappiness about some aspects of the convention, but I had no idea it was still viewed so negatively.

Kuruppuarachchi from UB2 gets up and hands over a sheaf of large pages to Wijay. "I wrote an account called 'The Rise and Decline of Farmer Organizations in Gal Oya'," he says. Wijay and the farmers discuss whether to read the whole report out loud. Since as it is fairly long, Kuruppuarachchi is asked to summarize it. He says we can take it with us to copy and translate. "Farmers in Gal Oya originally came from many different areas, so there was no cooperation among them. Some didn't even attend the funerals of their neighbors. Then the IOs came and they sacrificed a lot to bring us together. Some were even accused of being CIA agents. There were pressures to obstruct the IOs at first, from *mudalalis* [merchants]. But the idea of farmer organizations was accepted." Sarath and Dissanayake collect the bottles and go over to a pump to wash the glasses. Chuti, one of our first and best IOs, still teaching in Uhana, arrives at the school with her young son and sits on a school bench with the rest of us. Kuruppuarachchi continues: "Now there is a new generation, and some young people are not aware of the basic facts. One high-school graduate did not even know about the IOs or that she lives on the Left Bank of Gal Oya. We need more training for everyone. About 80 percent of the field channel groups were not functioning by the middle of 1984."

This is a disturbing figure. What was the reason? The first is "the farmer convention." Second, a few of the IOs were "not good." How so? A few had links with the officers and even with *mudalalis* and were not on the farmers' side. This is the first time I have heard such an allegation. Wijay tells me that some farmers blame one or two IOs for the unsatisfactory outcome of the farmer convention. Kalumahattea, who "betrayed" farmer interests is known as a *mudalali* here. Kuruppuarachchi continues: "Third, even today, this meeting was organized by the IOs, our friends here. Farmer-representatives cannot do this. The teachers wouldn't give us use of the school hall." Why not? I ask. Because of the emergency situation. There is a prohibition against use of this type of facility for meetings. IOs could get this place, but we couldn't. (Maybe this meeting is illegal) "Fourth, some farmers tend to form direct links with officers at the expense of our organizations." (This may also reflect unhappiness with the convention.)

Sumitipala from UB2 says the major reason for the decline is the convention: "We expected a lot from it." He adds that the rehabilitation project was not implemented well, that "90 percent of the construction was substandard. Also, now farming has become less profitable as costs have risen. This makes for dissatisfaction." He suggests that "foreign companies" are benefiting, referring to producers of agrochemicals, Wijay thinks.

I thank them for their criticisms and tell them that I was concerned about possible politicization of the farmer convention before it was held. "Unfortunately, we all can see problems more clearly after the fact. I hope we have all learned something from this experience." I then offer an alternative explanation for the decline of the field channel groups, which all agree dated from 1984.

By the time of the convention, the adequacy and reliability of irrigation deliveries was considerably improved. Physical rehabilitation had progressed far enough to provide more water control; the supply of water in the reservoir from the start of maha 1984 was ample owing to better rainfall. Moreover, farmer cooperation for managing water at D-channel and field channel levels had become fairly routine. There was thus less need for the field channel groups to meet regularly. Because they did not diversify their activities into other areas like agricultural extension or credit, groups had little to do once water management problems, by farmers' own account, had been mostly rectified. Kuruppuarachchi interjects that his account of the rise and decline of farmer organizations makes this point.

Making improvements in the crop insurance system, I suggest, would help revive interest in the farmer organizations. One farmer-rep asks how they can make changes in that program, since it is a national one and there are farmer organizations only in Gal Oya. Good point.

I reply that under the INMAS program there are farmer organizations elsewhere. Wijay and I tell them about the workshop we held at ARTI earlier this week with two dozen project managers. Ratnayake asks whether farmer-representatives from Gal Oya could be invited to speak to all project managers at their next workshop. "Not to argue, but to exchange views," he adds. Wijay says we will suggest this.

Sumitipala brings up an objection that for attending Project Committee meetings, officials are paid 250 rupees, whereas farmer-representatives get only 50 rupees. "It can cost us as much as 75 rupees to attend a day's meeting—50 rupees to hire someone to work on our farm in our place and 25 rupees for travel and meals. It is unfair to pay officials so much when they come to meetings on government time and in government vehicles, while farmers have to come on their own time and by private means." He explains that farmer-representatives have refused to accept this unequal payment. Using farmers' O&M fees to pay this would add insult to injury. How many farmers attend the meetings? I ask. Sometimes 100 percent, sometimes 60 to 75 percent. Do the organizations pay farmer-representatives' expenses for attending Area Council meetings? No. "We don't want to spend farmers' money on this. Representatives cover this cost themselves."

Sarath stands to say that there have been some good outcomes from the Project Committee, though farmers have had to be firm. They boycotted one meeting. One farmer-rep says that priorities for maintenance work are now set in consultation with farmers, which is very good. But when contracts are given to farmer organizations, the Irrigation Department tends to underestimate the cost, whereas if they are awarded to a technical assistant or his friends, the amount given is larger. I can see that there are still problems to be worked out in this area and suggest, "Farmers need to insist on disclosure of the details of all contracts and should refuse to take contracts that are underfunded." They say they are already doing this.

It is now after 5:00, and the last bus to Weeragoda will leave from Uhana in a few minutes. Given the security situation, buses do not travel after dark. We bring the meeting to a close. I tell them I hope it will not be a full year again before I can visit them. Kuruppuarachchi says that farmers are glad to share their experience with visitors from other countries, but they would like to be able themselves to visit counterparts elsewhere to learn about different ways of managing irrigation systems.

Evening Assessment

After picking up Munasinghe and Jinadasa at Mune's house, we pass the project manager's jeep on the road. Tilak has just arrived back from attending the workshop that Wijay and I ran at ARTI, and he says he was hoping to meet us for more discussions. We invite him to join us for supper at the circuit bungalow, where we plan an evening meeting with the IOs and Munasinghe. We want to take stock of the program and its deficiencies, to see how it can best be given new impetus and direction under the present difficulties. Mune gets out to ride with Tilak to make more room in our car. With Tilak is his successor in charge of crop insurance for Ampare district, who will also join us for supper. This will make the evening more interesting, since we need to talk about making the crop insurance program more satisfactory to farmers.

Upon arriving at the circuit bungalow, we sit down in a circle with cold drinks to begin our discussion. I can't resist telling our new acquaintance from the Crop Insurance Board what we have been hearing from farmers today. He explains some of the difficulties and constraints he must deal with, such as having very few staff members to cover all the claims in a whole district within a few weeks. He asks if we have any ideas about what to do, and we go over the suggestions from farmer-representatives. He agrees to consult with them and to come up with a plan for improving the program before the start of the next season. The project manager agrees that this should be a priority.

I ask about the strength of field channel groups now, and IOs say only 10 to 20 percent are still strong. This is broken down into 10 percent in Uhana, 15 percent in Weeragoda, and 25 percent in Gonagolla—about what I would expect. How many meet formally at least once before each planting season? 50 percent in Gonagolla and Uhana, 60 percent in Weeragoda. Kularatna arrives from where he works now at the head of the Left Bank system. He says the percentages are about the same in his area. How about the Area Council? It meets on the twenty-fourth of every month. Munasinghe notes proudly that the Paragahakele council has never failed to hold its monthly meeting since started almost three years ago. Quite a record.

What did the council discuss at its last meeting? Kule says: (1) A request for the Paddy Marketing Board to operate a buying store (warehouse) at Paragahakele so farmers don't have to take their paddy so far to sell it. This is now agreed. (2) Upgrading the secondary school at Paragahakele to senior status. This has been approved.

(3) Shramadana repair of the road between Himidurawa reservoir and Uhana. Four hundred farmers took part in this work with the Highway Department, improving six miles.

These are considerable accomplishments for just one month. I ask whether farmers can now sell their paddy directly to the Marketing Board rather than having to sell only through agents, one of the council's urgent requests at the meeting I attended in January 1986. This is now possible, I am told, a considerable boon to farmers.

What about other Area Councils? Uhana meets on the fifteenth of every month, except in April and September when farmers are having to get their crop planted quickly. At least 60 percent of the twenty-four members attend, at least one from each D-channel organization. Weeragoda does not have a fixed date, but the project manager says he is always notified in advance about the date so he can attend. Usually about half of the council members attend. This is said to be true for the Gonagolla Area Council too, which always meets on the tenth. The Area Councils also hold assembly meetings of all the farmer-representatives in their area—every three months in Weeragoda, twice a year (before each season) in Gonagolla, and once a year (in May or June) in Uhana. Farmers have evolved an appropriate schedule and structure of organization.

I ask about D-channel organizations, and Tilak says he tries to attend all of their meetings each month. Each of the eight DCOs in Uhana has a fixed date for its monthly meeting. All but two of the DCOs in Weeragoda meet monthly—these are M16 and Malwatte, where there is terrorist activity. (I understand why regular meetings are not held in M16 when Wijay tells me that Dissanayake, the chairman there, had to be relocated to a settlement scheme in the north for his own safety; as leader of a civil defense unit he shot two suspected terrorists, and his father-in-law was then killed in a retaliatory raid on Dissanayake's house.) In Gonagolla, some DCOs meet monthly, but all meet at least every two months, with 75 to 90 percent attendance. Tilak confirms the figures given us by IOs and farmer-reps.

I do some more checking. The Project Committee responsible for the Gal Oya Left Bank meets on the twenty-ninth of every month. There is a farmer vice-chairman and vice-secretary. About 90 percent of the twenty-six farmer-members attend. Even though they are not receiving any payment? I ask. The IOs groan and laugh, yes. Who are the members of the District Agriculture Committee and its Gal Oya management subcommittee? I recognize most of the names they cite. These are very knowledgeable and committed farmers.

The higher levels of organization in Gal Oya seem to be functioning as well as can be expected considering the violence and uncertainty surrounding the project. The IOs and project manager are attending as many meetings as possible to assist the organizations and to facilitate communication with officials as needed. We have to be concerned with restoring the vitality of the field channel groups, however, so that communication travels upward from and downward to the individual farmers through the structure now created of field channel groups, D-channel organizations, Area Councils, and the Project Committee.

The IOs say the Irrigation Department has so far given thirty-six contracts worth 380,000 rupees (about $10,000) to farmer organizations. This compares with 700,000 rupees paid by farmers so far in O&M fees. More contracts will be given. The dissatisfaction we have heard today from farmers about the way contracts are estimated is probably justified, the IOs say. The program needs to get more standardized estimates from the ID for all work.

I ask about the IOs' schedule of work now. They do not even have bicycles for getting around their large areas of responsibility. By the twenty-fifth of each month, they prepare an "advance program" for the coming month that is discussed with the project manager. They attend an average of eight D-channel organization meetings each month and at least five field channel group meetings. They collect statistics on fertilizer use, loans, and such from the Agrarian Service Centres and banks for the Irrigation Management Division, and they monitor the farmer organizations. They visit farmer-representatives who don't come to meetings and find out why not. They also conduct training programs, usually four each month with farmer-representatives and four with farmer groups.

We talk about how to make the training more effective, emphasizing problem-solving approaches. The IOs are giving or arranging instruction in water measurement and also various agricultural subjects like land leveling and crop insurance. They are not, however, doing any systematic evaluation of their training, as was done earlier. D-channel organizations are encouraged to undertake annual reviews. That was just started in 1986. The self-evaluation process developed last year for field channel groups has been tested. About 60 percent of the items were easy for farmers to deal with, but others need to be revised. The IOs agree that they should make this part of their regular duties, since we need to strengthen the field channel organizations.

One IO says they have been working under very difficult conditions. I tell him I understand this, but he says I don't know the full story. In his view, "people at ARTI, except for Munasinghe, are trying to destroy the program." He also excepts Wijay, who was away from ARTI for three years doing his Ph.D. studies at Cornell. I have heard this allegation before, though I didn't care to believe it. The precipitous withdrawal of bicycles and motorcycles last year certainly seemed like sabotage. I have no question that staff members like Nanda and Razaak have been strong supporters of the farmer organization program, but others may have been less sympathetic than they professed. The director who gave us good, if often cautious, support in the first years seems to have lost his enthusiasm, and his deputy, while publicly approving the effort, has said and done things in private that were unhelpful. This is sad to learn, because these persons have acted for years as friends with shared ideals. The IOs cannot have a complete picture of what goes on in Colombo, but others at ARTI have reported similar things to me. It is even more marvelous that the program persisted in spite of such tensions. Some staff members deprecate the IO program, saying it is not as successful as some persons believe (this is surely true).

It is getting late for supper even by Sri Lankan standards. Bandara tells us the food is ready, so we conclude our meeting. The project manager asks Wijay and me to keep giving him "feedback" on how he is doing, a sign of seriousness on his part. He must know that many people are not happy with his role, but I can tell him farmers say the relationship is improving.[2] I suggest that if he and his friend can improve the crop insurance program by working with farmer-representatives, it will be a big victory for everyone. His reputation suffers because of his previous association with that program, and making reforms in it would detach him from this liability. The IOs and Tilak seem to interact easily at supper, and this is encouraging. He gives several of them a ride home in his jeep when supper is over.

Return to Colombo

Next morning we get an early start. As we leave Gonagolla, we pass the jeep in which Munasinghe and four farmer-representatives are also

2. Later we learn that the project manager is a relative and close friend of the District Minister, which explains why everyone treats him so delicately. The minister may not know that this appointment has undermined the program that he has helped along in other important ways.

traveling to Colombo and exchange waves. The return drive through the "war zone" is more tense than the first time because it is still fairly early in the day, but there are no problems. Shortly after I get back to the small hotel near ARTI where I stay in Colombo, Munasinghe and the farmer-reps arrive, having also made good time on the road. The farmers are glad to find someone they know. The representative sent from Paragahakele, whom I have not met before, is particularly impressive. This shows that the program continues to produce good new leadership.

As we are having tea together and talking about the situation in Gal Oya, the delegation from Egypt arrives to meet us. We order more tea, and after I give an introduction to the farmer organization program, the discussion is turned over to the farmer-representatives. They show no hesitation in taking it over. Next morning the farmers participate in a large group meeting with the delegation and with government officials at ARTI, though most of the time is spent on matters of policy beyond their knowledge. After lunch, when the meeting is over, they leave to return to Gal Oya with Munasinghe. From the heartiness of their farewell handshakes, it is clear they are more appreciative of the experience than awed by it.

As their jeep drives away, there is both satisfaction and sadness. The program is still far short of its potential, owing to our own shortcomings and repeated government failure to understand it fully and to keep promises. But the program has progressed even in the past year, despite the many difficulties. The original strategy and philosophy seem validated by the results, but achieving these has depended on the caliber and commitment of farmers such at these—Appuhamy, Mudiyanse, Ranatunga, and Sumitipala—as well as on the capability and perseverance of IOs like Dissanayake, Jinadasa, Kularatna, and Sarath Silva, who remain in Gal Oya.

The Irrigation Management Department says that by the end of this month it will appoint a new cadre of Institutional Development Officers, permanently assigned to work with farmer organizations in major irrigation schemes. In addition to the IOs still with us, a number of our most committed former IOs have applied to be interviewed, willing to give up their easier and somewhat better paid jobs as teachers to resume the work that gave them such satisfaction through opportunities for both social service and personal growth. This permanent cadre was agreed to in principle by the Minister and Secretary back in 1982, but perhaps such an innovation in the way the government works must be expected to take a long time. We are still only partway along the road toward fully effective and institutionalized farmer participation in irrigation management.

Stocktaking

This was to be my last visit to Gal Oya for some time. I set aside several weeks to return in January 1988 and again in the summer of 1988, but the situation was judged unpropitious for a trip. The Irrigation Systems Management Project that was to build on our Gal Oya work got started later than originally planned. When I returned from my July 1987 trip to Sri Lanka, I met with the consultant team that had been recruited to assist the project. They were assembled in Washington, and I briefed them on my visit to Polonnaruwa and on what I thought we had learned from Gal Oya about farmer organization.

But the new project, like the Sri Lankan government, faced many difficulties in the field, and work proceeded slowly. The best thing the project was able to do, in my view, was to hire Sena Ganewatte to work with it full time as a local consultant. He knew the Gal Oya experience intimately and could help translate and extrapolate it with unusual commitment. For a while it was rumored that Wijay would also join the project as its agricultural economist, but that did not materialize. He was hired instead by the United Nations to oversee monitoring and evaluation in a project intended to improve the management of irrigated settlement schemes. He hoped he could introduce more participatory development through this work. It meant, however, that ARTI's irrigation group lost his leadership, though Wijay continued to assist its younger staff on his own time.

A compensating plus for the ARTI group was that our best friend within the government bureaucracy, Joe Alwis, formerly director of water resource development in the Ministry of Lands and then director of the Irrigation Management Division, became ARTI director in June 1987. I had arrived back in Colombo on the day he was inaugurated in the new post. In more detail than on my previous visit, I was told how the outgoing director and his deputy had been less than helpful for the farmer organization program, some thought even sabotaging it. Explanations of deeds and motives were too complicated for me to sort out. These individuals had previously gone out of their way at various times to maintain and strengthen the program. Why they would later undercut or simply neglect it, I could not fathom. Rather than try to assess blame, I tried to appreciate all who had furthered the program, many unexpectedly and some inconsistently, for whatever reasons.

Quite saddening was the conflict that arose between two staff members at ARTI who had been most intimately involved with the program. It was painful enough personally that I still prefer to speak of it only generally, but I cannot pass over it without mention. These were some of my closest friends, and they were now making dire accusations against each other, even to the point of threatening force. I thought perhaps such behavior was spurred by the growing insecurity and violence in the country at large. Fortunately, the program no longer depended on individuals, since the United States and Sri Lankan governments had written it into various project and policy plans.

Though creation of a permanent IO cadre took longer, once again, than promised, a cable arrived from Colombo in September 1987:

IMD SELECTED INSTITUTIONAL DEVELOPMENT OFFICERS. ALL ARE EX-GALOYA PEOPLE. I MUST SHARE THIS HAPPIEST NEWS WITH YOU. REGARDS.   WIJAYARATNA

This was indeed happy news. Why there was so much delay was not clear. Perhaps it was impossible for the civil service to accept these roles if designated as "organizers." When training and supervising IOs, we had emphasized that they should never think of themselves, or act, as "officers." But ultimately the bureaucracy won on this point. IOs could be made permanent only if they were called officers, it seemed.

We trusted that those appointed would retain the right spirit, because they were our stalwarts.[3] Sarath Silva remained in Gal Oya; P. B. and Priyantha continued working in Anuradhapura; four IDOs started working in Polonnaruwa. A succession of new IO batches were trained by ARTI to work under the IDOs during 1987 and 1988. Farmer organization work continued despite the insurgency waged in the north by Tamil militants and in the south by Sinhalese JVP subversives.[4]

3. A great injustice nearly occurred when D. S. and Jinadasa, two of our most experienced and dedicated IOs, were ruled ineligible because by civil service rules they were too old to join as new recruits. Munasinghe wrote that Wijay deserved "big credit for getting all the IOs as IDOs when there was so much pressure from various higher-ups. Jina is now at Lunugam Vehera and D. S. is at Padaviya. Everybody is happy about their performance except a few project managers through jealousy over their [D. S.'s and Jina's] efficiency."

4. In June 1987, when I visited the Kaudulla irrigation scheme in Polonnaruwa district, where USAID was starting the new project, the local hospital was full with a dozen seriously

Although I could not go back to Sri Lanka, I was able to start similar work in Nepal. Sena Ganewatte helped train there in March 1987 a group of organizers who began working with farmers in the Sirsia-Dudhaura system on the border with India. At the end of 1988, after spending three weeks with the project in Nepal, I phoned Sena to wish him a happy New Year, tell him about progress at Sirsia-Dudhaura, and find out how he and others were faring. By now the Gal Oya project area was in the thick of the conflict. Six weeks before, news had reached Ithaca of a massacre in Central Camp, a place where some of our Tamil IOs had lived and that I had often visited. Terrorists opened fire on people in the marketplace, killing twelve. That both Sinhalese and Tamils were shot showed how murderous and confused the struggle had become, lacking clear political purpose as Sri Lankans were being killed indiscriminately.

Sena reported that the farmer organizations in Gal Oya were "actually doing fairly okay," though he had not been there for many months. Nobody was traveling there. Even the mail was no longer reliable. But three new IOs had recently gone to help the IDOs there, bringing to five the total number assisting the farmer organizations. Senthinathan was still the deputy director of irrigation in Ampare. All the farmer organizations were being registered with the Ministry of Lands, finally gaining the legal recognition long sought by ARTI and by farmers themselves.

In Polonnaruwa the program was as good as or better than in Gal Oya, Sena thought. The Gal Oya farmer-representatives were "very special people," he said, "but Polonnaruwa has some very special people too, of the same caliber." Rehabilitation work at D-channel and field channel levels would no longer be done by private contractors, because of terrorist threats against them. "So the farmer organizations are doing excellent construction work. The Irrigation Department is very satisfied with it."

The program was about to experiment with using educated farmers as catalysts, Sena said, with a training program scheduled for March. "Volunteer" IO programs were operating in two of the irrigation schemes in Polonnaruwa. These were "very effective," according to Sena, with young high-school graduates doing organizing work, each on one thousand acres. This represented a welcome evolution of our

injured men, women, and children. Their settlement not far away had been attacked by "Tigers" the night before I arrived. The JVP, revising its campaign to overthrow the government, was attacking police and army posts in the south to capture weapons. For a detailed account of the rise and fall of the JVP, see Gunaratna (1990).

strategy. Such volunteers could probably not have won engineers' confidence as was needed in Gal Oya eight years before, but now with official acceptance of farmer organizations, this new approach might get the job done at less cost to the government.

The program's biggest problem stemmed from the political conflict tormenting Sri Lanka. Our plans to give farmers control over maintenance work and budgets were stalled because insurgents had forbidden farmers to pay O&M fees or to do any unpaid irrigation work. Apparently the JVP did not understand that shramadana was a traditional institution and that irrigation channels need more maintenance than the government can afford to pay for. So institutional innovations to improve resource mobilization and system maintenance were at a standstill. But organizational structures were now in place that farmers could use if allowed to proceed with activities that would benefit them and their country.

Return in 1989: Bad New, Good News

In July 1989 I was able to return, this time as a short term consultant for an Asian Development Bank project intended to strengthen the institutional capabilities of the Irrigation Management Division. The specific task was to improve the performance evaluation of project managers, IDOs, and IOs. The work was supposed to begin in January, but the level of violence continuing between the presidential election in December and the parliamentary elections in February caused several months' delay.

It was a bizarre good news/bad news situation. The good news was the caliber and commitment of political and administrative leadership in the irrigation sector. The former District Minister for Ampare, P. Dayaratne, who had been so supportive and appreciative of our efforts, was now Minister of Lands and Irrigation. One of his first pronouncements as minister was to endorse the devolution of management responsibility to farmers. His confidence in this approach was based on his experience over eight years with farmer groups in Gal Oya. But even before he took office, the cabinet approved a policy in December 1988 that made "participatory irrigation management" government policy.[5]

5. Joint Cabinet Memorandum: Participatory Management of Irrigation Systems, no. 942 (16-88-12-01). This was a result of the ARTI-IIMI workshop referred to in footnote 8 of

Also important, the minister appointed the former Government Agent in Ampare from 1980 to 1985, D. M. Ariyaratne, to head the Irrigation Management Division. His strong backing for the farmer organizations once he saw their constructive potential has been reported several times in preceding chapters. Munasinghe wrote me that the minister, when told I might be returning with the Asian Development Bank project, had said he would give whatever help was needed.

The new Secretary of Lands was the former Land Commissioner who had been willing to take the IO cadre into his department when we were seeking a home for it in 1984, also good news, and the new State Secretary for Irrigation was K. D. P. Perera, the engineer who helped us in the May 1981 showdown reported in chapter 2, who later became director of IMD and then director of irrigation, all the time being more favorable toward a larger farmer role than were most of his engineer peers.

Most gratifying was the appointment as director of irrigation of Godfrey de Silva, the engineer who personally introduced farmer participation in the Minipe irrigation scheme before we started in Gal Oya. This created a network of understanding and friendship that could give more permanent effect to the goals of our program. The "bad news," however, was the conditions under which they were trying to promote developmental change. By the time I arrived, the average daily death toll in Sri Lanka was about fifty, higher than in Lebanon and climbing. In the five weeks after I left, about five thousand persons were killed, more than twice the number lost in Northern Ireland over twenty years.[6]

---

chapter 8. The policy included transfer of ownership of the irrigation network below the D-channel level to farmer organizations "when they are found ready to take on that responsibility." It also canceled O&M fees for farmers, who would manage the O&M of their distributary systems, pledging the government to continue managing the main system and bear what amounted to about half the total maintenance cost.

6. *India Today*, September 30, 1989, 14. The government's armed forces, the Tiger secessionists, an Indian peacekeeping force, and the JVP (operating much like the New People's Army in the Philippines and sometimes like the Sendero Luminoso in Peru) were locked in a four-way conflict, with noncombatants being killed more often than soldiers. The most militant Tamil group, the LTTE, had agreed to a cease-fire with the government to put pressure on India to withdraw its forces. Practically all Sri Lankans agreed on getting the Indian troops out. The JVP was periodically shutting down the country by declaring islandwide strikes to protest the Indian presence, threatening to kill anyone not observing the strike. The government security forces were responding with curfews that brought all activity except their own to a halt. JVP hit squads were assassinating pro-government figures, politicians, military and police officers, journalists, even occasionally Buddhist priests, while "suspected subversives" were being rounded up by the hundreds each day, many never returning. Paramilitary death squads like those in El Salvador and Guatemala took revenge on the families of JVP supporters or "suspects" in retaliation for acts against security officers and their fam-

In this environment, there were nevertheless signs of progress. New IOs were working in eight irrigation schemes under the direction of former Gal Oya IOs. The government was experimenting with turning over operation and maintenance responsibilities to D-channel organizations, including transfer of some O&M budget to the DCOs. Initial results were good, and this was now to be accelerated. It was actually becoming a necessity, as Irrigation Department officers faced difficulties working in the field. During the 1989 maha season, farmer organizations in Polonnaruwa patterned after those in Gal Oya were doing de facto main system management of major irrigation schemes, opening and closing reservoir sluices because officials were afraid to go and do this work.

Although most other government employees did not venture forth from their offices, the project managers, IDOs, and IOs were going about their work without interference or threat from the JVP insurgents.[7] Because most private contractors were intimidated by the terrorists, farmer organizations were doing more and more of the ID's construction and maintenance under contracts. These could now be up to 250,000 rupees, ten times the limit permitted two years before.

Since road travel to Amparc was considered dangerous, I was advised to fly there on one of the twice-weekly flights of the Sri Lanka Air Force available to the public. (When filling out an application for a ticket, I was surprised to be asked to give "address of NOK" and had to ask what it meant. Next of kin.) Senthinathan met me at the Uhana airstrip and drove me to his office, where Ratnayake from Weeragoda and Narangoda and tall Punchibanda from Gonagolla were waiting. The project manager, Tilakaratna, joined us for a long discussion and lunch together. The level of field channel organizational activity was reported to be now less than in 1987, but the functioning of the other

ilies. *India Today*, March 31, 1991, p. 87, estimated the number of persons killed in the counterterror at from 50,000 to 60,000. This estimate may be too high, but in the United States, a comparable figure at just a quarter of that rate would approximate 225,000.

7. According to ISMP report in November 1988, a letter purporting to be from the JVP was circulated warning farmers to stay away from the new organizations because they were linked to the government. It was thought this would kill the program—until another letter circulated the next week denouncing the first one as bogus. The latter communication said the JVP had nothing against the organizations and farmers should feel free to participate in them. (Also, it said the JVP would find out who sent around the first letter and "take appropriate action.") It was suspected that the first letter came from someone in the Irrigation Department who opposed a larger farmer role in management. At the time of my visit, four IOs had been taken into custody by security forces, and only one had been released. Since the JVP had infiltrated even the army and police, nobody could be sure the IO cadre did not contain any JVP members. If so, however, at least the insurgents knew from inside that our farmer organization work was nonpartisan and intended to empower farmers.

tiers was as good or better. Farmer-representatives were able to set priorities for maintenance and approving budgets. The Project Committee was still chaired by Tilakaratna, but he usually turned the chair over to a farmer-representative. There was full farmer participation on the Project Committee, though they still objected to receiving one-fifth as much as officials were paid for attending. Although Gonagolla farmers were no longer as active as before, Uhana farmers were now taking more responsibility, Senthi and the farmers agreed.

Sadly, Punchibanda had had to drop out of organizational work, he said, after receiving an anonymous death threat against himself and his family, ostensibly from the JVP. Senthi and others doubted it was a genuine letter, because Punchibanda, even if outspoken and controversial, was not pro-government. But nobody thought it wise for him to test the letter's authenticity. Narangoda was planning to leave Gal Oya soon, since his wife had been given a teaching job in Kegalle. He would be missed, but the organizations no longer depended on his leadership. His friends agreed that Narangoda would be moving to a more dangerous area, since Ampare was one of the most peaceful places in the country.

Technical irrigation performance remained good. This season, only two-thirds of the Left Bank was being cultivated because the main reservoir was only one-third full at the start of the yala season. But this was quite an accomplishment, since water issues were holding steady at 5.5 acre-feet per acre, indicating much greater efficiency than before the project. Even in this water-short year, farmers said there were no water distribution problems, and they had no complaints once it was decided which areas would receive water.

There had been controversy over this. When the government gave tail-end (Tamil) areas an equal share, this aroused some strong political pressures in the head end. These did not become partisan, however. I was surprised to learn that a farmer-rep who had been secretary of the Weeragoda Area Council was elected to Parliament in February on the opposition SLFP ticket. The farmer organizations were still regarded as nonpartisan, however. I hoped this would continue to be true.

The visit was as poignant as it was brief. Farmer and administrative leadership was changing and yet continuing. The program was far from perfect, but that it survived was a wonder. Most of the things we had worked for were in place and functioning, and indeed, the Gal Oya model was being extended throughout the country now with political and bureaucratic support from above. JVP attacks had halted progress in some ways and yet paradoxically opened up other possibilities for expanded farmer responsibility. The incumbent government lacked effectiveness and legitimacy in many respects, but nevertheless it had put in place some remarkable leadership in the irrigation sector. When K. D. P. Perera resigned as state secretary for irrigation, he was succeeded by Godfrey de Silva. Certainly we could not have launched a farmer participation program as successfully under 1989 conditions, if only because the conflict and insecurity preoccupied everyone and shortened time horizons. Yet the dialectic of peaceful and violent periods was not something we could control. We needed to understand and work through them. My efforts to develop explanations that went beyond immediate events and interests had already started some time before. There was evident need for concepts and principles of social science that transcended, at the same time they related to, observed phenomena.

Return in 1990: Institutional and Policy Support

In June of the next year, I was asked to help launch a new USAID funded Irrigation Management Policy Support Activity, intended to give effect to the policy of participatory irrigation management adopted at the end of 1988 and supported by the new Minister and Secretary of Lands. As if closing the circle, the person heading this effort was Godfrey de Silva, recently retired from government service as State Secretary of Irrigation. He had initiated an experiment with farmer organization in the Minipe irrigation scheme twelve years before and had been a continuing adviser and supporter of the efforts in Gal Oya. This new project was assisted by the International Irrigation Management Institute, specifically by its director for field operations in Sri Lanka, Doug Merrey. Doug had observed and aided the IO program while a consultant in Ampare in 1981–82, and he had been instrumental in keeping participatory irrigation management on the agenda since.

The government was more stable on one front—the JVP insurrection had faded after its leader was captured and killed in November—but the war with Tamil separatists exploded shortly before my arrival. The LTTE had been holding peace talks with the government and, it seemed, was getting most of its demands acceded to. Then suddenly on June 11 it attacked a string of police stations along the east coast and took away six hundred policemen, who had been instructed not to fight back. The bodies of three hundred were found shortly thereafter, and the other three hundred were also almost certainly dead. Three days later, LTTE gunmen crashed into a meeting of a rival Tamil separatist party in Madras, India, and killed fourteen of its top leaders, including the secretary-general and a member of the Sri Lankan parliament. Other Tamil groups now feared they would be similarly eliminated or subordinated. These events solidified Sinhalese, Tamil, and Muslim support for military action against the LTTE. Government officials now stated repeatedly that this was a struggle not between Sinhalese and Tamils but of Sri Lankans against an ambitious, murderous minority of a minority, denounced on all sides as fascist.

Unfortunately, Ampare was now caught up in the conflagration. In retaliation for the killing of policemen from their towns, small mobs in Ampare and Inginiyagala lynched several dozen Tamils living peaceably in their midst, apparently instigated by local police angered by the murder of their colleagues. A USAID technical assistance team that included a Tamil engineer working on the Right Bank managed to leave just before the violence broke out. Because of the unsettled conditions, it was best that Senthinathan stay at his home in Kalmunai near the coast with his family, since even he was probably at risk in Ampare town (though not among farmers in the countryside).

My assignment was to work only in Colombo, but in any case it was considered unwise to visit Ampare. At least ten thousand families in the district were in refugee camps. Munasinghe visited me at my hotel to brief me on the situation, and I spoke several times by phone with Senthinathan. There were two IDOs still working with the farmer organizations, now assisted by ten IOs. Fifteen more Sinhala-speaking IOs had been recruited and trained to help strengthen the field channel groups on the Left Bank, and twenty-five Tamil-speaking IOs were supposed to start work on the Right Bank. (These plans were, naturally, being reconsidered in light of recent events.)

Operation and maintenance of twenty D-channels had been handed over to DCOs, and turnover was planned for twenty-five more at the end of the yala season. How were the twenty working? "Generally all right," Senthi said, with two exceptions, where political jockeying had disrupted them. Munasinghe thought that about half were working well. Specifically, G3 D-channel had been handed over and was working very well without Narangoda.

Was new leadership coming forward? Yes, plenty, was Senthi's reply, though there were some problems now on UB2. Where turnover has been completed, maintenance plans are made according to farmer priorities, and in all areas farmer-representatives review and approve the maintenance budgets. This was good to hear, but I also learned that the Irrigation Management Division was giving free food-for-work to pay farmers for channel maintenance, work previously done as a farmer contribution. It was sad to see the IMD undercutting the self-reliant action encouraged by our program since its inception.

The Right Bank looked as if it would be more of a challenge than the Left Bank, not because it is mostly Tamil but because there are more absentee owners and more tenants and laborers doing the cultivation. Also, many communities are heterogeneous, with farmers and fishermen living together and having no occupational solidarity. Senthi said they would probably have to proceed somewhat differently in organizing farmers on the Right Bank, and I agreed. While one might speak of a Gal Oya model, it is not to be treated as a blueprint. A learning process approach means we should plan for experimentation and modification in new circumstances, I stressed with Senthi.

When I met Sena Ganewatte, still working with the USAID project in Polonnaruwa, he was full of enthusiasm for the progress there, reporting that the IDOs and IOs were doing wonderful work. All of the D channels had been handed over in Minneriya, and 80 percent in both Kaudulla and Giritale. The large Parakrama Samudra scheme was lagging behind at 25 percent, but it was doing other impressive things like negotiating credit for farmers on reasonable terms from the banking system. The organizations hoped eventually to establish a farmers' bank in Polonnaruwa. "It will come," Sena said. Project-level farmer organizations were being set up alongside the Project Committees made up of officials and farmers. Sena said there were many really good farmer-representatives, "just like in Gal Oya, maybe even some better."

The farmer organizations at Kaudulla were trying to purchase facilities to store their rice crop at the end of the season, when prices are at a minimum, to hold it several months and gain a better price for their members. They were also negotiating a bank loan to set up their own rice mills, to retain more of the value of their production. IOs whose appointments terminated at the end of this year could become managers for the D-channel organizations.

One Kaudulla D-channel organization had won a prize as the best DCO in the country. It was at the tail end, an hour's drive from the town, where water is really scarce. It got a full yala crop for the first time through full cooperation of all members using their water with utmost efficiency. They had gotten an abandoned Land Commission building from the Government Agent and fixed it up beautifully. They were also negotiating bank loans to install six wells for groups of ten to fifteen farmers each, to grow high-value crops in the off-season. Their accomplishments are "really heartwarming," Sena said. In the previous maha season, he added, farmer organizations at Minneriya, where Kule was doing fine work as the IDO, the water issue was reduced by 25 percent, from 4.45 acre-feet per acre (already below the national norm) to 3.3 acre-feet, another real accomplishment.

Farmers were doing most of their D-channel maintenance work by shramadana, so the money they got for this from the government could be put into their development funds, which 80 percent of the DCOs now operate, to make group investments and individual loans. The ID was giving construction and maintenance contracts to FOs for up to 250,000 rupees with little hesitation. These are only for the labor costs, since the ID provided materials, so logistical problems and chances for corruption could be reduced. Still, Sena recognized, this is an area where more training and oversight are needed, to ensure that technical standards are high and that the organizations do not founder because of mismanaging money or materials.

The most interesting initiative Sena told me about was "mortgage release." Started by one DCO, this had been taken up by ten others and was spreading. Many farmers who had borrowed money from moneylenders had lost de facto ownership of their land when they could not repay at the exorbitant rates of interest charged. With the support of the Government Agent, DCOs were recalculating the loans at a bank rate of interest. Since this is much lower, by such reckoning most of the loans had been already satisfied. So the farmer organizations simply declare these farmers free of further financial obligations to their creditors. The latter have to accept this, since mortgaging is illegal in the first place and they face a united body of farmers, backed by the GA's authority.

So while armed battles were going on in the north and east of the country, we could see institutional and policy support growing for participatory irrigation management, buoyed by the initiative and creativity exhibited by farmers, organizers, and officials. When I left in mid-July, I was asked to return at the end of the year to help further with the institutionalization of this mode of development introduced a decade earlier. It was now part of official thinking and planning (IIMI 1990).

When I returned in December, among other activities I participated in was a workshop organized at ARTI by the IMPSA secretariat with farmer-representatives from twenty-four irrigation schemes spread across all of Sri Lanka. Their views were being solicited as part of IMPSA's formulation of irrigation sector strategy. Of the twenty-seven representatives attending, all but four wore sarongs, an indication that they were "real" farmers and not ashamed of this. The two representatives from Gal Oya were farmers I had not met before, M. A. Gunapala, chairman of the Uhana Area Council, and R. Karunaratne, its secretary and also secretary of the D-channel organization for UB9 and 10. It was good to see new leadership coming forward, though I was sorry not to be meeting again the farmers who had become friends over the years.

At the end of the opening session, I was invited to talk to the group. I said I was there mostly to listen but was very pleased to see this kind of meeting taking place, unthinkable when we started ten years ago in Gal Oya. I recalled for them Narangoda's statement in 1986 that the farmers there would not consider their organizations a "success" until they were spread across Sri Lanka.

Karunaratne and Gunapala responded, expressing their thanks for what Cornell and ARTI had done, saying that their farmer-organizations were "90 percent." (This was translated for me as 98 percent, which worried me because it would be an obvious exaggeration; the farmer-reps later restated that the situation was "90 percent advanced compared to 1980.")

When we adjourned for dinner, Karunaratne and Gunapala stayed behind so I could get some details on the situation in Gal Oya. Recently a delegation of FRs had been sent to Thailand to observe farmer organizations there; three had gone from Gal Oya. (The former District Minister for Ampare, now Minister for Lands and Irrigation, told me later that these three had been chosen by their fellow farmer-reps, and were all members of the opposition party, but this was acceptable because they were good FRs.)

One of the farmer-reps, from UB15, had been instrumental in setting up fourteen women's organizations, sponsored by the FOs to increase household incomes. The women started with mushroom growing and had moved on to jam making and textile weaving. When the People's Bank refused them credit at one point, farmers put up 150 rupees each as start-up capital. The weaving had made a profit of 10,000 rupees this year, but there were still problems with it.

The main farmer efforts now were to achieve "turnover" of all responsibility for operation and maintenance at the distributary channel level and below to D-channel organizations. Twelve of the fifty-four

DCOs now had full operation and maintenance responsibility and another ten were due to assume responsibility this month. (One problem was that the document governing turnover, brought from Colombo, was in English and needed to be translated into Sinhala.) How many weak DCOs are there, I asked? They said four or five, ten percent.

Recalling old Ratnayake's concern about the declining strength of field channel groups, expressed in the imagery of a tall tree without roots, I asked about them. The FRs agreed that most groups did not meet regularly, but there was now a date set by which all groups should meet each month, preceding DCO and Area Council meetings so that problems could be channeled upward. The higher level organizations set their agendas, listing items to be discussed. If a field channel group fails to hold its regular meeting, it is expected to discuss the items on the agenda for the meeting it missed. This sounded a bit bureaucratic, but it is systematic and should provide linkage upward and downward.

We have a rule, said Gunapala, that all matters for the DCO must come through the field channel group, not from individuals. The same applies for the Area Council, which receives items from the DCOs. These meet every month, quite regularly. How often did field channel groups meet? Every two or three months was the reply. When I expressed concern that this could mean farmer-reps are speaking only for themselves, not their fellow farmers, Karunaratne suggested that I discuss this with the organizations on my next visit to Gal Oya.

They said the Project Committee was operating well, now with a farmer as chairman. The Project Manager no longer directs the meetings, instead coming and mostly listening, they added. Members of the committee are taking initiative to address priority problems of farmers and have asked the GA to give them some land on which to construct buildings for their Area Councils, which has been agreed.

Because we got to the canteen twenty minutes after the others, there was not much rice and curry left. So Vidanapathirana, who had returned from the Sudan where he had been a UN volunteer and was now an economist with IMPSA, and I took Karunaratne and Gunapala to a small restaurant near ARTI for some supper.

My first question after we had ordered was, How about politics? Are they intruding on the farmer organizations? I was told this is not a problem. I asked for some confirmation and was told about a plan the farmers devised to get one D-channel extended so it could irrigate an additional fifty acres of land. They invited the four MPs from their area to come, two each from the UNP and SLFP, and asked them each to contribute equal amounts from their discretionary development budgets toward the construction. Both parties got credit and neither could claim any political advantage. Also, when the farmer organizations were invited to send representatives to the *Gramodaya Mandalayas,* the Project Committee decided that no FOs should participate in these local councils because the government had made them rather partisan. This is a bold move.

What about Kalumahattea and Waratenne, our old adversaries-turned-cooperators? Both have died, I was told. How about old Ratnayake from Gonagolla? He was not in good health, but "okay." Ratnayake from Weeragoda, I was glad to learn, had been a member of the delegation to Thailand, surely a thrill and a reward for him.

What about the budgets for operation and maintenance? This had been a controversial matter as ID personnel tried to keep control. The Irrigation Department now made an allocation to each D-channel area, and the DCO could set priorities for the funds' use, making the money go further by contributing their own labor as needed. The farmer-representatives praised the cooperation of Senthinathan and Wijayadasa.

What was their biggest problem? Marketing was still the biggest problem. The Paddy Marketing Board was not buying the rice they harvested, so they had to sell to local merchants, who paid five rupees per kilo and then resold it for eight rupees a few months later. The farmer organizations were trying to develop a scheme where they could hold onto their paddy after harvest until the price went up. Their other problem was still credit. I asked what the interest rate was now, and they said at least twenty percent per month. Vida, who was translating, exclaimed that the official bank rate was twenty percent *per year.* So farmers still were not getting their fair share of income.

As we left the restaurant, I asked about the cooperation of government officials. This was generally good, they said. They now get good cooperation from the Irrigation Department. How about the project manager? He is cooperating too.

On our way back to the ARTI hostel where the farmers were staying overnight, I asked about water management, the task for which we had started this whole enterprise. This is fine, the representatives said. They had a new system where each field channel group designates two farmers to be the water distributors for each season, and they take turns doing the work, cooperating with the ID in operating gates and making sure everyone gets water. "Everyone is satisfied," said Karunaratne.

Whether or how much he might be exaggerating I could not know, but their discussion was frank and often critical of their organizations' shortcomings. Is Gal Oya a "success" yet? Some ARTI staff members, perhaps jealous of the recognition our work has gotten, have been criticizing the program, saying it has many failings. But they did not know what Gal Oya was like in 1980 and have not spent enough time there to know it now. The situation is not yet what it should be or could have become with better ideas and more effort. Weaknesses there are, and they may increase if farmers do not make continuous efforts to strengthen their collective capacities. But it appears that what we set out to accomplish for and with the farmers in Gal Oya has been largely attained.

Reflections

Seeing such accomplishments amid disorder gave impetus for developing different conceptions of causation and change than I had brought to Sri Lanka a dozen years before. Our learning from Gal Oya would not be complete if it dealt only with the insights we had gained about possibilities and means for participatory development. More important were ways of improving how social scientists think more generally about their subjects and tasks. At some point, I came to see that social science could be enriched and expanded by adding "post-Newtonian" dimensions.

The discussion that follows in part 2 is a summary of ideas and explanations that attempts to give coherence to the dynamics and outcomes observed in Gal Oya. To keep the presentation from becoming too abstract, it is grounded in the personal experience of coming to terms with the unexpected progress made in promoting participatory development there. The resulting framework for understanding social opportunities brings together concerns normally addressed within the purview of philosophy and physics with the practical problems dealt with in our several social sciences. That social scientists need to be more self-conscious about how they approach "reality" is one of the main lessons learned from working in Gal Oya. Reality is more like a river than a rock.

Having most of the summer of 1988 free, without the usual trips to South Asia I had been making for ten years, I was able to concentrate on writing this book. What can be seen from this case study is how much potential exists in rural communities, as well as in government bureaucracies, for promoting "bottom-up" development. Although this approach has become increasingly accepted in principle as both ethically and practically desirable, it is still the exception rather than the rule.

One reason is that approaches to development are so often viewed dichotomously and treated as mutually exclusive. Paradoxical though it may seem, "top-down" efforts are usually needed to introduce, sustain, and institutionalize "bottom-up" development. We are commonly constrained by thinking in "either-or" terms—the more of one, the less of the other—when *both* are needed in a positive-sum way to achieve our purposes.

Such a realization might be regarded as idiosyncratic, either as a quirk of the Gal Oya case or as something unique to development. But the more I examined the experience, the more it appeared that the ways of thinking I had brought to Sri Lanka were themselves misleading and constraining. "Rigorous" and "analytical" approaches to problem solving, although they have their merits, can blind one to possibilities. They stereotype judgments about people and impose notions of narrow, sequential causation where more diffuse, concurrent influences and effects need to be understood and utilized, with more expansive conceptions of human nature and motivations.

When I discussed our experience in Sri Lanka with others, new axioms and explanations kept recurring that appear to have validity for social relations and social dynamics more generally. These attempt to move the concrete learning from Gal Oya onto a higher plane. I increasingly think one cannot achieve the goal of participatory development without some major reorientation of thinking.[8] In short, the mechanistic, deterministic ideas we use to plan and implement initiatives of economic and social change can themselves impede the objectives sought. We need to reconsider the mental underpinnings of our actions. The metaphors employed in part 2 will be less concrete than rocks and rivers as thoughts turn to things like *conjunctions and sums*. But the results should enrich our capacity for both analysis and practice.

---

8. So do some others who work on participatory development: Anisur Rahman and Matthias Stiefl, who have respectively directed research programs on this subject for the International Labor Organization and the United Nations Research Institute for Social Development. Discussing with them the ideas presented in part 2 encouraged me to pursue these formulations systematically, since Rahman and Stiefl had experience in promoting participatory development each in a dozen countries or more. See, for example, Rahman's introduction to Tilakaratna (1987). Also, I found that evaluations of success in rural development programs in India (Clements 1986) and Indonesia (Craig and Poerbo 1988) were drawing on the same emerging bodies of theory from different disciplines as I was.

PART TWO

EXPLANATIONS

It is said that if you want to know reality, you must try to change it.
Henry Volken, Ajoy Kumar, and Sara Kaithathara,
Learning from the Rural Poor

10

Of Conjunctions and Sums:
Conceiving Possibilities

Relativity eliminated the Newtonian illusion of absolute space and time; quantum theory eliminated the Newtonian dream of a controllable measurement process; and chaos eliminates the Laplacian fantasy of deterministic predictability.

Anonymous physicist, quoted in James Gleick,
Chaos: Making a New Science

The changes observed in Gal Oya were in many ways kaleidoscopic, concurrently encouraging and discouraging. At the same time, certain patterns could be seen that reflected persisting values and relations. Order coexisted with disorder. Many constraints in the physical and social realms had to be taken seriously, yet others turned out to be amenable to some transformation or transcendence by imaginative action—the constructive kind that explores various human, material, and other possibilities. Such efforts would be misdirected if the universe were as mechanistic and determined as that proposed by Newton or Laplace.[1] But our experience in Gal Oya, once we were immersed in efforts to improve irrigation with farmer participation, made reality look ever more like a river and less like a rock.

River metaphors continually offered insights. A river proceeds within constraints—its banks—but continually perturbed by outside influences, some quite remote, like rainfall and all the factors that

---

1. Laplace, an early nineteenth-century philosopher-mathematician, achieved a kind of immortality with his assertion: "We ought to regard the present state of the universe as the effect of its previous state and as the cause of the one which is to follow, . . . a mind which could comprehend all the forces by which nature is animated and the respective situations of the beings who compose it . . . [would] submit these data to analysis . . . [and] would embrace in the same formula the movements of the greatest bodies of the universe and those of the lightest atom; for it, nothing would be uncertain, and the future, as the past, would be present to its eyes" (cited by Wolf 1981, 43).

affect runoff. Indeed, the variation itself varies, as volume and rates of flow interact with natural obstructions and with human uses. A river presents considerable but not complete predictability. Although long-term and seasonal patterns can be identified, these are continually interrupted by unique events. In fact, a river changes itself as the action of its water alters its riverbed and banks. Predictions are possible, as are beneficial uses for humans. So making plans and investments can increase our gains. Yet a potential for destruction coexists with the river's beneficence.

A river's resemblance to the ebb and flow of social life is suggestive but limited. Many currents within our existence are well known and predictable, but the banks that contain and direct them remain indistinct, and we do not know the sea into which they flow. Still, river metaphors for understanding our human situation are more suitable than the mechanistic models now prevailing in most social science, whose imagery was inconsistent with what could be seen and done in Gal Oya.

Making sense of this experience was itself an evolving process, punctuated by insights and occasional misplaced hopes. Understanding this reality, with its many surprises, required some simplification and some reduction of the immense complexity into intelligible concepts and patterns. This reassessment led to a growing appreciation of the ever diversifying, contingent nature of physical and social phenomena. Explanations framed in terms of necessary and sufficient causation no longer seemed either necessary or sufficient. But as a social scientist, I persisted in the expectation that some patterning should be possible.

The problem of reconciling simplification with complexity was paralleled by the difficulty we had of accommodating engagement and detachment in our work in Gal Oya. Our assignment there was considered "action research," something sounding legitimate and appropriate but really quite ambiguous. The agricultural engineer in our Cornell group, Gil Levine, warned in a friendly way that as academics we needed to guard against becoming so involved in the project that we lost our objectivity. Our task had been defined as designing and assessing models for introducing farmer organizations with the goal of formulating recommendations that had a sound social scientific basis. If we became partisans of certain ideas or approaches, or if we got caught up in wishful thinking, our usefulness would be compromised, not just as social scientists but as persons who wished to help improve the situation in Gal Oya.

Neutral objectivity by itself offered no solution, however. This was not some kind of laboratory experiment we were embarked on. We were not just testing hypotheses. People's well-being was at stake. It was hard to be indifferent to whether proposals were getting implemented, whether the (apparently) best course of action was being followed, whether personnel were appointed on schedule, whether proper explanations were being given to farmers. There was reason to try to help the process along, not just to plan and then monitor and evaluate it. But would such activism spoil the effort? We might end up not producing the valid increments to knowledge expected from us. Could detachment coexist with engagement? Were these really as mutually exclusive as people assume objectivity and subjectivity must be?

Some resolution of these quandaries was found in a book that Gil recommended as good general reading, not because it had anything to do with Gal Oya—*The Dancing Wu Li Masters: An Overview of the New Physics,* by Gary Zukav (1979). It made quantum mechanics, which had previously seemed difficult to grasp, quite intelligible. The physics was not really very new, since these theories for understanding atomic and subatomic phenomena had been developed in the first quarter of this century. What *was* new was the growing interest in quantum physics. Popular books on it have multiplied in recent years, making it accessible to nonphysicists who previously knew only classical physics—Newton's encompassing theories, augmented by the laws of electromagnetics and thermodynamics. Almost everyone knows the law of "equal and opposite reactions," which frames all relationships in perfect zero-sum terms, but this does not characterize all physical reality.

Concepts used in the "new" physics for thinking about the behavior of subatomic particles, such as uncertainty and complementarity, appeared marvelously isomorphic to what we were seeing in Gal Oya. In many ways the social realm could be better understood by exploring and extrapolating from what physicists thought they knew about material realities. All understanding, it can be argued, in some way comes, after all, by analogy or metaphor.[2] McCloskey's suggestion that economists no less than poets think and communicate through metaphor must infuriate some economists. "To say that markets can be represented by supply and demand 'curves' is no less a metaphor

2. The use of metaphors has too often been deprecated and depreciated in social science. Lakoff and Johnson (1980, 3) correctly argue that "metaphor is pervasive in everyday life, not just in language but in thought and action." This applies also to formal scientific analysis such as physics (Jones 1982).

than to say that the west wind is the 'breath of autumn's being' " (Mc-Closkey 1983, 502). One of the most frequent and persuasive justifications invoked for a free-market economy, for example, relies not on evidence but on allusion: Adam Smith's brilliant figure of speech, "the invisible hand."

The distinction between objectivity and subjectivity, so stoutly made by many social scientists, has been reconsidered by physicists. Once they got to the atomic level, where Newton's principles no longer worked, they found intricate connections between the observer and what is observed. In fact, many "objective" phenomena are inextricably influenced by "subjective" factors. The doctrine that the world is made up of objects whose existence is independent of human consciousness is contrary to the theories of quantum mechanics and to facts established by experiment.[3]

Whereas physicists have reconsidered their assumptions about what is "real," social scientists, in the name of empiricism, have tended to insist on ontological premises that reflect an unwarranted disregard for metaphysical analysis, too readily equated with mysticism. Unexamined assumptions are made that "what you see is what you have" without an appreciation that comprehending what we see depends cognitively on assumptions. Any knowledge contains unavoidable subjective elements or dimensions, since there is no cognition without concepts and no recognition without preconceptions.

Appreciating "Both-And" and "Either-Or" Thinking

Upon returning from my sabbatical year at the Agrarian Research and Training Institute in Sri Lanka (1978–79), I had the good fortune to be appointed to the Social Science Research Council's South Asia Committee.[4] Though I was more a comparativist than an area specialist on South Asia, I could pass as the latter, having studied Sinhala and

3. See d'Espagnat (1979) and Rae (1986). A contributor to the "new" physics, David Bohm, says of the process of physical investigation: "Our knowledge of the universe is derived from this act of participation which involves ourselves, our senses, the instruments used in experiments, and the ways we communicate and choose to describe nature. This knowledge is therefore both subjective and objective in nature" (Bohm and Peat 1987, 55). The findings of quantum physics do not eliminate or transcend the subject-object duality, but they transform our understanding of the relation between subject and object, establishing connections where we had thought the two could be completely separated.

4. The committee is actually the Joint Committee on South Asia, sponsored by the American Council of Learned Societies together with the SSRC.

some Hindi. It was a wonderful learning experience to participate twice a year in meetings with anthropologists, historians, economists, and humanists of various disciplines.

After I had been involved in Gal Oya for several years, the South Asia Committee undertook to assess its previous ten years of support for research focused on the theme of "South Asian conceptual systems" (described in Heginbotham 1977). Workshops and conferences had been held on subjects as varied as Indian myths and folklore, South Asian monetary systems, and political and cultural concepts of order and chaos in that part of the world. The purpose had been to understand any fundamentally different ways of thinking about the world South Asians might have. To what extent, and with what effect, might South Asians perceive, evaluate, and act according to ideas, assumptions, and meanings that diverge from those prevailing among American researchers?

Committee members each reviewed several completed projects to see what could be synthesized from different proceedings and publications. My assignment was to look at multidisciplinary examinations of the Sanskritic concept of *karma* and the Islamic concept of *adab* (O'Flaherty 1980; Keyes and Daniel 1983; Metcalf 1984). Both concepts embody complex philosophies and metaphysics concerning mankind's place and duties in the world, and both present a Western reader with discomforting ambiguities. What should one think, for example, of a worldview that muddles the distinction between such obviously opposed principles as free will and determinism (Daniel 1983)? How can one reject "the distinction between internal and external, between inward and outward" or defend a statement like "the relation between physical and spiritual self is closer than one might expect (Metcalf 1984, 10–11)? To my way of thinking, such ambiguity was not agreeable.[5]

Ever since starting to work in South Asia, like other Westerners I had been frustrated by what was stereotyped as an "Eastern"

---

5. I was reminded of my perplexity when first reading the classical Indian epic the *Bhagavad-Gita*, in which the Hindu god Krishna persuades the hero Arjuna to enter into battle because it is his duty, not because his is the right side. Krishna counseled indifference toward *both* worldly virtue and vice, treating as equivalent what appeared to me to be opposites or negations (Miller 1986, 36). Arjuna asks Krishna whether the life of discipline and engagement is better than the life of philosophy and detachment that the god has been praising. "Simpletons separate philosophy and discipline [action], but the learned do not; applying one correctly, a man finds the fruit of both," Krishna replies. "Men of discipline reach the same place that philosophers attain; he really sees who sees philosophy and discipline to be one" (57).

orientation—declining to make the kinds of rigorous distinctions and classifications we pride ourselves on in Western countries. Sri Lankan, Indian, or Nepali friends would often follow my line of argument until I thought I had gotten them to agree on A rather than B, then they would evade my conclusion by saying, "Ah, but there is something to be said for B too"—which there usually was. I was trying to force "either-or" choices, while they continually saw in "both-and" terms what I was regarding as alternatives.

Reading through the SSRC conference papers, it became apparent that both-and thinking is essentially what game theorists in social science consider *positive-sum,* whereas either-or reasoning is what they call *zero-sum* (or constant-sum). In the latter situation, the parts always add up to a fixed amount, so that having more of one thing necessarily means having less of another. In positive-sum situations there are no such limits, and the whole can be more than the sum of its parts, capturing synergistic effects. Both-and reasoning appeared to me more valid once presented in this "rigorous" terminology of contemporary social science. Either-or formulations direct thinking toward zero-sum solutions, whereas both-and modes of thought point toward positive-sum outcomes. Can one say that either kind of thinking is right and the other wrong? Not really. Both can be valid concurrently. One need not always choose between presumed opposites, as seen from quantum physicists' understanding of complementarity, which appreciates that alternative explanations can be valid even for the same subject at the same time, as discussed in chapter 11.

My report to the SSRC committee suggested that our exploration of South Asian conceptual systems supported the committee's hypothesis that there are characteristically different ways of understanding the world. Though it is a simplification to say this, South Asians, indeed many non-Westerners, tend to employ both-and conceptualizations, while persons in the Western tradition prefer to view things in either-or terms. Indeed, as if to confirm my proposition, I was myself proposing an either-or classification with mutually exclusive categories. Making this distinction produced certain insights, however, even if overstated.

From our experience in Gal Oya, it was apparent that one need not choose, and indeed should avoid choosing, one abstract principle to the exclusion of another. Rather, it was better to draw on *both,* though not necessarily always equally, in a positive-sum way. I suggested to the committee that social scientists and humanists in both East and West could beneficially utilize both principles, though I would argue for the general superiority of both-and—i.e., positive-sum—thinking.[6]

Paradoxically but certainly, both-and logic accords validity to *both* both-and *and* either-or ways of thinking about problems. This may sound complicated, and one may balk at it on first hearing, but the principle is in fact quite simple and sound once one recognizes and reins in one's propensity to classify things in mutually exclusive categories, as if these categories were themselves somehow real or true. The capacity to construct and utilize categories is not to be discarded or forgotten, only to be kept within useful limits. To give up *all* either-or modes of analysis would itself be drawing an unwarranted kind of either-or conclusion, inconsistent with the insights of both-and thinking.

Learning to reorient my thinking was not easy. Like most academics, I have a penchant for analyzing things into categories, the more tightly defined the better. My discomfort was eased, however, by this conclusion: It is possible and desirable to be analytical *tactically* and yet to be synthesizing *strategically*. Analysis—breaking whole things up into constituent components—can be beneficial, but it needs to be directed toward some goal; it is not an end in itself. Synthesis, on the other hand—putting pieces together into some meaningful whole—exemplifies both-and logic and is ultimately more important. But it coexists with and depends on analysis, which employs either-or ways of thinking. We can benefit by learning to think in both both-and and either-or terms. These can be contrasted as, respectively, *binocular* and *binary* ways of looking at the world. The latter may give clarity from its simplicity but the former gives focus and depth of vision.

Regarding analysis as a tactic and synthesis as a strategy, which is itself an analytical distinction, leads us to reject holism as a hegemonic and exclusive principle in science, medicine, philosophy, or any other field. There are many valid objections to any preoccupation with analysis, since separating things that are directly or indirectly related can obscure more than it reveals. But this caution should not lead us to

6. Heracleitus was probably the first explicit proponent of both-and thinking: "We both step and do not step into the same rivers" (Barnes 1979, 66). Williams (1988), writing on domestic household labor relationships, employs the same both-and versus either-or distinction introduced here, similarly prompted by ideas of Einstein, Heisenberg, and others. He proposes a similar shift from classical to quantum physics thinking, viewing reality as multiple, constructed, and holistic rather than as single, tangible, and fragmentable; regarding causation in terms of mutual simultaneous shaping rather than temporally precedent or simultaneous causes having deterministic effects; and seeing knower and known as interactive and inseparable rather than as independent (1988, 124–38).

treat all things as being indistinguishable, or to believe that no meaningful differences exist to be accounted for. Although the underlying connectedness of phenomena may be underestimated by the analytical propensities of mainstream physical and social sciences, we will not get very far by not making distinctions.

Holistic approaches in various fields of inquiry may produce insights by compensating for the limitations of reductionist analyses, but by themselves they are no solution. A both-and orientation brings together holistic "systems" thinking with treatments of reality that focus on differences and take subsystems seriously. We should not opt for just one approach—analysis *or* synthesis—any more than we can reasonably choose to think only inductively *or* only deductively. A both-and approach includes either-or analysis, while at the same time it considers phenomena and relationships that are not reducible to numbers, pure types, or airtight compartments. Analytical methods can produce, in the language of economics, plausible intermediate products but not final ones.

The Power of Positive-Sum Thinking

The changes in Gal Oya could be made more comprehensible by exploring a distinction now used widely in social science—between zero-sum and positive-sum alternatives, with negative-sum outcomes as a third possibility. These concepts offer analytical avenues that lead to fruitful syntheses of knowledge, as seen in the following chapters. Thinking in positive-sum terms (not the same thing as just "thinking positively") can help to enlarge streams of benefit, which is especially valuable when resources are a scarce as they are in less-developed countries.[7]

7. The term *positive-sum* was not used initially by the founders of "game theory," from which it derives. See von Neumann and Morgenstern (1953), Luce and Raiffa (1957), and Rappoport (1969). Given the priority attached to self-styled rigor in social science, they considered zero-sum situations at length and then treated "non-zero-sum" alternatives almost as an afterthought. Von Neumann and Morgenstern acknowledge that this makes their analysis less realistic and useful. "The zero-sum restriction weakens the connection between games and economic problems quite considerably . . . [by emphasizing] problems of apportionment to the detriment of the problems of 'productivity' proper" (1953, 504). Their limited treatment of non-zero-sum games focused, unfortunately, on the difference that changing the number of players made rather than on the expansion of potential payoffs (504–86). Ordeshook's comprehensive review of game theory (1986) lists ten subheadings under "zero-sum" in the index, but there is no entry for "positive-sum." Non-zero-sum analyses (still not called positive-sum) he considers on only four pages out of five hundred.

In a zero-sum situation the total net value is not changed, only re-distributed, as the gains of some persons equal other persons' losses. In positive-sum situations, by contrast, some can gain without others' losing (or without their losing as much as others gain), so that total welfare is increased. The opposite is a negative-sum situation, where everyone loses (or the gains of some are more than offset by others' losses, as in warfare), with the result that total net benefits diminish.[8]

Real-life situations and outcomes can be classified and understood in these terms. Economic, social, or political activities can simply produce a transfer or reallocation of resources (zero-sum); or they can "expand the pie" (positive-sum); or they may have the consequence of making the pie smaller (negative-sum), whether this was intended or not. Because these alternatives sound so concrete, it is easy to forget that they are only analytical, and that the real world does not exist in such neat categories. Situations can have both zero-sum and positive-sum aspects simultaneously, or they may change from one into the other. Their nature and effects depend substantially on the values and orientations of the persons involved, since in thinking about sums, we are dealing with phenomena that are fluid like rivers rather than solid like rocks. Examples from the relatively "hard" social science of economics make this clearer.

Situations Simultaneously Zero-Sum and Positive-Sum

Economists talk about their science as helping allocate scarce—meaning limited—resources. At any point in time, there is a fixed amount of goods and services to be allocated (zero-sum). But if that

8. Morgenstern defines zero-sum games as ones where "as in games played for entertainment . . . the gains of some are exactly balanced by the losses of others." He continues, "In other instances, the sum of all payments may be a constant (different from zero) or may be variable; in these cases all players may gain or lose. *Applications of game theory to economic or political problems require the study of these games*, since in a purchase, for example, both sides gain. An economy is normally productive so that the gains outweigh any losses, whereas in a war, both sides may lose" (1968, 63; italics added).

Note that fifteen years after introducing game theory, Morgenstern still did not use the designations "positive-sum" and "negative-sum." Though he pointed out their significance, his analysis ignored them. Anthropologists have employed the concept of zero-sum thinking when debating whether peasants in "traditional" societies presume there is only "limited good," so that anybody's good fortune must come at another's expense, leading them to resist change and reducing cooperation among them. In more "modern" cultures, it was argued, people conceive of good things as amenable to being expanded without someone else losing (Foster 1965). Zero-sum thinking is, however, ratified by the "modern" values of individualism and competition as contrasted with the "traditional" values of collective identity and cooperation.

allocation provides enough incentive to persons to bring into production some previously underutilized or "slack" resources, or if part of the output is saved and invested to increase productive capacity, the amount of goods and services produced should grow. So whether certain resources being allocated are zero-sum or positive-sum depends on whether they are regarded as a *stock* or as a *flow*. They can have both aspects at the same time, much as, in physics, the electrons surrounding atomic nuclei appear either as particles or as waves, depending on how they are measured, and thus possess both aspects concurrently. Are economic resources stocks or flows? Zero-sum or positive-sum? The view one chooses should reflect the problem being addressed, since neither characterization is true in any absolute sense. Either may be more appropriate within a certain frame of reference, and both can be concurrently valid.

Situations Objectively Changing from One to the Other

The distinction between objective and subjective realities is more equivocal than usually implied, but it remains useful if applied carefully. To consider questions of resource allocation further, according to neoclassical economic theory, production inputs are supposed to be allocated among alternative uses so as to equalize the marginal costs and marginal benefits of production and exchange. This means that at any point in time, a vast array of zero-sum transactions are taking place, where what is given up practically equals what is received.[9]

But, if some or many people think that the compensation they receive for their contribution of resources to production and exchange is unfair, or if the prices of goods and services purchases are thought to be too high for the satisfaction they give, or if the dissatisfactions associated with earning the income needed to procure goods and services are judged too great, people will *diminish* their subsequent contributions to production. This change will depress future flows and stocks

9. According to the same theory, positive-sum benefits accrue from these zero-sum exchanges because of something called consumer surplus. This arises because people making exchanges value what they are getting more than what they give up. The price at which transactions occur is the one that "clears the market," meaning it covers the cost of producing what the nth buyer will pay for that particular good or service. All consumers who are willing to pay more than the market equilibrium price for that good or service enjoy a windfall benefit. This is another example where zero-sum and positive-sum situations coexist.

of goods and services, so zero-sum transactions can become negative-sum, even when each transaction was analytically equal and thus zero-sum. (Actually, since real-world exchange is often less free than economic models assume, exchanges may not have really been equal.) If people were pleased with their transactions, however, they can expand the volume of production and distribution, thereby creating positive-sum consequences. We can know the dynamic outcome of such collective processes only over time.[10]

Situations Subjectively Changing from One to the Other

Perhaps most interesting is how zero-sum exchanges can be converted into positive-sum exchanges by changing the value that people attribute to each other's well-being. Economists presume that people's preferences are sovereign, and they consider as rational any decisions that seek to maximize the attainment of such preferences. To simplify calculations, it is assumed that people wish to maximize only their own well-being, though nothing prevents them from wanting to enhance others' welfare. (This latter possibility is handled by assuming that people will seek to maximize their income and are then free to spend this money however they like, for themselves or for others if they so desire.)

The various things people desire and want to maximize can be summarized in what economists call utility functions. These, constructed separately for individuals and then often aggregated, assume that people seek only their *own* well-being. Independent utility functions further presume that everyone is indifferent to others' welfare. But people can and often do want others to be better off, or at least not harmed, thereby creating an *interdependence* of utility functions; that is, persons consider that their well-being depends at least partly on that of others (Hochman and Rodgers 1969). This unfortunately complicates economists' efforts to determine what is rational behavior, because it is hard enough to assess what people want for themselves, let alone for others.

If goods and services are considered limited at some point and are allocated among individuals who value only their own well-being, this

10. An example of zero-sum transactions becoming negative-sum would be where each member of a crowd, "wishing to see better a passing parade, rises to his tiptoes, with the result that no one sees better but everyone is uncomfortable" (cited in Streeten 1987, 6).

is a classic zero-sum situation, where one persons' gain comes at another person's expense. But what if someone chooses to value others'
well-being *in addition to his or her own?* This is permitted by economic theory, but it is ignored for simplicity's sake and even implicitly
discouraged as being vaguely "irrational."

In fact, whenever we include others' welfare in our own calculation
of utility, what was a zero-sum situation becomes positive-sum. Those
who care about others consider themselves better off whenever the
valued other persons receive benefits. Goods and services thus can
produce multiple benefits to the extent that we choose to value each
others' well-being. The value of "scarce" commodities can therefore
be enlarged without increasing their volume. Why has this understanding not been recognized and approved as a means of getting more
benefit from limited goods and services? It is difficult to put monetary
value on such satisfaction, but it is no less real for being derived
from others' good fortune. Contemporary social science exalts individualistic calculations of benefit and disdains considering collective
welfare, making an ideology of what was initially justified as a methodological assumption.[11]

When we started working in Gal Oya, we followed the conventional
approach and tried to figure out how the individual and mostly material self-interest of farmers at the head and tail of channels could be
accommodated, to get them to use water more efficiently and equitably. Their quick acceptance of our suggestions to save and share water
surprised us in light of their thirty-year history of conflict over water
and their previous indifference to each other's needs.

In irrigation, a fixed (zero-sum) volume of water can often yield
positive-sum increases in total production if distributed more evenly or
in a more timely way. Some farmer effort is needed to clean channels
and rotate water deliveries, so the added output should be considered
net of these other costs. But in Gal Oya we could see from the crops
growing in fields that had been barren in previous dry seasons that
zero-sum and positive-sum dynamics can interact. Once farmers
started valuing others' well-being in addition to their own, many new
possibilities opened up, even when there was a smaller (negative-sum)
water supply to distribute during the 1981 and 1982 yala seasons.

---

11. There is an important body of literature on "public choice" that tries to reconcile the
pursuit of public goods and collective well-being with individualistic rational-choice assumptions. It has been a challenging field of deductive debate, but it has run aground on its assumptions. See Ingram and Scaff (1987).

People's subjective willingness to contribute to each other's improvement produced measurable, objective consequences in terms of higher water use efficiency and larger harvests. Our efforts to analyze, predict, and improve situations in either-or terms proved inadequate. In this and other ways, we found both-and principles more beneficial. Once I started thinking through this newfound appreciation of "conjunctions and sums," I was in a position to suggest in theoretical terms some explanations of what had occurred in Gal Oya and to suggest strategies for improvement elsewhere. These were consistent with many insights that researchers in other disciplines have been gaining into the problems they were studying along similar conceptual lines. We were not giving up all conventional ways of thinking, since that would itself be inconsistent with the both-and principle, but we were enlarging and enriching our mental armory with more expansive, less rigid concepts.

Beyond Mechanistic Models

Social science presently tends to regard people individually and in groups much like objects in Newton's celestial mechanics, considering how they are moved by external forces. These produce at least potentially predictable behavior in a social universe that however inadequately it may be understood at present, is thought to operate according to some general laws immune to individual influence. Personal values and particularistic attachments—loyalty, affection, solidarity, a sense of justice, creativity, or pride—are regarded as epiphenomenal, as aberrations that interfere with the orbits we would normally follow, pulled by the otherwise steady influences of larger bodies and forces upon us.

In place of gravitation, social scientists consider the invariant force of motivation, which is comprehended about as well as physicists understand gravity—not very well. Personal and individual factors get excluded from theoretical consideration and are treated as random or residual variations, because they are regarded as idiosyncratic. They do not receive the attention they deserve as things that hold social enterprises together and make them succeed. Yet whether they are strong forces or weak ones, they should not be ignored.

Social scientists know that people are not inanimate objects like planets and stars with no will or energy of their own. But will and energy get treated as mediating factors, affecting people's responses to outside stimuli, rather than as independent factors in their own right.

Once preexisting preferences or habits are known, it is thought that one can predict how persons will act and that their behavior can be manipulated by external means. Behaviorism in psychology (e.g., Skinner 1971) may be an extreme extension of Newtonian thinking into the human realm, but the influence of this kind of thinking is found throughout the social science disciplines.

Hirschman (1977) has detailed how some of the earliest contributions to social science in the seventeenth and eighteenth centuries drew on ideas from natural science formulated by Newton, Galileo, Copernicus, and others. The invention of marvelous new machines gave rise to mechanistic notions about human nature and about the proper institutions to regulate human behavior and fulfill human needs. The idea of a clockwork universe, completely regular and fully explainable in material terms, created corresponding images of humans.[12]

Not all social science has been so mechanistic. Marxian analysis was animated by concern with the power and productivity of human potential. Marx opened his essay "The Eighteenth Brumaire of Louis Bonaparte" by declaring existentially: "Man makes his own history, . . . but he does not make it under conditions chosen by himself, but under such as he finds close at hand." Marx's dialectical method interprets the emergent outcomes of historical processes as syntheses between the contradictory principles of thesis and antithesis, a both-and perspective. Unfortunately, his rejection of Hegel's concern with the role of ideas in history led to an either-or emphasis on material factors. The resulting mode of analysis has too frequently led to a mechanistic understanding of social processes, framed in reductionistic materialist terms with a predetermined end point.

Social scientists in this century have sought to emulate the natural sciences more than the humanities. They have written often about power, but seldom about *energy*, revealing the influence of old rather than new physics on their thinking. Most analyses of power have been in the Newtonian tradition, regarding people as objects acting on

---

12. Hirschman says: "The advances of mathematics and celestial mechanics held out the hope that laws of motion might be discovered for men's actions, just as for falling bodies and planets" (1977, 13). He notes that Hobbes based his theory of human action on Galileo, while Spinoza sought to "consider human actions and appetites just as if I were considering lines, planes or bodies" (13–14). Helvetius proclaimed: "As the physical world is ruled by the laws of movement, so is the moral universe ruled by laws of interest" (43). Early political economists regarded the economy as working with "the delicacy of a watch," while the movements of economic men proceeded "with the uniformity of a machine" (93). See also Prigogine and Stengers (1984, 27–77) and Gleick (1987, 12–15).

one another, with predictions based on the simplifying premises of closed systems.[13]

As the physicist quoted in the epigraph to this chapter states, Einstein's theory of relativity made into variables the previously fixed parameters of time and space that had anchored Newtonian versions of the world about us. About the same time, the advent of quantum physics challenged the view that things could be known with certainty; everything is better regarded as probabilistic, and not necessarily with fixed probabilities. Even so, most social scientists have persisted with a mechanistic view of our social universe that is derived from old-fashioned concepts of the physical one.

This is not to say that Newtonian physics is obsolete or should be discarded. Quantum mechanics coexists with celestial mechanics. The two bodies of theory offer different laws because they address different kinds of relationships, at different levels of analysis.[14] Classical physics, which also includes the laws of electromagnetics and thermodynamics, has provided the theoretical and practical foundations for remarkable material accomplishments, building huge skyscrapers and bridges and putting astronauts on the moon. Without them we would not have our precious computers, word processors, or laser printers. We obviously benefit greatly from classical physical theories.

But much of the social universe functions in a way that makes "quantum" concepts more appropriate. Analyses constructed on the premise of closed systems suggest that the best we can hope for is equilibrium, and that the future will culminate in entropy, the loss of all energy and order and the cessation of existence. The possibilities contained in open systems, by contrast, should make us less certain about this. Rather than be preoccupied with equilibrium and an entropic fate, contemporary physicists and other natural scientists have begun

13. This literature is reviewed in Uphoff (1990). Newtonian influences are clearly seen in the anthology by Bell, Edwards, and Wagner (1969), which contains that literature's most "scientific" contributions. My own initial analysis of power was in that same tradition (Ilchman and Uphoff 1969, 50–51), but, our resource-exchange analysis by treating authority and legitimacy as positive-sum rather than zero-sum resources is consistent with the view taken here of power. Adams's study (1975) is one of the few treatments of energy in social science. His evolutionary perspective seeks to convert zero-sum into positive-sum dynamics, and his position is essentially monistic, combining material and mental factors, "yet [he writes] I have to resort methodologically to dualism." His effort to reduce all power relations to control over technology unfortunately weakens his analysis.

14. On this matter of levels, William B. Munro proposed in a presidential address to the American Political Science Association over sixty years ago: "By analogy from the new physics, [political science] should also turn part of its attention from the large-scale and visible mechanisms of politics to the invisible and hitherto much neglected forces by which the individual citizen is fundamentally actuated and controlled" (1928, 10).

grappling with the creativity of natural processes and with the potentials of energy (e.g., Jantsch 1980; Prigogine 1980; Prigogine and Stengers 1984; Gleick 1987). Energy exists throughout the physical universe in practically infinite supply, though only a small part is accessible for use.

Less well understood is what Hirschman has characterized as *social energy* (1984, 42–57). It is a "quantum" concept, since it seeks to break out of equilibrium thinking and regards people as moving not like planets in fixed paths around a sun but more like electrons around a nucleus, capable of shifting orbits and gaining or giving up energy as they do so.

One of the facts of social life is that persons and organizations invariably operate well below their potential. Individuals seldom approach—and certainly cannot sustain for very long—their maximum possible levels of effort, creativity, and cooperation. At best we can sustain effective outputs maybe 70 to 80 percent of what we are capable of achieving, and such personal shortfalls are compounded collectively. How many organizations manage to function on average at much above 30 to 40 percent of their potential?

These numbers are only estimates rather than actual measurements. Potential, by definition, is hard to measure, because one cannot know its limits until they are exceeded. But, the percentages suggested are accepted by practically everyone I have discussed them with because they are consistent with experience and observation. While working in Gal Oya, we always operated well below 100 percent effectiveness, never capitalizing on all of the talents, ideas, and material resources available or attainable. Knowing the program from the inside, we were painfully aware of what could and should have been achieved with more time and forethought, some additional resources, more consultation, or better ways of working. But the institutional organizers and ARTI staff performed, under difficult conditions, at times up to 80 or 90 percent of their potential, according to my estimation. Such intense effort probably contributed to some of the turnover in the program, though most IOs who left insisted they would rather be working in Gal Oya than in less taxing jobs. Most government personnel on average achieved barely 50 percent of what could be expected of them.

Using outside standards of comparison (what was usually attained) rather than assessing them by our own internal criteria (what could possibly be attained), IOs were performing superbly. And many farmers like Narangoda and Punchibanda, whose possibilities had been previously disparaged and unfulfilled, suddenly moved from the low to the high end of their scales of potential. Organizers and farmers managed to sustain collective efforts in the range of 60 to 70 percent, compared with 20 to 30 percent in official institutions. Such differentials made our program very impressive, as the Government Agents for Ampare and Batticaloa attested. The levels of energy and innovation demonstrated, while certainly below a hypothetical maximum of 100 percent, were out of the ordinary and had both synergistic and cumulative effects. In fact, they energized government personnel in Gal Oya so that officials' performance also improved dramatically, without an increase in salaries and even some loss in material rewards.

What accounts for such conversion of financial and other inputs into useful work? Motivation is the standard social science explanation, but following Hirschman, one might better think in terms of what can be called energization. What explains the circumstances in which individuals and groups become energized to produce results and achieve goals? Usually we look for how to overcome the equivalent of individual or collective "friction," which impedes forward movement. Social scientists focus on the manipulation of incentives, the social equivalent of gravitational pull. Better explanations would consider how our program, unlike others, drew on large reserves of personal and group energy, manifested in unexpected effort and creativity.

The output of people and organizations is only roughly related to the inputs they receive. There is usually some positive correlation between inputs and outputs, but because the relationship is not linear, it is much lower and less predictable than mechanistic models associated with input-output terminology imply.[15] Persons with equivalent inputs—salary, years of education, office space—perform at widely varying levels, and the same is true of organizations. Responses to and transformations of inputs are not predictable according to classical physical laws because human enterprises resemble machines only by analogy, and then imperfectly. Hydraulic and organic metaphors suggesting less tightly linked relations of cause and effect and greater variability in outcomes are more revealing.

15. In his analysis of institutional development, Israel (1987, 64) criticizes the common assumption of a linear relation between incentives and performance, suggesting that the "incentive function" is somehow curvilinear, but also that its shape for different combinations of effects remains unexplored. Leibenstein (1976) previously argued the impossibility of specifying the process of transforming inputs into outputs, paralleling some of Heisenberg's argument about uncertainty (chap. 11).

This view is supported by chaos theory, which is emerging in physics and other natural sciences. It explains how small causes can have large effects, as when small influences on weather in one part of the globe produce significant weather system changes elsewhere.[16] This is contrary to Newtonian notions that friction and entropy always diminish rather than accelerate processes. Moreover, this new body of theory shows that nonlinear relationships are common and natural, and that outcomes do not have to occur every time or in just the same way to be "caused." The world looks less and less like a grand machine.

Open versus Closed Systems: Reconciling Rigor and Relevance

An appreciation of chaos theory moves us away from deterministic models without abandoning the presumption that there are causes and effects to be analyzed and understood. It encourages us to think of systems as being more open than closed and as evolving rather than being fixed. If this is true for physical, chemical, biological, meteorological, and other natural phenomena, why not for social relations?

If one presumes that any system of relationships under consideration is closed, this makes it zero-sum, whereas assuming that it is or can be open directs thinking toward positive-sum outcomes. Is the specific system actually open or closed? Any system is likely to be closed in some respects and open in others. Perhaps some systems operate or exist essentially as closed systems while others are fundamentally open, but none are completely closed unless artificially so constructed. A river can be regarded as a closed system, but weather that is ultimately global makes it open. Even a rock, if viewed as a system, is open to outside influences, however minute or slow.[17]

---

16. This is known as the butterfly effect—"the notion that a butterfly stirring the air today in Peking can transform storm systems next month in New York" (Gleick 1987, 8). In chapter 14, I examine the implications of chaos theories for social science.

17. Conversely to the well-known saying, "the more things change, the more they stay the same," it is also true that even when things appear to be staying the same, they can be changing. To take an example from international relations rather than from Gal Oya, an intercontinental ballistic missile that is the subject of disarmament negotiations is not the same thing from year to year. What it *is* is changed by the development of other offensive and defensive weapons systems, by an increase or a decrease in tensions between the superpowers, whether secrecy about the missile's capacities has been breached, and so on. Thus something as solid and material as a missile is an open system, subject to all the changefulness predicted by a Heraclitean worldview.

So whether a system is regarded as open or closed is an analytical question. One can treat it as either. But more insight comes from considering it as both open and closed, though in different ways, examining the implications of each and looking for positive-sum possibilities that can be discerned from each frame of reference.

Analysis is done by constructing either-or categories. For the sake of rigor, the more mutually exclusive the alternatives are, the better. To think analytically, one must set aside the diverse characteristics exhibited by all but the most elementary things. Complex phenomena can be treated as simple realities and classified according to one prominent feature, or at most a few. I have myself been using such distinctions as solid versus fluid, Western versus non-Western, and inductive versus deductive. These categories represent in one-dimensional terms, objects, cultures, and thought processes that have many aspects. Things put in any category, while similar in one respect, will vary along other dimensions.

There is of course value in analysis, but as suggested already, only if undertaken tactically—as a means to larger ends—not as an end in itself. Analysis, if done carefully, can illuminate reality more than it distorts it. As long as the fact of simplification is not forgotten and resulting conclusions are fruitful, analysis is worth the effort and price. In the social sciences, we need both analysis and synthesis. But since analysis logically comes first, we spend most of our time on it. Having invested effort and ego, it is tempting to reify our categories, to cling to them and exalt them.

We have for good reason come to value rigor as the foremost quality of good analysis. Ambiguous or incomplete categories lead to dispute rather than agreement, but rigor assures only greater possibilities of replicability and consensus, not necessarily more valid and valuable conclusions. Unfortunately, there appears to be an inverse relationship between rigor and relevance in most social science work. This may be because rigor always requires some reductionism, since certain aspects of phenomena are necessarily excluded by any classification and measurement. Moreover, their changing nature tends to be ignored because taking this into account greatly complicates analysis.

An important question for social scientists is the nature of this inverse relation—whether it is either-or or both-and. If rigor and relevance are mutually exclusive, we have a transformation curve as in figure 1a. This is either-or and negative-sum in the extreme. If, on the other hand, there is a linear, zero-sum relationship, the curve looks like figure 1b—the more of one, the less of the other. With imagination, it

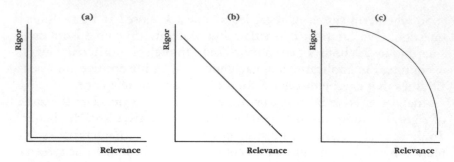

Figure 1.

may be possible to make this relationship convex, as in figure 1c, where we give up some of one to gain more of the other in a net positive-sum situation. Which relationship prevails depends on how we conceive of rigor and how far we take our reductionism. Some reductionism can be beneficial if, to echo the proposition stated above, it is tactical rather than strategic.[18]

If the argument stopped here, readers could dismiss this effort as itself only analytical. But constructive syntheses can be derived from combining both-and and positive-sum thinking. Analysis, to be rigorous, must always be in some way reductionist; rigor is possible only when dealing with closed systems, since open systems by definition encompass influences, however remote or changing, that are beyond the scope of the analysis.

Certain characteristics go together. Rigor is achieved by engaging in reductionist either-or analysis, applied to a closed system with zerosum relations. (Indeed, my analytical argument here exemplifies such thinking.) More relevant kinds of understanding require as a rule more holistic both-and modes of thought, which assume more open systems that exhibit positive-sum dynamics. The logic of both-and thinking supports, paradoxically, attempting rigorous analysis, with all the simplification and classification this approach implies. Such reasoning is to be employed as a tactic, however, to help us move strategically toward syntheses. Analysis is a path—not the complete or the only road—to understanding. We should accept reductionist thought to the extent that it expands our intellectual horizons.

18. Indeed, this critique of analysis is itself reductionist. There seems to be no avoiding the paradoxical position of using reductionist logic to expose and critique the negative effects of reductionism in social science. To extend an idea suggested by Gunnar Myrdal (1957), even heretics must use the language of orthodoxy to express their heresies.

Probabilities and Possibilities

The general nature of this discussion may make it appear somewhat remote from Gal Oya, but the questions and concepts are ones that arose during the very specific events related in part 1. The situation in the field disturbed not only the categories with which we approached it, but also our penchant for categorizing and for passing judgments. The more we got immersed in action, however, the more we saw the coexistence of strengths and weaknesses in people, the difficulty of devising any "perfect" plan, the limitations of every idea, and the extreme complexity of human motivation.

To the extent that we treated things according to abstract categories (as if their "essence" was homogeneous), and especially when we judged people, plans, ideas, or motivations to be wholly unworthy, we were reducing the possibilities for getting positive contributions, which were always needed. We were led to begin using categories more cautiously, more tactically. We needed to make judgments, of course, but we learned to keep them tentative, being continually surprised by support from unanticipated sources and by some unexpected disappointments. Regarding ambiguous situations (or persons) in positive terms created expectations of a mutually beneficial outcome and tended to make such outcomes more likely. We were humbled by discovering that we sometimes made the "right" decisions for the "wrong" reasons, just as "right" reasons could produce "wrong" results.

The first premise to be set aside by experience in the field was the concept of determinism in human affairs. We could construct plans of action much as engineers design machines, specifying what we thought were correct inputs to achieve certain objectives within some predictable time period, but this was soon seen as illusory. At first we thought miscalculations were due to our newness to such an undertaking and to our own inadequacies. But it became clear that nobody had much control over what was going on, though practically everyone had or could have some influence, however small.

There is a saying in military lore, that no battle plan survives its first contact with the enemy. The people in Gal Oya—the farmers, engineers, and administrators—were of course not our enemies. Yet there were so many actors, with different ideas and interests, that the complex, dynamic situation we grappled with was more like a river than a machine. Any idea of a clockwork universe to be understood and tinkered with mechanistically became ludicrous. Conversely, the quantum mechanics notion that everything is a matter of probabilities—that there are no certainties—was helpful. Deterministic constructs had little utility in Gal Oya, and even fixed probability distributions, not varying over time and needing only to be mapped, were not realistic.

At the same time, in the midst of flux there also appeared to be some regularities. It made no sense to adopt the antithetical worldview of randomness. Value orientations, as analyzed in chapter 12, could be identified, and predictions could be made from them. It was possible to establish some order despite disorder once we gave up either-or thinking, colored by zero-sum presumptions about individual and collective action.

It was important to appreciate that no factor could produce desired outcomes with certainty or even a high probability. But certain kinds of initiative and rationale could improve the odds for achieving such outcomes. All relationships could best be understood in terms of probabilities that were themselves subject to change. The likelihood that a certain head-end farmer would accept a water-sharing rotation might be only 0.2 initially, giving a probability of 0.8 that he would not cooperate, but this did not necessarily mean he would be obstructive. At the start, it was unlikely he would join with others, but if those others talked to him, or if they went ahead and set a good example, or if an official visited and encouraged him to set a good example himself, the probabilities could be altered. Even without such influences, there was some chance—one in five—that he would cooperate, but it is better, if possible, to change the odds in favor of program goals rather than to gamble on luck or whatever external factors might affect his choice. As long as there was some possibility (even a small one) of getting a desirable outcome, one could try to capitalize on this and enhance it, changing and beating the initial odds.

This relation between probabilities and possibilities had perplexed me ever since I took a graduate seminar with Warren Ilchman at Berkeley. He startled the class by declaring that he wasn't particularly interested in the central tendencies that most social scientists are drawn to—the mean, median, and mode of statistical distributions. He was more interested in deviant cases, those that are not average or typical, specifically those above the mean in a desired direction. In development, it is axiomatic that the most common situations or characteristics at present are not the most desirable ones. In our quest to achieve more beneficial conditions—the objective of development—deviant cases are more instructive than the mean, median, or mode.

To be sure, one wants to know what the most probable outcomes are. They represent baselines to work from and indicate what is most likely to result if nothing new of different is done. The challenge in development work is how to make possible outcomes that are deemed desirable somehow more probable. This view departs in two important ways from the standard mode of scientific inquiry that presumes a certain unchanging truth exists "out there," to be discovered by some combination of induction and deduction.

First, the view here assumes that the task of inquiry is arriving at instrumental knowledge. This is not to deny the possibility of eternal verities, but truth is regarded as an amalgam of contingent and constant relationships. Factors that vary are harder to pin down than the steady influences, but they are probably more important for accomplishing results. Rather than just seek to know the workings of the world around us, an effort is made to *change* those workings, appreciating that we will get to know reality better as a result of trying to change it.[19]

Second, such a view regards value judgments as an intrinsic and unavoidable part of the enterprise. There may be such a thing as abstract, value-free knowledge, but it is not the only kind of knowledge or even the most important. Consistent with both-and thinking, there is place and need for both absolute and relative kinds of knowledge. In development work, what is contingent and value-led is certainly crucial, the views of absolutists (antirelativists) notwithstanding.

Which brings us to considering possibilities as a special kind of probability. How can one produce more desired outcomes by putting (nudging) an appropriate configuration of circumstances into place? One evening during our first year in Gal Oya, Ed Vander Velde and I had a long discussion. With acute insight, he identified so many obstacles and difficulties ahead of us. If all his estimates of probability were correct, and they probably were, the odds of success were so slim that we might as well quit. The director of irrigation took a dim view of farmer participation; his deputy director in Ampare was obstructing our program; the engineering staff was mostly opposed to farmer involvement; we were not getting the administrative backup we needed from ARTI, and so forth. But there were possibilities, 20 percent, 10 percent, sometimes only 5 percent, that the director would change his

19. This idea, with which part 2 was introduced, has been stated by Smith (1987, 165) as follows: "Action science practitioners suggest that many of the mechanisms which are responsible for the status quo are invisible until they are activated by genuine attempts at change."

view, that we could get the deputy director to cooperate or that he might be replaced, that some engineers would go out and talk with farmers, that ARTI's support could be improved. If any of these occurred, it would improve the odds for the other supporting actions we needed.

If our task as social scientists was just to make the most certain and safest predictions, we should have declared the Gal Oya project's goals unattainable and withdrawn. Or we could have warned that the project was unlikely to succeed and watched while it faltered and failed, never knowing how much of its demise was due to our prediction. The role of social scientists in development has too often been either to make dire predictions—such as "rural elites will capture all the benefits" or "bureaucrats will sabotage the effort"—or to make glossy estimates of success that later prove wrong while someone else gets blamed for faulty implementation. A detached posture of objectivity at least permits one to have made the right prediction if pessimism is justified. If the enterprise succeeds, people are likely to be so pleased that they forget the gloomy prognosis. Enough qualifications can always be made in a forecast to protect posteriors and reputations.

Ed had taken a leave of absence from his teaching position in Michigan to work on our project, and he would not get much satisfaction from saying, "I told you so" if nothing positive came from our efforts. Both of us believed that participatory development was possible if not probable. Given an opportunity to change and beat the odds, we were not about to give up. Wijay, Lucky, Sena, Munasinghe, Joe Alwis, the IOs, and others saw this as one of the rare chances they might ever have to make a difference in how their country developed. They felt strongly that the conditions under which Gal Oya residents lived and worked needed to be improved, and they were satisfied that the idea of participatory irrigation management was sound, encouraged by what we knew of Philippine experience. So we embarked together on the effort reported in part 1.

The strategy had to take account of both probabilities and possibilities at the same time. We should not delude ourselves or others about the points of resistance or the large areas of uncertainty. We had to be realistic, acting according to our best estimates of what was most likely to happen if certain things were done. But at the same time, we could try to alter the constraints on what was probable. Without being self-deluding, we had to identify what was both desirable and possible and to work toward its realization. We got enough positive reinforcement from this approach that it became our mode of operation.

We reconciled the tension between being simultaneously detached and engaged, being both objective and subjective, by being neutral with regard to *means* and maintaining a self-critical attitude as best we could while being firmly committed to the *goal* of participatory and egalitarian development. In this both-and way we derived energy from each orientation while trying to avoid the converse debilitating influence of being indifferent to results but blindly attached to certain means.

We had to learn to keep both probabilities and possibilities in view simultaneously. In our Gal Oya work, we needed to keep one eye focused on near-run probabilities while the other eye kept longer-run possibilities always in view. Successful vision merges the respective images, combining realism and imagination. We chose to be neither pessimistic nor optimistic but rather both at the same time.[20]

Four Fallacies, the Greatest of Which Is Reductionism

This resolution of what seemed like a contradiction sounded foolish at first. Wasn't it true that you can't eat your cake and have it too? (This aphorism has been repeated by millions of parents to inculcate "realism," that is, zero-sum thinking, in their children.) I found some justification in the concept of paradox, that certain things might only *appear* to be contradictory and mutually exclusive—something the ancient Greeks appreciated.[21] Can optimism and pessimism be applied to the same person or the same object of evaluation? It makes sense to have both attitudes operating side by side when looking at the future. The both-and principle justifies accommodating both attitudes of mind, but one should not expect to be completely at ease mentally; living with some tension between alternative principles is personally energizing.[22]

20. This way of understanding our strategy came from my having an anomalous eye condition, anisotropia, where one eye is near-sighted and the other far-sighted. When using the two eyes together, thanks to the brain's ability to accommodate, I have functionally normal vision, and eyeglasses are more of a nuisance than a help. Anisotropic contact lenses are now often prescribed for persons who need their vision corrected, to give them concurrent possibilities for seeing clearly things that are near and far away.

21. Hirschman, an adept discerner and analyst of paradoxes, has lamented the failure of social scientists and policymakers "to understand the contradictory character of every social process" (1971, 353).

22. F. Scott Fitzgerald has written: "The test of a first-rate intelligence is the ability to hold two opposed ideas in mind at the same time and still retain the ability to function" (cited in Peters and Waterman [1982, 89]).

Reflecting on our experience in Gal Oya (in good reductionist fashion), I identified four analytical orientations that are common in contemporary social science and result in much fallacious thinking and action, at least when taken to extremes. They therefore warrant challenge and revision. These four fallacies are *reductionism,* greatly simplifying complex phenomena or relationships, casting them into simple either-or terms (Hofstadter 1979); *individualism,* treating social or collective phenomena as if they were only reflections of personal interests; *materialism,* denying the reality and importance of nonmaterial factors; and *mechanism,* regarding things as if they were machines.

To give up these ways of thinking completely would itself violate the logic of both-and reasoning. Such logic justifies their retention but challenges their hegemony within social science. They should not be accorded any excessive or exclusive credence, however, because they distort conclusions when their use denies the validity of their opposites: holistic synthesis, collective interests and phenomena, nonmaterial realities, and hydraulic, organic, and other nonmechanistic models. If one has to choose, these four alternatives are generally more valid than the restrictive conceptions they enlarge upon, but both sets should be used for more satisfactory results. The four fallacies identified here are not so much wrong as inadequate, not broad enough to carry the large intellectual load assigned to them.

Reductionism has become a dominant intellectual strategy in most scientific disciplines, driven by methods of analysis and particularly dichotomization.[23] Its limitations do not mean it should be given up entirely, since reductionism can produce valuable insights for building up knowledge, provided it is employed more as a tactic than as a strategy.

Individualism as a methodological assumption has more limitations than presently acknowledged; but it can be a powerful and productive force— when not endorsed and promoted to the neglect or violation of social interests.

23. Gleick writes: "In these days of Einstein's relativity and Heisenberg's uncertainty, Laplace [see note 1 above] seems almost buffoon-like in his optimism, but much of modern science has pursued his dream. Implicitly, the mission of many 20th century scientists—biologists, neurologists, economists—has been to break their universes down into the simplest atoms that will obey scientific rules. In all these sciences, a kind of Newtonian determinism has been brought to bear" (1987, 14). Gould (1985, 378), writing as a biologist, regrets "our vulgar tendency to take complex issues, with solutions at neither extreme of a continuum of possibilities, and break them into dichotomies, assigning one group to one pole and the other to an opposite end, with no acknowledgment of subtleties or intermediate positions."

Material factors and needs are tangible and compelling, not to be replaced by nonmaterial factors any more than nonmaterial phenomena should be regarded as just a reflection of material things.

Mechanistic models are too rigid and certain to reflect real-world relations in their entirety, but they can produce insightful approximations.

These orientations become fallacious when extreme or hegemonic. They have validity as part of larger schemes of understanding. Actually, individualism, materialism, and mechanism represent particular kinds or extremes of reductionism. So the main quarrel is with reductionism as a dominating mode of thought.

I contend here that a social science shaped by Newtonian concepts of mechanistic cause and effect, of abstract impersonal forces acting at a distance, of linear relationships where all parts are proportional and interchangeable, is bound to be misleading. Elegant, parsimonious conclusions can be derived from such assumptions, but they are not helpful for promoting the kinds of developmental changes that were needed and introduced in Gal Oya.

We need a social science oriented to the puzzles and potentials of energy rather than to the stable state of equilibrium or the dismal prognosis of entropy. Even if our fate is ultimately zero-sum or negative-sum, the scope for positive-sum analysis and action should not be narrowed by the way we conceptualize the social universe. A post-Newtonian social science looks beyond reductionist either-or/zero-sum thinking to tap the social energy to be found in collective action and nonmaterial realities. The methods and assumptions of positivist social science do not do justice to values, ideas, and motive forces like human solidarity. Because these have very real consequences—indeed, some inspiring possibilities—they deserve more attention than received within reductionist frameworks for modeling our social universe.

11

Relativity, Complementarity, and Uncertainty

With Einstein's theory of relativity and with quantum theory, traditional concepts of motion, matter and causality change . . . the Newtonian concept of absolute space and time, a holdover from earlier Aristotelian notions, was finally found to be incompatible with Einstein's relativistic ideas.

David Bohm and David Peat, *Science, Order, and Creativity*

So often during our work in Gal Oya we wondered what was true and what was not, whether indeed there could be such a thing as truth, absolute and unchanging. Within a framework of deductive logic, indisputable conclusions were possible. But in the field, truth resembled the blind men's elephant—not one like the tame creatures on display in the Colombo zoo, but more like the wild tuskers that occasionally entertained us outside the Ampare circuit bungalow. One's first response when confronted with an elusive, changing set of "facts" from which to construct an understanding of what is true and real—when these do not fit into stable, certain patterns—is to get more and, one hopes, better facts. But this often makes truth appear even more convoluted. One comes to appreciate that facts are contextual or relative.[1] One searches, as Pascal did, for that proverbial "bedrock" of truth on which to stand and build securely, but this is found only by those who

---

1. Nietzsche's explication of Heracleitus's worldview is relevant here: "When the other philosophers rejected the evidence of the senses because these showed multiplicity and change, he [Heracleitus] rejected their view, which presented all things as permanent and homogeneous. . . . Heracleitus is absolutely correct, that 'being' is an empty fiction [meaning 'becoming' is the permanent state of existence]. The 'apparent' world is the only one there is; the 'true world' is a fabrication. There are no eternal facts, just as there are no eternal truths" (Nietzsche 1955, 2:957–58; my translation).

fabricate it by deduction.[2] The empirical world refuses to stand still and eludes precise specification.

An image of reality that came to mind was, not surprisingly, associated with rocks and rivers. The search for truth is like building a bridge across a river. We start from one bank, our present base of understanding, and work toward an imperfectly perceived one on the other side. If we are lucky and the riverbed is firm, we can drive piles down to what we think is solid rock to support the bridge of understanding we are constructing to reach the other side. Actually, a physical bridge can be built on a real river bottom of gravel and mud if enough footings are put down and they are appropriately designed. Some weight can be carried across bridges with pretty insubstantial foundations. It is fortunate when we can put down footings of knowledge on relatively solid rock.[3]

Most of the time in the social world, we are building bridges of comprehension toward ill-defined shores across muddy, murky river bottoms. Can such efforts succeed? Yes, if we recognize that we seldom have rock-bottom foundations and therefore must devise epistemological strategies that are suited to a world that is ontologically uncertain. Given the paradoxical nature of this both-and world, we must remain uncertain even about uncertainty. Often we can presume a reasonable degree of factualness about some knowledge. The river bottom is not all sand and mud or simply flowing water. But neither is it all rock if we only dig down far enough. It is a both-and river bottom we are trying to traverse, so we might as well get accustomed to the heterogeneity of the materials from which we must set about constructing serviceable knowledge.

Einstein's Relativistic Train: Appreciating Frames of Reference

A first step toward post-Newtonian social science is to extrapolate Einstein's concept of relativity in physics to the social universe. No

---

2. Blaise Pascal, a seventeenth-century philosopher-mathematician, oriented by a Newtonian worldview, wrote despairingly: "We burn with desire to find solid ground and an ultimate sure foundation whereon to build a tower reaching to the Infinite. But our whole groundwork cracks, and the earth opens to abysses" (*Encyclopedia Britannica*, 1979 ed., 7:74). We can be more confident if our task is to build bridges of understanding across various gaps rather than to "reach the infinite."

3. What appears to be solid in the physical world is mostly space. In a rock, a minuscule amount of matter is bound together by massive fields of energy. If an atom were the size of a stadium, the nucleus that contains most of the atom's substance would be about as big as a football.

mathematics is required, only an appreciation of how the both-and nature of reality is illuminated by his basic insight about "frames of reference." Even time and space are relative, understandable only within contexts of meaning and evaluation that we ourselves construct. I cannot claim to understand all the intricacies of relativity theory; I still apprehend only its essentials, aspiring to become not a physicist but rather a more sophisticated social scientist. But from a thought experiment discussed below, readers can see how the concept of relativity can contribute to the social sciences much as it has contributed to the physical ones.

During my first year in Sri Lanka, I started writing a long case study of the relation between that country's local institutions and its rural development as part of a larger comparative study of development experience in sixteen countries from East to West Asia (Uphoff 1982–83). In preparation for my own fieldwork at village level, I had read everything on local institutions in the Agrarian Research and Training Institute library. Since what I knew about Sri Lanka was mostly book learning, however, I wanted a "reality check" on my conclusions. A colleague at ARTI, Ranjit Wanigaratne, who had more than ten years of experience studying rural local institutions firsthand, agreed to collaborate on this evaluation so it would be tempered by an inside view to complement my outsider's view.

Working together was enjoyable but not always easy. I was impressed by the contributions of rural local institutions to Sri Lanka's remarkable development record.[4] Ranjit, however, pointed out many shortcomings and deficiencies in the cooperatives, the Cultivation Committees, the Rural Development Societies. We hammered out wordings that could do justice to our respective evaluations, but we were vaguely unsatisfied with the compromises. Which of us was right? Would determining this make one of us wrong? Neither was willing to abandon his point of view, so we both adjusted and carefully crafted our conclusions.

Before finishing the study, while traveling on a train in India, I hap-

---

4. Gross national product growth had not been spectacular, but real income had been doubled in twenty years. Life expectancy, nearly seventy years, was approaching European levels while per capita income was still around $200. Literacy rates were very high, with women almost as well educated as men and with rural literacy (84 percent) almost as high as that in urban areas (89 percent). The rate of population growth had fallen from 3.2 percent to 1.7 percent between 1950 and 1975, and infant mortality had been cut by two-thirds over three decades. Moreover, income distribution had become more equal. The rate of growth in agriculture had been one of the best in Asia (Uphoff and Wanigaratne 1982, 486–99).

pened to read in a popular magazine an article on Einstein's theory of relativity, commemorating the centenary of his birth. The article made the theory intelligible by presenting the following thought experiment, coincidentally about a train.

Imagine a train two million miles long traveling through space at a speed faster than light. While nothing is known to travel faster than light, in a thought experiment, anything can be assumed. Suppose a sleeping conductor wakes up and notices that it is dark, so he switches on a powerful light in the middle of the train, one million miles from either end.

If you were a passenger *inside* the train and could look in both directions, you would see *both* ends of the train illuminated at the *same* time, since light travels at a constant speed (186,000 miles per second) in all directions.

But if you happened to be *outside* the train on a planet or star, you would see light coming out the rear window of the train *instantly* (actually, faster than the speed of light), but would *never* see light coming out of the front window. Why? Because the train was traveling faster than the speed of light, and light inside could never catch up with the moving front of the train.

How can both of these observations be simultaneously true? Because each has a *different frame of reference*. According to this formulation, two apparently mutually exclusive conditions can coexist (the front end of the train being both lighted up and dark) from different perspectives. Recall the illusion one has of moving when sitting in a stationary train while another train passes by. Depending on whether one's frame of reference is the standing train or the moving one, one gets the sensation of traveling or of standing still.

Where there is some agreed frame of reference, such as the ground, it is easier to decide which interpretation of the sensory data is "true." (Consider how oblivious we are to spinning and hurtling through space at unimaginable speed because we regard the earth's surface as our frame of reference.) In our social relationships there is seldom any single, consensual framework of reality to refer to, because most "facts" are social constructions rather than things having an autonomous existence. In any case, the *meaning* of social phenomena is an essential part of their existence, and meaning must be a common creation, existing both in our individual consciousnesses and in some shared set of collective interpretations. So there is no solid ground beneath us to provide a single frame of reference for social phenomena, determining in effect who is moving and who is not, or who is right and who is not. Conclusions and evaluations depend on what frame of reference is employed or assumed.

> An irrigation system operating at 50 percent "efficiency" is performing poorly compared to the hypothetical ideal of 100 percent, but may be considered praiseworthy if one knows that U.S. Bureau of Reclamation systems in the Western United States average around 40 percent.
> The current conflict in Sri Lanka looks one way if regarded as an ethnic struggle (the Tamils, taken together, constitute 18 percent of the population of Sri Lanka, and consider themselves discriminated against by the majority Sinhalese) or as political one (armed secessionist Tamil groups have killed more Tamils than Sinhalese since 1977, seeking to coerce fellow Tamil-speakers into supporting a separate state). The conflict also looks different if one knows that the main Tamil population supporting secession has been an advantaged, rather than a disadvantaged minority.

It is thus necessary to be conscious of and explicit about one's frame of reference with regard even to facts. Gaining insight into social relativity, understanding that the social universe presents no absolute reality, helped Ranjit and me to understand where and why we differed in our evaluation of local institutions and their role in Sri Lanka's rural development. His frame of reference was the country itself, with all its needs and possibilities, its capabilities and constraints. He was keenly aware of how much poorer its performance had been than what was needed and might have been achieved. My frame of reference, since I had edited the other case studies, was all of Asia. I viewed Sri Lanka in terms of what had been attempted and accomplished elsewhere, and I saw more success than failure. To Ranjit the glass of institutional performance was half empty, whereas from my comparative perspective it was at least half full. We could *both* be right.

This should not lead to uncritical and eclectic acceptance of any and all views that may be expressed as true. That would take relativism to an unwarranted extreme. A both-and view expects some things, if not all, to have a fairly high degree of consensus and certainty and to be reasonably regarded as "true." But not all frames of reference are equally valid or, more important, relevant. One needs to assess *them*, not just facts, when trying to sort out differing perceptions and judgments. Social relativity does not mean that nothing can be considered incorrect. Rather, it calls our attention to the frames of reference we and others employ, usually implicitly and unconsciously, when perceiving and reasoning. Conclusions should be drawn carefully and consistently within any given framework. Certain perspectives may be judged and agreed on as irrelevant or of lesser significance. But the principle means that multiple conclusions about the same phenomena may be tenable. Several positions that are inconsistent within a single frame of reference can reasonably be held if more than one frame is applied, according to mutually agreed criteria such as utility and coherence.

In Gal Oya it became evident that many clashing frames of reference coexisted in a single area. Which should be the prevailing one? That of engineers? Farmers? Administrators? Politicians? IOs? ARTI? Cornell? Even *within* each group there were divergent understandings and interpretations of what was true, of what was desirable, indeed, of what was happening. Those groups were most effective that had the most collective coherence in vocabulary, assumptions, and methods of reasoning.

All of us begin by crediting the frame of reference we have constructed for ourselves as largely if not completely valid. To give no credence to our own understanding will immobilize us. But we should not assume that ours is the only understanding, or the only valid one, this being a both-and world. Where there is widespread agreement on something, we can approach something like socially validated truth. When nobody concurs, we have complete relativism. Dissenters from a consensus may be correct within their particular frame of reference, which should itself be examined and debated. Relativity itself remains a matter of degree.

The example of our effort to reconcile "objectivity" and "subjectivity" while doing organizing work in Gal Oya was noted in the previous chapter. The precedent of physicists' probing the subatomic world was helpful. They discovered that electrons, thought to be the ultimate irreducible physical reality, did not have a single mode of existence. Einstein had made important theoretical contributions to the theory of light by treating them as *particles*, infinitesimally small bodies of matter, whereas Bohr and other quantum physicists had experimental evidence that electrons acted like *waves*, like bundles of energy. Which face electrons showed the researcher depended on the question being investigated and on the devices being used for measurement. Why wasn't something as elementary as an electron either matter *or* energy? How could it be both simultaneously, since these are understood as quite different things? Physicists have learned to accept and live with this ambiguity. Why shouldn't we try to be both objective and subjective concurrently? Rather than think in a zero-sum way (the more of one, the less of the other), one can draw on the best of both as much as possible, in a positive-sum way.

Is nothing absolute? By both-and reasoning, some things should be. Yet even something as concrete and quantitative as measurement, it turns out, is relative because it depends on the instrument used. One of the insights from fractal geometry that has contributed to chaos theory is that something as "objective" as the length of the coastline of Great Britain varies severalfold according to the length of the ruler used. Widely differing lengths are produced by using measuring sticks one kilometer, one meter, one centimeter, or one millimeter long, because there is so much convolution of the coastline. To be "absolutely" accurate in one's measurement, one would have to measure around not just pebbles but also grains of sand (Gleick 1987, 94–96). But then of course the length would also differ between high and low tides. What is the real coastline? Even measurements are unavoidably relative.

Taking frames of reference seriously helps us to see how contradictions may be compatible and reconcilable. Which is the preeminent reality—the individual or society? The answer depends on what problem is being addressed. Sometimes one starts from the individual perspective, at other times from the societal end of the continuum. The individual is shaped by and reflects society in crucial, even undetermined ways, while society is made up always of individuals. Who can say which is the true reality—society or the individual—and which is then derivative? Many apparent oppositions get understood in a different light once we drop the assumption that absolute truth and immutable characteristics exist and our task is to discover them. The world itself contains contradictory phenomena and relationships. Absolute and immutable qualities and measures turn out to be creations of the mind more than of the world we observe and within which we act.

Essences versus Emergent Properties: An Existential Perspective

An important question in development studies, but also in social science, is whether there are *essences*, such as "tradition" or "modernity," according to which we can assess things' significance and predict consequences, or whether we should think more in terms of contingent, emergent properties. The Greek philosopher Plato, taking a view in opposition to Heracleitus, argued that the tangible things of the observable world were illusions because they were transient or impermanent. According to his ontological premises, only something eternal and unchanging could be true. This led him and those who followed him to look for ideal forms that lasted forever and transcended human existence—abstract concepts like goodness, beauty, and virtue.

Plato's presuppositions led to an essentialist view of reality, which understands things by characterizing them in unvarying, homogeneous, and mutually exclusive terms. His intellectual rival Aristotle, who laid the conceptual foundations for existential philosophy, rejected this mode of deductive reasoning, considering as more real the palpable existence of things rather than some abstract essence. He preferred a more inductive approach, developing instrumental, purposive categories according to which similar things could be grouped and dissimilar things set apart. The distinctions made were not simply matters of observation but could be decided only with reference to some problem or objective (frame of reference).

Unfortunately, the dispute between Plato and Aristotle got cast in either-or terms, so that existence and essence, material qualities and ideas, were regarded as alternatives. An extreme essentialist position maintains that my teacup does not "exist," though I can grasp it and use it—that only the idea of a teacup is real because teacups come and go but the idea inspiring them lasts forever. The contrary materialist position is similarly unrealistic, insisting in reductionist fashion that my teacup is only a piece of ceramic. It is more reasonable to regard the cup as being concurrently, and as importantly, an idea. Both-and thinkers have no difficulty regarding a teacup as a material object and at the same time, an idea, indeed fully *both*.

In "essentialist" social science, complex phenomena are summarized and dealt with according to some characteristic regarded as their essence, assuming that what they are is immutable, not shaped or altered by the choices and actions of the people associated with them. This approach contrasts with what I have elsewhere called "existential" social science.[5] In the essentialist tradition, we find debates in the

5. These alternative approaches to social science were sketched in Uphoff and Ilchman (1972, 14–17). The existential alternative is not a literal extrapolation from the philosophies having this designation, but a number of themes are common to both: the need to make choices, the existence of many possibilities, narrowing the gap between objectivity and subjectivity, reconciling free will and determinism, appreciating indeterminacy and the ubiquitous process of becoming, giving more weight to the existence of things than to their essence.

literature over whether one-party states are more or less prone to
stagnation than multiparty states. Will military regimes govern more
efficiently than civilian regimes because the former are more technocratic? When gross typologies proved embarrassing, they can be refined to compare progressive military regimes with conservative ones,
competitive with noncompetitive single-party systems, tutelary democracies with immobilist democracies.[6] Particularly dubious generalizations have been made by contrasting the essences of "tradition"
and "modernity" (e.g., Lerner 1958; Pye 1962; Apter 1965). Conclusions based on such reasoning are now largely rejected, but more because the resulting descriptions and predictions were seen to be wrong
than because social scientists recognized their ontological and epistemological shortcomings.

Both Newtonian and essentialist worldviews look for inherent, fixed
qualities, in contrast to the more Heraclitean or existential concern
with contingent, emergent properties (Popper 1972, 194). Within a
Newtonian framework, people are acted upon and respond to stimuli;
their characteristics are seen as determining their behavior. Their personal values and goals are significant only as they affect larger inanimate patterns of necessary and sufficient causation. Context is
considered important if outside forces are seen as determinant, but
analysis includes only factors within the (closed) system that is being
considered. There should be few surprises in a mechanistic, deterministic social world.

There were plenty of surprises in Gal Oya, of course: the "selfish"
head-end farmers now sharing water, the "lazy" officials now
working hard in the field, the cooperation between Sinhalese and
Tamils. After several years, it was evident that we were indeed *doing*
existential social science, something I had written about deductively
before I had some experience with it. Now I got a feel for what it
meant to step outside the assumed clockwork universe—to be dealing
with values and goals (my own and others'), to be focusing on the
backdrop and the stage setting as well as on the actors center stage,
to be assessing and making choices along with others. A whole stream
of outcomes flowed like a river, often sluggishly, sometimes rapidly,

---

There also are associations with the philosophical school of phenomenology, as discussed in
chapter 14.

6. Such generalizations found their way into many earlier volumes on comparative politics
such as Janowitz (1962), LaPalombara and Weiner (1966), and Almond (1970).

with currents and eddies, blockages and backwaters. We did not assume that we were or could be value-free. This was engaged social science.[7]

At the same time, we needed to maintain objectivity, taking a critical look at events and problems. Self-deception would hurt us and our goals. Rather than try to gain as much as we could within the prevailing system of roles and values with its observed probabilities, we were trying to change them, converting parameters into variables where this approach would further our objectives.

It was important to appreciate how much in flux the social universe was. There may be some immovable obstacles, but rivers find ways around them. It was not possible to say what was the essence of people or problems, because they were always in some ways changing, if only because as their context changed, so did they. I was accustomed to working in an analytical, reductionist mode, breaking things deductively into smaller and smaller parts, categorizing, checking for consistency, and analyzing further. But gradually the idea of emergent properties began to make more sense. One can break almost anything into component parts, but that does not make the whole the same as, or nothing more than, its parts.[8]

An atom is not the same thing as its constituent protons, neutrons, electrons, and all the other particles, muons, gluons, and quarks. It is all of these, but it is also more than these. A molecule has properties different from those of its component atoms. A living cell is made up of molecules, but how they are structured is all-important, producing effects not obtainable from different arrangements of the same molecules. An organ in the body is composed of cells, but how can one

---

7. "Science is not achieved by distancing oneself from the world; as generations of scientists know, the greatest conceptual and methodological challenges come from engagement with the world" (Whyte, Greenwood, and Lazes 1989, 515). Whyte and his colleagues report on action research experience with the Xerox Corporation in the United States and the Mondragon worker-managed cooperatives in Spain. "Engaged" research methodology helped Xerox achieve up to 40 percent savings in annual production costs. The cases show it is "possible to pursue truth and solutions to concrete problems simultaneously," the authors say.

8. In the realm of biology, Stephen Jay Gould says: "As levels of complexity mount along the hierarchy of atom, molecule, gene, tissue, organism, and population, new properties arise as results of interactions and interconnections emerging at each new level. A higher level cannot be fully explained by taking it apart into component elements and rendering their properties in the absence of these interactions." He regards these emergent principles, needed to encompass life's complexity, as "additional to and consistent with, the physics and chemistry of atoms and molecules" (1985, 380) He notes that in reductionist studies the living thing disappears and a mere agglomeration of parts remains (387).

explain the workings of a heart or a brain just in terms of its cells?[9] Functions emerge from organs that are beyond the capacity of any cell or set of cells, and of course persons are more than the sum of their organs. Is there any more reason to think that groups of people, organizations, or nations are merely collections of persons, that social wholes are no more than sums of their parts?[10]

This perspective underscores the relevance of relativity with frames of reference as a focus and unit of analysis. Each level considered, itself a frame of reference, presents its own emergent qualities, which reductionism ignores by looking instead at subordinate levels. Lower-level frames of reference can be revealing and may suggest to us the mechanisms whereby certain outcomes are obtained. But the purposes and potentials of these outcomes cannot be adequately understood by studying their components.

An essentialist social science approach classifies things in either-or terms, compared with its existential alternative, which explores both-and possibilities, not only across alternatives but up and down the different levels (though not down to organs, cells, or atoms, which were discussed above only to illustrate a larger point—we are talking here about social science, not biology). Individuals, groups, and higher-level institutions up to the state are all regarded as potential sources of energy and support, each linked to adjacent levels, not reducible to or wholly explained by the others. Social scientists, like the physicists studying the behavior of atoms, are part of the process of analyzing

9. Discussing the hierarchy of organization within the human nervous system, Arthur Koestler describes how "functions at the higher levels do not deal directly with the ultimate structural units, such as neurons or motor units, but operate by activating lower patterns that have their own relatively autonomous structural unit. . . . The structure of the input does not produce the structure of the output, but merely modifies intrinsic nervous activities that have a structural organization of their own. This has been proved by observations and experiment" (1964, 434). Each level of operation within the heart is controlled by the next higher level, and yet each has a rhythm of its own. Similar self-regulating capacities of other organs enable them to function "as autonomous sub-wholes, while at the same time they function as parts of a multi-levelled hierarchy, [each level] subject to the control of still higher levels" (437–38). Koestler's challenge of reductionist premises and methods applies similarly for social science.

10. The general properties of "open hierarchical systems" are analyzed insightfully by Koestler (1967, 341–53). He proposes the analytical concept of the *holon* (pp. 45–58) to represent the two-faced nature of phenomena that exist autonomously in some respects while they function subordinately as parts of larger structures. Cells exist discretely and at the same time they function as parts of organs, much as people lead their own lives while being parts of communities or nation-states. Structures, whether biological or social, present emergent properties that differ from the characteristics of their parts. Contemporary scientific experiments show how structural organization can emerge spontaneously out of randomness (Coveney and Highfield 1990, 168–69).

and acting. If we conceive and perform our roles in terms of detachment and objectivity, we will have a different impact on the situation than if we are engaged.

The Reality of Subjectivity: Linking Three Realms

Accepting the principle of social relativity might seem to lead only to subjective evaluations, but the dichotomy between objectivity and subjectivity is itself misleading. Knowledge is generated and maintained in individual, subjective mental processes, while being expanded and evolved in shared, collective ones. Practically, all knowledge is embedded in at least some people's consciousness, but sharing and consensually validating knowledge gives it some objective status.[11] Like truth, knowledge is neither just subjective nor only objective; it has qualities of both simultaneously.

The weight of scholarly opinion has stressed objectivity and disparaged subjectivity. During our first several years in Gal Oya, I was continually apologizing, to myself if nobody else, for the extent to which subjective factors had to be introduced into our analyses and strategies. As good social scientists, we felt we should be dealing more with objective matters. But gradually the omnipresence and inextricability of subjective phenomena persuaded us that it was wrong to dichotomize and discount them. Accepting both-and ways of thinking justified operating in both modes concurrently and interdependently.

A connection between objectivity and subjectivity was found in a book by John Eccles and Daniel Robinson (1984), which expanded on the typical juxtaposition of two alternatives by detailing a continuum composed of three realms.[12] We generally recognize that there is a difference between the brain and the mind, the one constructed of gray

11. The parallel analysis by Berger and Luckmann (1966) of "society as objective reality" and "society as subjective reality" offers demonstration of both-and thinking. Jürgen Habermas and some other contemporary philosophers are thinking along lines quite similar to those mapped out here, even if much of their language and argument is more complex than that evolved out of practical experience like ours in Gal Oya. An example of Habermas's formulations relevant here is: "Every instance of problem solving and every interpretation depend on a web of myriad presuppositions. Since this web is *holistic and particularistic at the same time*, it can never be grasped by an abstract general analysis" (1987, 305; emphasis added).

12. Eccles and Robinson were amplifying on analysis done previously (1977) by Eccles, a Nobel laureate in brain physiology, with Karl Popper, a renowned contributor to philosophy and social science. In earlier work Eccles (1970) and Popper (1972) referred to these as three separate but related "worlds."

matter and the other eluding any material definition, though it obviously has some material basis. Eccles's research on the brain satisfied him, as well as his collaborators, that we should think in terms of three realms that interact and even overlap but that can and should be distinguished conceptually.

The first is the physical realm of *matter and energy*, within which the brain exists and operates. The brain perceives other parts of this first realm and acts as a liaison between realm 1 and a second, the realm of *consciousness, or mental states*. It is associated with the mind, which engages in thinking, feeling, remembering, imagining, and so forth. It is itself informed by a third realm, that of *knowledge*, which becomes objective for being shared by many consciousnesses (realm 2). Realm 3 is manifested in what is known as culture, a collective extension of the mind, which is itself an extension of the brain. Without the cumulative, shared knowledge that culture represents—its concepts, rules of reasoning, and so forth—we could not perceive and understand the physical world (realm 1).

Contrary to the assumptions of naive realism, one cannot make any sense of a world "out there" without some prior ideas about that world. These ideas are partly self-generated but are largely learned from the social world around us. To Westerners, Indian music sounds on first hearing like monotonous cacophony, because we lack South Asians' concepts of scales, rhythm, and harmony. To an utterly untrained eye or ear, sensory inputs cannot be comprehended without the conceptual apparatus needed to decipher them.[13]

In thinking, we make distinctions, through analysis, at the same time as we construct coherent patterns, through synthesis, from what was other wise unintelligible. We associate similar things and set unlike things apart through comparison, while attaching meaning and values to various things through evaluation. These functions of this second realm make for unavoidable subjectivity even when dealing with "objective" facts. The realm of consciousness is not autonomous. It is suspended between the physical realm of matter and energy and the ideational realm of concepts and meanings. It depends on the material bases of the first realm and on the symbolic templates of the third.

Rather than try to choose between objectivity and subjectivity, we should appreciate their concurrent and interdependent existence. It is

13. That preconceptions are needed for both conceptions and perceptions is increasingly supported by experimental evidence (Boff et al. 1986). A popularized extension of this realization is found in Wayne Dyer's book *You'll See It When You Believe It* (1989).

often useful to make rigorous, analytical either-or distinctions, but all things considered, their connectedness is likely to be more significant. Indeed, Eccles and his coauthors direct our attention to something beyond objectivity (realm 1) and subjectivity (realm 2)—that is, to knowledge (realm 3), a product of the interaction between the first two realms.

Whether one realm is more important than the other is not a meaningful question because of its zero-sum, either-or presumptions. The second realm is in many ways the most fascinating and the most difficult to explore. Eccles and Robinson regard the brain as a liaison at the interface between realms 1 and 2, with the outer senses performing perception and the inner senses performing cognition, both directed by what can be called volition, variously identified with the will, the psyche, the self or the soul. These constructs appear unscientific because they cannot be observed or measured, but this does not make what they represent any less significant. For decades, perception has been eagerly studied by psychologists because it is researchable. In recent years there has been an overdue growth of interest in cognition (see Gardner 1985; Bruner 1986). Unfortunately volition, which is the ultimate in subjectivity, remains off limits for most scientists.

The three realms can be regarded, respectively, as the domains of things, thoughts, and ideas. The last can of course be embodied in material things like printed words or stone monuments glorifying ideals like national independence or personal bravery. The rest of realm 3 exists in the repertoire of concepts inherited, refined, and passed on from generation to generation. The term idealism has gotten a bad name in recent times, represented according to either-or logic as the opposite or negation of realism and thus as unrealistic. But realm 3 is not to be dismissed as nonexistent or trivial.[14]

Both the "ideal" and the "actual" aspects of things constitute valid frames of reference. One of the most common fallacies in social science is to compare the ideal of one thing with the actual condition of something else. If ideal decentralized development is contrasted with the actual way centralized bureaucracies work, the former is obviously superior. But if I compare typical decentralization with its many limitations and disappointments and centralized administration that

14. Vincent Ostrom has written: "The human capacity to comprehend the meaning of experience and to articulate this meaning in a system of communication depends upon the development of semiautonomous symbolic systems ... [that] comprise the elements of a culture, and include such phenomena as language, numbers, law and politics, ritual and belief, art and skill, science and technology" (1964, 85).

performs as it is supposed to, the latter will be more attractive. Apart from the fact that one should not expect to choose either as better for all circumstances and tasks, it is manifestly foolish to judge the ideal form of one thing against the actual form of another, though contrasting ideal forms with other ideals or actual with actual can be fruitful. Ideal forms, to the extent they are widely shared, take on a reality that is very important, shaping behavior by invisible means. They provide common frames of reference that help us make evaluations and choose courses of action.[15] As discussed in chapter 13, ideals represent a special category of ideas, and both have energizing power.

Escher's Hands: Different Realities Grounded in Each Other

There remained something disquieting about the both-and concept of synthesis. It implies that one should never fully accept or completely reject either extreme in a dualistic or polarized scheme. This needs some elaboration. It is consistent with a both-and worldview to maintain that in some situations both-and thinking is appropriate, whereas in others it is not. Some opposites are mutually exclusive by definition, such as hot and cold or light and darkness. Certain relationships are zero-sum by their nature and do not hold out positive-sum possibilities. But the concern here, which derived from our experience in Gal Oya, is with discovering where *self-imposed* either-or assumptions make our actions and evaluations *unnecessarily* zero-sum. One wants to be able to take advantage of every opportunity available for furthering people's development by overcoming what only appear to be real constraints.

The South Asia Committee of the Social Science Research Council in July 1984 held a conference titled "Symbolic and Material Dimensions of Agrarian Change in Sri Lanka," a topic I had proposed in connection with our interest in South Asian conceptual systems. How might Sri Lankan and Western social scientists reconcile, if they could, the main lines of disagreement about recent changes in that country? One school of thought drew largely on cultural and social anthropology, whereas the other got most of its formulations from the moorings of Marxist theory. The predominantly rural nature of Sri Lankan society had begun to change. Economic development was shifting employ-

15. This conclusion reconciled me to some of Plato's philosophizing, which I had rejected in favor of Aristotelian premises before I appreciated the significance of both-and logic.

ment and population from rural to urban areas. The intrusion of tourism and both foreign trade and foreign aid were changing consumption patterns and life-styles. Literacy rates were approaching 90 percent, and the status of women was the highest in South Asia.

Did what Sri Lankans think about themselves and each other shape these patterns of change, or were the changes themselves more strongly altering Sri Lankans' values and conceptions? The conference goal was to assess, in effect, whether the perspective of "culture" or that of "class relations" produces more insight. In the terms of Eccles and Popper, this match was realm 3 versus realm 1.

It was very difficult to get participants to address the large epistemological and ontological issues involved. Most remained preoccupied with what they thought was true about "their" villages. Rather than force a simple dichotomous choice, I asked participants to present what they thought was the most defensible position along a continuum between symbolic and materialistic extremes. But this, especially in a South Asian setting, was resisted. A Sri Lankan anthropologist now teaching at Princeton University, Gananath Obeyesekere, cast aside the question, saying it made no sense to him. "Both symbolic and material phenomena are so grounded in each other that although they appear different, they are not separable." Someone reminded us of the "drawing hands" sketched by M. C. Escher, the recursive artist featured in Douglas Hofstadter's popular book *Gödel, Escher, Bach* (1979). This image of realities reciprocally growing out of each other offered a graphic solution to the question I had posed (see next page).

Not all dualities can be expressed in this way, but many oppositions such as those probed in this book appear more comprehensible and dynamic in this light:

objectivity and subjectivity	abstract and concrete
induction and deduction	probable and possible
energy and matter	absolute and relative
analysis and synthesis	ideal and actual
tactics and strategy	interior and exterior
individual and society	quantitative and qualitative
similarities and differences	free will and determinism

These can be understood as being "grounded in" each other. Only with energy and matter do we have a formula like Einstein's $e = mc^2$ that states the relationship of interdependence mathematically. But productive tension and convertibility can be seen in each of these po-

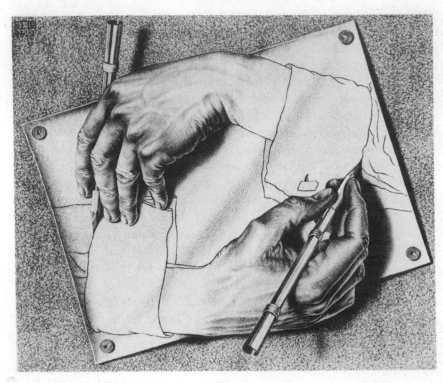

M. C. Escher, *Drawing Hands.* © 1948 M. C. Escher/Cordon Art, Baarn, Holland.

larities. They are not the thesis and antithesis in a dialectical relationship that produces a superior synthesis. Rather, they are alternative, concurrent realities of the sort Heracleitus wrote about in his theory of the unity of opposites. Even things as apparently mutually exclusive as order and randomness are related since randomness can be mathematically expressed as "order of infinite degree" (Bohm and Peat 1987, 121–28). Anthropologists use what they call "emic" and "etic" frames of analysis concurrently. One represents reality as understood by persons living within a community or society, while the other constructs reality from the vantage point of an outside observer. The interpretations and conclusions drawn from each perspective are valid even though they differ and even when they are derived from observing the same situation and phenomena (Pike 1967, 1, 8–28; Goodenough 1970, 104–12; Headland, Pike and Harris, 1990).

There are, of course, different kinds of oppositions to be reckoned with, grounded in each other in various ways. The most definite category includes opposites like dark and light or heat and cold that are physically distinct, since the presence of one means the absence of the other. These appear to be differences in kind but they are fundamentally still matters of degree—the more of one, the less of the other. This makes them inextricably related, because the significance of one derives not just from its own qualities but also from the absence of the effects of the other.

Unfortunately, such archetypal polarities, which are indeed mutually exclusive, have given the mistaken impression that all opposites have this quality. Most do not, being instead contrasts, complementarities, or alternatives rather than negations. Social circumstances like war and peace or inflation and deflation, for example, though technically one is the absence of the other, have emergent properties and second-order effects so that peace is not just "no war," and war is more than just the absence of peace.

Some opposites are issues of form, such as solid and liquid, or matter and energy, where the alternatives are convertible one into the other, at least in principle. Forms may be quite distinct but closely related, such as blocks of ice and snowflakes (crystals), which are both water in different solid forms. In such cases we are dealing more with qualitative than quantitative differences, though even matters of degree manifest both kinds of difference.

Other forms of opposition are matters of contingence such as interior and exterior or life and death, where each depends on the existence or possibility of the other. Still others are matters of contrast. These are more abstract and derive their meaning from their opposites—for example, objective and subjective, inductive and deductive, and indeed, abstract and concrete. And still others represent different principles, like private and public or linear and nonlinear.

Efforts to come up with a satisfactory classification of types of opposites, something Heracleitus tried to do 2,500 years ago (Hussey 1972, 45–50), have continually been undercut by the nonexclusiveness of categories. This very exercise shows the limits of either-or analysis.[16] Yet the exercise itself confirms that what are regarded analytically as opposites invariably have some connection in significant

16. The limits of either-or analysis have been reached in computer science, where "fuzzy sets" give more power in applying logic to complex problems than do the strictly binary categorizations with which digital computing started out (Zadeh 1987; Wang 1983).

physical or intellectual ways. Highlighting differences through analysis can contribute to larger schemes of synthesis, which show how presumed opposites are grounded in each other.[17]

Complementarity and Uncertainty: Borrowing from Bohr and Heisenberg

Niels Bohr, one of the founders of quantum physics, proposed the principle of complementarity to reconcile the contradiction involved in electrons' behaving in some instances like particles and in others like waves. Rather than viewing the wave and particle theories of light as excluding one another, this approach presents both concepts as necessary to achieve a complete description. "Complementarity meant the universe can never be described in a single, clear picture but must be apprehended through overlapping, complementary, and sometimes paradoxical views" (Briggs and Peat 1984, 53). Though formulated to deal with subatomic phenomena, the principle of complementarity finds application at the macro level with biologists and other natural scientists, and it has been proposed as relevant to social scientists as well.[18]

17. A paper presented at a U.N. Centre for Regional Development symposium in Nagoya where I also made a presentation in November 1991 cited an old Chinese proverb, almost as old as Heracleitus's thinking, that is relevant here: Each one of the many kinds of opposites acts as if it could get along without its other. But nature treats opposites impartially, dealing with each of every pair of opposites with the same indifference. And the intelligent man will regard opposites in the same manner. Wang Huijiong, "Globalization and Regional Development," 26.

18. See, for example, Pattee (1978). Ilya Prigogine, Nobel Prize winner for his work on the thermodynamics of nonequilibrium systems, suggests that different languages and points of view about a system "may be complementary. They all deal with the same reality, but it is impossible to reduce them to one single description. . . . The real lesson to be learned from the principle of complementarity, a lesson that can perhaps be transferred to other fields of knowledge, consists in emphasizing the wealth of reality, which overflows any single language, any single logical structure. Each language can express only part of reality" (Prigogine and Stengers 1984, 225).

The principle of complementarity is applied to contemporary public law by Erich Jantsch (my interpolations show where concepts of relativity also come in): "We are used to distinguishing unambiguously between opposites, especially between 'good' and 'bad' or between 'good' and 'evil.' In such a view, to act ethically means simply to further the 'good' and to suppress the 'evil.' But in multilevel ethics [which involve different frames of reference], values may become reversed from level to level. The reduction [of complex issues] to a single semantic level [to a single frame of reference] is responsible for the confusion reigning in the implementation of minority rights in America, for example. In order to increase minority participation in academic professions, admissions standards are lowered for minority students. In order to restore or achieve equality in the long run, equality is violated in the short run. But instead of making this two-level [two frames of reference] policy explicit in the law,

That supposed mutually exclusive qualities can coexist is readily seen from the following two sentences, both of which we know to be true. The human body is extremely vulnerable; people can die from small biochemical imbalances or from a small cut in a major artery. At the same time, the human body is incredibly resilient; we have myriad examples of people who have survived devastating injuries or diseases. Zero-sum analytical thinking suggests that one or the other must be wrong. The principle of complementarity helps us expand our way of thinking to accept and appreciate both "truths" for their respective validities.

Some oppositions are intrinsic, as I noted already—for example, heat and cold or darkness and light. They have been extended, without critical thinking, as "models" for practically all opposites or contrasts. Fortunately, we now have physical scientists calling into question such apparent dichotomies as that between observer and observed or between subjectivity and objectivity. Even truth and falsehood or love and hate, long regarded as simply excluding one another, may coexist, according to psychologists. The line between animate and inanimate forms gets more interesting and fuzzy as virologists pursue their research. Ethicists as well as theologians continue to wonder about their ultimate dichotomy—whether "evil" is simply the absence of "good" or whether contrary principles of motivation exist, each in some sense needing the other. The relation between free will and determinism also looks different in the context of complementarity, where two competing principles coexist.

My purpose is not to take social science into the realms of biology or cosmology but to suggest that the perspective of complementarity opens up theoretical and practical possibilities for development. Zero-sum categorizations that are mutually exclusive have tactical value, but only that. As strategies, they invite narrow, static conclusions, forgoing opportunities for insight and action and depriving us of some benefits that could be obtained from available resources, whose scarcity is at least partly a consequence of our thinking.

Analysis is made more complex by the fact that things appear not only in twos but also in threes or larger sets. For example, the

the sloppy fiction of equality as an absolute, constitutional right is upheld. No wonder that the US Supreme Court decision in the case of Alan Bakke vs. Regents of the University of California met with such a confused reaction—both sides claiming victory, others becoming deeply concerned about the future of minority participation. . . . In a multilevel, evolving reality, opposites vanish ultimately. . . . In process thinking, there is only complementarity in which the opposites include each other" (1980, 272–74).

distinction elaborated in chapter 10 among positive-sum, zero-sum, and negative-sum dynamics is threefold.[19] Complementarity is thus not restricted to polar opposites but rather refers to the interactive nature of different aspects of what are ultimately larger complex realities. These overlapping and often paradoxical aspects of phenomena can be made mutually exclusive by definition and deduction, but they remain functionally intertwined. Any single view of them is invariably partial, made possible—but at the same time limited—by the frame of reference used. Physicists have discovered this in studying electrons at the micro level and conjecturing about speeds in the solar system beyond that of light. In between is the world that social scientists seek to understand and contribute to.[20]

Accompanying this appreciation of complementarity as a general principle is the concept of *uncertainty*, which is generalizable from Heisenberg's proposition of quantum physics, that many if not all things are inherently unknowable. The very act of trying to measure or ascertain them affects the phenomena being considered. This particular caution applies in social science perhaps even more than in physics. It does not mean that approximate measurements and determinations should not be made just because they will not be exact. In a world of uncertainty, all knowledge is to some extent an approximation. Thus our statements are most apt if probabilistic, modified with appropriate qualifications.[21]

19. Physicists find matter existing in one of three states—solid, liquid, or gas—with possibilities for being transformed from one to another. Economists classify inputs not into two categories but rather into three sets of factors of production—land, labor, and capital. Freudian psychologists study the interaction among ego, superego, and id. Sociologists have considered three ideal types of authority: traditional, charismatic, and legal-rational (Weber 1947) and of compliance: coercive, utilitarian, and normative (Etzioni 1961). This last distinction parallels that made by economists among types of systems (threat, exchange, and integrative—Boulding 1989), incentives (condign, compensatory, and conditioned—Galbraith 1983), and strategies (exit, voice, and loyalty—Hirschman 1970).

20. In his APSA presidential address, William Munro said: "A revolution so amazing [as quantum theory and the doctrine of relativity] concerning the physical world must inevitably carry its echoes into other fields of human knowledge. . . . By no jugglery of words can we keep Mind and Matter and Motion in watertight compartments" (1982, 2).

21. The concept of uncertainty is applied by Guillermo O'Donnell and Philippe Schmitter in their analysis of "transitions from authoritarian rule," saying that the problem of prediction arises from "underdetermined" situations, where there are insufficient structural or behavioral parameters to guide outcomes (1986, 3). Their analysis does not reject the long-run causal impact of structural factors but gives weight to the choices and actions of individuals and groups. Because of a "high degree of indeterminacy embedded in situations . . . unexpected events (*fortuna*), insufficient information, hurried and audacious choices, confusion about motives and interests, plasticity, and even indefinition of political identities, as well as the talents of specific individuals, are frequently decisive in determining outcomes" (1986, 5).

Even when we know that our understanding is imprecise or that relationships are contingent, the language we use for the sake of brevity commonly obscures this, asserting more certainty than is warranted. Many misunderstandings and disputes stem from overstatements that fail to reflect our uncertain state of knowledge. When uncertainty is due to our own inadequacies, we should seek to remedy them, but we should constantly bear in mind that there are inherent limits to what can be known.[22]

Reflecting on the Gal Oya experience, I see that many of the surprises resulted from the ambi-valence of phenomena and the occasional reversal of opposites. The continuing dynamic of changefulness became very real to us, though we did not construct any metaphysical theory to account for this. Many insights could be gained from the ideas of quantum physics and other disciplines. That positive and negative electrical charges, for example—as opposite as physical phenomena can be—are so interdependent and cannot exist without the other, like Escher's drawing hands, supported the both-and reasoning that seemed more congenial in Sri Lanka than in the United States.

While learning to resist the allure of reductionist thinking, we were revising our understanding of the logic of collective action. Efforts to make sense of behavior at the interpersonal level may seem far from the broad conceptual questions addressed in this chapter, but their implications for social science become clear when we explore the limitations that arise from reducing all human motivation and action to individualistic and materialistic bases. The ways of thinking described here and in the preceding chapter have practical application and theoretical power for dealing with personal and social behavior, as will be seen in the next two chapters.

22. Max Weber expressed this in Heraclitean metaphor: "The stream of immeasurable events flows unendingly towards eternity. The cultural problems which move men form themselves ever anew and in different colors, and the boundaries of that area in the infinite stream of concrete events which acquires meaning and significance for us. . . . are constantly subject to change. The intellectual contexts from which it is viewed and scientifically analyzed shift" (1949, 80).

The Rehabilitation of Altruism and Cooperation

The conventional economic model not only predicts (correctly) the existence of problems with free riders but also predicts (incorrectly) such severe problems that no society we know could function if its members actually behaved as the conventional model implies they will.

Howard Margolis, *Selfishness, Altruism, and Rationality*

One of the most prized oppositions in social science is between presumed rational and irrational behavior. It has led to a wealth of analyses and predictions regarded as rigorous by their producers, who assume for the sake of analysis that people can be best understood as self-interest-maximizing individuals. The suggestion made previously applies especially here, that analysis should be regarded as a tactic rather than a strategy and that judgments and conclusions are best derived from both-and syntheses.

Although useful insights can come from simplifying models of motivation and behavior, no inference is justified that these are the only kinds of motivation and behavior to be expected and valued or that other kinds are therefore irrational or spurious. What was introduced as a methodological assumption has become within the social sciences and within our culture at large a prevailing belief, that people *are* essentially selfish and individualistic (by definition) and that policies and institutions should therefore be based on this presumption.

The parsimony and elegance of this model of human nature makes more open-ended things like generosity, compassion, a sense of justice, or hope for a better world less attractive to social scientists. Such things get regarded as epiphenomenal and are not taken seriously or integrated in development planning and programs. Assumed to seek and act upon personal advantage only, people are thought not to care

326

about others' well-being. To be sure, there is abundant evidence that individual self-interest is a powerful and pervasive motivating force, so the assumptions that "rational actor" analyses are based on are not capricious or unfounded. They are grounded in much readily observable behavior.

Indeed, if one had to choose a single premise by which to analyze and predict people's actions, the principle of self-interest maximization arguably gives the best description of probabilities. But what about possibilities? As discussed in the next chapter, expectations constitute one of the major influences on behavior, creating self-fulfilling prophesies. The assumption of selfish motivation on the part of individuals tends to produce such outcomes, thereby reducing the influence of collective interests and social ideals. Opposite assumptions contribute to converse outcomes. Such relationships are probabilistic, like everything else in this relativistic, quantum, chaotic world, and the probabilities themselves are not fixed. Just because expectations do not invariably and directly cause certain behavior does not mean, however, that they should be neglected. Until we depart from the limiting assumptions of Newtonian psychology, where stimuli and responses are mechanistically linked and treated as essentially individual, we will not adequately appreciate the subtle influences of common interests and shared ideas. It is important to see the social world, like the natural one, as having both-and possibilities that create positive-sum dynamics, alongside outcomes that are shaped by zero-sum thinking and either-or categorizations.

The Tragedy of Regarding Collective Action as Illogical

The strongest case against basing social organization or public policy on presumptions of altruism and cooperation is supported by two influential contributions to social science: Mancur Olson's book *The Logic of Collective Action* (1965) and Garrett Hardin's essay "The Tragedy of the Commons" (1968). These analyses have had the unfortunate effect of confirming the reductionist and coercive predispositions of many policymakers and analysts. But their arguments have some validity, and they are valuable for having pointed out how individual self-interest can undermine the effectiveness of voluntary organizations or community natural resource management.

A growing literature has begun challenging the Olson-Hardin argument on empirical and theoretical grounds, however.[1] Both the thesis and the critique will be outlined because both are relevant to the experience in Gal Oya. As Howard Margolis points out, the "rational actor" analysis correctly predicts problems of free riding, but it incorrectly presumes that they will suppress cooperative and altruistic behavior, without which no society can function. Bad money may drive out good money in economics, but selfish behavior need not displace all other-regarding action in society.

In Olson's analysis, the prospect of free riding—where some individuals avoid paying the costs of collective activity but obtain its benefits—is held like the sword of Damocles over voluntary organizations such as the water user groups in Gal Oya. Public goods are benefits that can accrue also to persons who did not contribute to creating them, such as increased flow of water to all farmers along a channel that has been cleaned by group action if the group has no authority to withhold water from anyone. When free-riding individuals can get benefits without contributing their fair share of the costs of creating them, this causes resentment among those who have borne the cost of benefits that others are also receiving. Collective action is at risk when people do not want to contribute to establishing or maintaining an enterprise that benefits "undeserving" persons.

An example would be a trade union whose success in collective bargaining raises the wages of all workers, whether or not they are members and pay dues, so long as nonmembers cannot be excluded from getting benefits. If nonmembers can gain at others' expense, it is presumed that as rational actors they will do so. Existing organizations will become defunct because they cannot mobilize enough resources to maintain themselves when members opt out and free ride if permitted to do so. More important and more unfortunate, organizations that could be beneficial will not come into existence because everyone will wait for others to bear the costs of their establishment and operation. This is what we feared would happen in Gal Oya.

This "tragic" outcome can be averted only under certain conditions: (a) if the organization is sufficiently small, free riding can be deterred

1. Though these two analyses address somewhat different tasks of collective action, voluntary organization, and natural resource management, they are treated together because they derive from the same game-theoretical assumptions, essentially those of the "prisoner's dilemma." This analytical model is well explained in Dawkins (1989, 189–201). A theoretically and empirically powerful critique of this model is presented in Ostrom (1990). Hampton (1987) shows that many collective action problems involve coordination (positive-sum) rather than conflict (zero-sum) dilemmas which the PD model assumes.

because it is more easily detected and group sanctions can be invoked more effectively against it; or (*b*) if the organization offers certain benefits valued by potential members that can be withheld, such side-payments may induce beneficiaries to disavow free riding and contribute to organizational maintenance. If neither condition applies—if the task requires large-scale organization and the benefits created are indivisible and nonexcludable—voluntary efforts to produce public goods will not be viable, according to Olson's analysis. To get the benefits of collective action, then, coercive authority must be invoked to make membership and resource contributions compulsory.

Hardin offers similar assumptions and conclusions. The "tragedy of the commons" arises when persons overuse a "common property resource" like forest or pastureland to the point of destroying it. If many herders are grazing their livestock on a commonly held pasture, for example, each derives the full benefit from adding another beast to his herd but bears only a fraction of the cost, since each additional animal subtracts only marginally from the grass and water available for all the others. Thus all herders have an incentive to increase their herd size, to the point where the pasture is overgrazed and all their animals (and they) suffer. They are acting on the same premise as Olson's free-riding nonmembers, deriving benefits at the expense of others, likewise to their own ultimate disadvantage. If individuals do what they think is in their own interest without regard to its consequences for others, this can prove detrimental for everyone.[2]

Various objections can be raised to the conclusions and predictions put forward by Olson and Hardin. A great deal of collective action certainly occurs, as we observed in Sri Lanka and as Hirschman (1984) found in his survey of the Latin American experience. Moreover, common property resources are not necessarily mismanaged.[3] The theses of Olson and Hardin have been widely attended to because they model a considerable range of outcomes with theoretically attractive

2. This is what economists call the fallacy of composition. A standard example is that although individuals can make themselves better off by saving some of their income and putting it in the bank, if *all* individuals do this and reduce their current expenditures, the economy will collapse and everyone will be worse off. Thus what is good for the individual may not be good for everybody. This also demonstrates what were referred to previously as emergent properties. Actions at one level of analysis (in one frame of reference) can have different consequences at another level (within a different framework).

3. Since Hardin used rangeland management examples for his argument, the empirical counterattack is most relevant in this area. See Netting (1976), Gilles and Jamtgaard (1982), and Sandford (1983). Theoretical limitations on the Olson-Hardin thesis are spelled out by Ostrom (1990) in her comparative analysis of collective action experience in various countries.

simplicity and rigor. Rather than refuting their arguments in an either-or way, I propose an alternative understanding based on different premises about human motivation and choice. Collective action and common property management may not be as illogical or irrational as Olson and Hardin suggest. To abandon voluntarism and self-management because reductionist social science tells us these courses are impossible or at least improbable would itself be a tragedy, especially in developing countries where financial and natural resource bases are so slim and where such motivations could produce varied benefits.

Structural Explanations: Channeling Potentials

How can one explain the emergence of so much collective action among Gal Oya farmers with so little free riding? Why was water as common property managed in a more equitable and productive way once organizers started introducing such opportunities for groups? Some individuals shirked, and sometimes old patterns of self-aggrandizing behavior persisted, but overall a system of cooperation was put in place more quickly and easily than we expected. Explanations, I suggest, fall into two general categories—structural and cognitive.[4] The two do not represent alternative, closed systems but rather are interactive theoretical constructs producing reinforcing insights into the same phenomena.

Structural explanations assess the effects that situations and contexts have on behavior. They focus particularly on the influences of roles, created by the common expectations associated with certain statuses or positions. These provide capacities and establish interests that are generalizable to any persons in those particular roles. Such explanations try to predict "typical" behavior. How would *an average or modal person* be likely to act within the structure of a given situation, subject to the incentives and sanctions, the communication patterns

4. There is no fully satisfactory term to describe this second approach. It could be called behavioral but it would be misleading to characterize any one social science approach as behavioral, because all social science is concerned with explaining behavior. If not averse to neologisms, we could characterize the alternative approach as *ideational*, to encompass all the products of thought—ideas, norms, values, perceptions, evaluations, and so on. The term *cognitive* is chosen here because this represents a growing movement in the social and behavioral sciences, seeking to deal with what goes on in real people's heads and what happens among them as a consequence. This alternative could be called volitional analysis, but that is an uncommon term.

and information flows, the possibilities for action and the authority relationships associated with that role?

What can be regarded as cognitive explanations, on the other hand, emphasize the influence of ideas, particularly values and ideals. This approach focuses on what individuals bring to situations of choice, particularly to roles, in terms of their personal orientations, qualities, and objectives. How would *a particular person* be likely to perceive, interpret, and evaluate things and consequently to act in situations generally, other things like the structure of incentives, communication, and authority being equal? This is the converse of a structural explanation. The alternative approach considers what kind of impact individuals with their own consciousness are likely to have on any situation, while structural analyses ascertain the impact of situations on individuals, disallowing such idiosyncratic influences as values, friendships, energy levels, and memories.

The two modes of explanation are different ways of sorting out influences on behavior. They appear contrasting, yet we know that structures and choices are linked. Not only are the expectations that structure roles rooted in people's norms and values, we know that situations reinforce certain values and not others, thereby contributing to some degree of convergence between interests and ideals over time. This is a both-and world, so both analytical frameworks have validity, though they should also be transcended (Sheth 1987). The two approaches can be contrasted in the following simple way:

	Structural Approach	Cognitive Approach
Explains behavior of:	Average/typical/modal person	Particular person
In terms of:	Situations and contexts	Ideas, values, and ideals
Focusing on:	Roles and interests	Expectations and norms
Concerned with:	Capabilities and constraints in situation	Capacities and orientations of individual

As structurally oriented social scientists, we started out in Gal Oya trying to design roles and incentives that would channel the behavior of *any* person—the average, typical, or modal farmer in Gal Oya—in short, an abstraction. The continued emergence of causal factors that were highly individual and personal, however, forced us to take account of such factors in our analysis and action, though not to the exclusion of structural considerations. We found time and again that individual differences and personal values were crucial when trying to achieve or explain change, whereas generalizable interests remained the most pervasive, tangible influences shaping people's "equilibrium" actions.

This led to an interesting conclusion, reductionist but revealing, that drew on economists' distinction between average and marginal relationships. Whereas persisting or average patterns of behavior are best explained by *structural* variables, innovative changes, those at the margin, which are what we seek in development, are more often accounted for by more *cognitive,* individual, or normative factors. We could not understand or accomplish pathbreaking performance in Gal Oya just by varying incentives or information for typical persons. It was highly motivated, atypical individuals who broke the mold. But such performance was not sustainable or generalizable unless it was channeled into and reinforced by structures having predictable roles, incentives, and communication patterns, which shape the choices and actions of most people.

After initially being ambivalent toward exceptional or exemplary behavior because it was considered aberrant, we came to appreciate it. While encouraging it, we tried to institutionalize it as much as possible into "normal" relationships so that what at the time was *marginal* would over time raise the *average.* Whenever behavior at the margin was less constructive than the average, this lowered overall performance.

Irrigation lends itself to structural analysis because one can make many predictions about behavior based simply on people's situations and roles. There are ubiquitous locational differences within every system between head-enders and tail-enders. Water distribution creates foreseeable incentives among users. There is likely to be some competition, even conflict, among users within any given command area if the supply of water from a common source is inadequate to meet all the demand.

At the same time, there are incentives for cooperation to increase that supply, if possible, thereby reducing conflict and enhancing productivity, converting a zero-sum situation to a positive-sum one by collective action.[5] Farmers on different field channels who may clash over

5. An experiment with improved water management by the International Rice Research Institute in the Philippines found that total production could be increased by 39 percent through more equal distribution, with increases ranging from 8 percent at the head to 137 percent at the tail. All sections showed yield increases, averaging 13 percent, so nobody lost, but the lower part got more benefit from expanding the area irrigated (thanks to upstream water saving). The total number of acres irrigated increased by 24 percent, but in the tail the area under irrigation went up by 117 percent (Wickham and Valera 1978, 61–75). Farmers

the distribution of water among their channels have reason to cooperate when it comes to getting more water supply into the distributary channel that serves all their respective field channels.

For years in Gal Oya, farmers at the field channel level had not been cooperating. They seldom cleaned and repaired channels so that everyone would get water more reliably and efficiently. Leaving this task to individual action was ineffective because farmers at the head of the channel had incentives not to participate in group efforts to improve the distribution system. When a channel is silted up, relatively more water flows into head-end fields, to the detriment of farmers downstream. If head-end farmers do not clean their sections of the channel, there is little reason for those at the tail to clear the sections closer to their fields. Moreover, whenever the flow of water is slowed because of silt and weeds, distributing it requires more time (more labor), and the volume of water available to farmers gets reduced as water is lost through seepage down through the soil into the water table. Lack of collective action thus produces a negative-sum outcome.

Conversely, cleaning channels and rotating water deliveries among fields and among channels reduces seepage losses, makes the flow faster and larger, and gives crops a better water supply. The same gross amount of water can serve all farmers better, in what becomes a visible positive-sum situation. True, they must invest some labor in channel maintenance, water distribution, and conflict management, but they can save labor time if they can wet their fields more quickly and do not have to stand guard over their meager and uncertain supply around the clock (or go steal water at night, at some personal risk).

The institutional organizers introduced two structural innovations that changed the situation in Gal Oya: field channel organization and the farmer-representative role. Farmers' willingness to embark on these initiatives was encouraged by the fact that two new elements were being introduced into an unsatisfactory status quo: rehabilitation of physical irrigation structures was promised, and organizers came with some kind of official backing. Farmers' self-help efforts would be paralleled by government investment, but it was up to the farmers to forge cooperative relationships through the new social structures.

---

in the Bear River irrigation system in Utah have more water trading in some channels than others, depending, water attendants say, on "how nice people are" to one another. Where there is a high level of mutual trust and much trading, farmers can reduce their labor inputs by almost 60 percent without a significant decrease in yields (Steiner 1990).

Field Channel Groups

The groups proceeded with no formal authority to force anybody to cooperate, only with the social authority generated from discussions and decisions on common problems. To have all the farmers along a channel meet face to face fairly regularly created a public forum where before there had been only private communication.[6] Tail-enders in Gal Oya had an opportunity to present their difficulties directly to those who are contributing to their problems. Head-enders who were wasting water by not closing the pipes into their fields could not evade responsibility for this when it was discussed in a group.

When approached previously as individuals by individuals, head-enders could be unresponsive, make excuses, blame others, or make promises they did not intend to keep. ("Go talk to Gunadasa, he is wasting water too. When he does something, maybe I will cooperate too.") But farmers who acted selfishly in private found it more difficult to do so now in public. The well-known influence of social pressures was mobilized in the service of more efficient and equitable water distribution. Irrigation, it should be noted, offers less opportunity for "exit" (Hirschman 1970) than do other tasks of rural development and resource management. Also, if anyone takes more than his share of water, his flourishing crop is there for all to see.

Simply forming the organizations, even informally, created new flows of information, legitimacy, and status that affected behavior. If some farmers refused to cooperate, the groups sometimes just worked around them, or they could threaten to cut off water from any difficult farmer who interfered with their efforts. They had no legal authority to do this, but at field level water was practically under farmer control.

Enough benefit and goodwill was generated by the improvements being made in the system that some amount of free riding was simply ignored. Most farmers were not deterred by someone else's getting benefits from collective action when these were greater than the cost they incurred. As seen in chapters 3 and 4, the desire to get benefits from better management of a common resource overcame the obstacles that skeptics or opponents of the program placed in its way at the outset.

6. The same structural dynamic is reported in the United States, where providing "a forum for face-to-face discussions about their common problems" led to positive-sum thinking and changed patterns of water use in California's West Basin Water Association (Ostrom 1990, 114–24). The importance of having reasonably small base-level groups, for reasons similar to those in Gal Oya, is seen in California as well (189–90).

Farmer-Representatives

Oversight and implementation of group activities and decisions was entrusted to the farmer-representatives. These individuals, chosen by consensus, were unpaid and had no formal or legal authority, yet the social authority they wielded was impressive. The method for selecting FRs contributed to getting more public-spirited leaders than those who had acted in the name of farmers in the past. By starting with ad hoc work before any explicit organization or formal leadership had been agreed on, those farmers who had the most interest and skill in improving irrigation were the ones who took initiative for the group and distinguished themselves in the eyes of their peers.

To be sure, if a long-established leader in a particular community fitted this description, he was likely to take the lead, and then work usually proceeded even more quickly. For the most part, however, the persons who moved into FR roles represented a new cohort, not the more influential, prosperous, or partisan farmers with ties to government officials. A few farmer-politicians accepted the program and became FRs, but they were held accountable to majority interests and ideals.

Because the farmer-representative was chosen by consensus, all farmers on his channel had publicly consented to his acting and speaking on their behalf. If an electoral process had been used, there would have been some persons who had voted against the FR, and they would feel differently toward him (and he toward them). Secret balloting can protect expression of minority sentiment and in principle can reveal "truer" opinions, but at field channel level, enforcing decisions reached by any narrow or contested majority vote would have been nearly impossible. The rule of consensus created greater authority than elections could. That FRs served without pay enhanced their legitimacy.[7]

The approach described here did not work in all cases, as we saw in part 1. But in most instances where FRs did not work effectively, they were replaced, an indication that the role and the organization had become valued by the water users. It was very encouraging for my theorizing about this experience to learn that a colleague in Thailand had used a similar approach in his country with similar results.[8]

---

7. In Polonnaruwa district, where inequality of landholdings is greater, farmer-representatives are often chosen through elections because of the fear that rich farmers will dominate the selection process. Groups are offered the option of consensus or secret ballot.

8. In an effort to mobilize community self-help capacities in rural Thailand, faculty and

The field channel groups and FR roles are considered structural innovations because their effects did not depend on individual characteristics or values. New patterns of communication, incentives, and decision making produced reasonably similar kinds of behavior among all those who filled the roles and participated in the organizations, much as the role of institutional organizer mobilized and channeled both energy and ideas from young people into the construction of these organizations. Performance in these two roles did not depend on unusual idealism or special traits of personality. Expectations emanating from ARTI, USAID, Cornell, the government, engineers, local officials, and especially from farmers, other FRs, and the cadre of IOs, created both roles and held them in place, suspended in a web of social interactions and ideas. Still, as farmers and IOs frequently stated to me, it was important that new values become ascendant, spurring and reinforcing new flows of information, economic incentives, status, and authority. Structural changes by themselves were not regarded by the key actors in this drama as sufficient.

Normative Orientations as Probability Distributions

Instead of focusing on external structural influences on behavior, one can examine internal normative factors, recognizing that neither is absolute or autonomous. In some ways, economists assign most causal influence not to structures but to values. People's choices are regarded as reflecting their objective function of utilities or the interests that they seek to maximize. The speed with which Gal Oya farmers accepted participatory and egalitarian water management, however, contradicted this view. The IOs did not have sufficient time to make so profound an impact on values, nor were they trained, experienced, or confident enough to achieve such a "Copernican revolution" in prevailing norms.

The theory of cognitive dissonance (Festinger 1957) holds that people tend to change their values to conform to and justify their behavior, rather than alter behavior to give expression to their values. There is something to this. Some farmers became proponents of fair water

students from Thammasat University developed a similar method of house-to-house visits by "catalysts" to identify and encourage potential leaders (Rabibhadana 1983). Small base groups were formed, with informal procedures and selection of leaders by consensus. Organizers asked questions rather than instructing or preaching to the people.

distribution probably only after feeling social pressures to cooperate in the new program discussed, devised, and operated by their water user group, especially once the new approach produced welcome results. We increasingly came to see people's behavior as variable, as something probabilistic, rather than as something fixed, predetermined by some hierarchically ranked set of interests. The connection between norms and actions is a two-way street, with any adequate explanation of the changes in Gal Oya combining both structural and normative factors. Norms respond to situations, as I argue below. Yet situations are structured by persons who are value driven. So who can say which is prior?

The structural factors discussed above are consistent with Olson's stipulation that voluntary collective action will be more sustainable in small groups, but, even so, there was some free riding, which is thought to deter the cooperation required to produce shared benefits, except we saw more than individual calculations of net benefit. The logic for collective action changes when people attach some value to others' getting benefits. Normative influences can have evident impact, but they have to be activated and are more effective when reinforced by new structural relationships.

Change of values is seldom the issue, because people usually have such a diverse set or repertoire to draw on already.[9] Although we hold different values with greater or lesser intensity, at some level we accept virtually all the norms found in our society as valid, in varying circumstances. Because these values frequently conflict, there are no invariant orderings of values. What matters is not which values one has—we all have many—but which values are activated and applied in a given situation.[10]

---

9. Richard Neustadt (1960, 97) wrote years ago about "that marvelous rummage bag sometimes called our 'belief system' [that] is crammed with alternate and contradictory images . . . ready for instant use," which can provide a justification for almost anything one wants or needs. Cultures provide many diverse and apparently inconsistent principles or values that can be invoked. Be bold: "Nothing ventured, nothing gained." Be cautious: "An ounce of prevention is worth a pound of cure." Herbert Simon (1945, 20–36), at the time thinking in Newtonian fashion, considered such ambivalence "a fatal defect in the current principles of [public] administration that, like proverbs, they occur in pairs. For almost every principle one can find an equally plausible and acceptable contradictory principle."

10. Arthur Maass reports a psychological experiment that shows clearly this contextual aspect of values. Some years ago, a large number of Catholic college students were asked about their attitudes toward birth control after being randomly divided into two groups: "One group met in a small room where they were made aware of their common religious membership. The other group met in a large auditorium along with hundreds of other students of many religions, where no effort was made to establish awareness of common religious beliefs. Although all of the students were instructed to respond with their 'own

The effect of context is seen in Stanley Milgram's famous experiment (1974) where volunteers proved willing to administer apparently torturous electric shocks to "subjects" when instructed to do so. The value of not hurting other human beings could be overridden by the norm of deference to authority. The more authoritative the figure giving the instructions appeared to be, the more willing people were to inflict pain. This does not mean that values are of no consequence. People seldom engage in the kind of behavior that Milgram's experiment showed could be evoked under certain conditions, largely because humanistic values inhibit this.

Values get presented as if they were absolute, but we know from our own experience that they are relative, to the situation as well as to one another. Are we selfish *or* generous? We may tend to be one more than the other most of the time, but we are all capable of *both* selfishness and generosity, in different contexts or frames of reference. Value dispositions are like probability distributions.[11] For some people these are "normally" distributed, with equal chances of selfishness or generosity, as in figure 2b. In others they are skewed toward selfishness (fig. 2a) or generosity (fig. 2c). So long as there is even a small chance that one will be generous (or selfish), one may act this way in spite of or because of what else is going on in one's environment.

If I have a general tendency to be selfish (a) but others around me are being generous, that is likely to affect my behavior, over time shifting my probability distribution to (b) or maybe even toward (c). If I tend to be generous but everyone else is acting selfishly, for a while I may resist letting my value distribution move from (c) to (b), but "bad experience" could shift it all the way to (a).

Suppose we learn about an earthquake somewhere, and I would be initially disposed to contribute five dollars for relief—my initial typi-

---

personal opinions,' there was a significant difference between the replies of the Catholic group that were aware of their common religious membership and the unaware group. The former approximated more closely the traditional Catholic position against birth control. The question was the same for both groups, but individual responses depended on the institutional environment in which the question was asked" (Maass 1983, 23–24). What the students thought *they* believed, on a very sharply defined subject, differed according to what they thought *others* around them believed, even when the answers were private. This question of how behavior and even normative orientations are shaped by "behavioral settings" is analyzed with extensive experimental data and innovative social psychological theory by Schoggen (1989). What IOs did in Gal Oya was to reshape "behavioral settings."

11. This conceptualization is prompted partly by the idea of "wave functions" in quantum physics (quantum wave functions, colloquially called "qwiffs"), which are thought to represent the underlying nature of physical reality. See Wolf (1981, 169–91). This conception will be discussed again in chapter 14, since such a probabilistic view of the universe is important for constructing a theory of social change and stability.

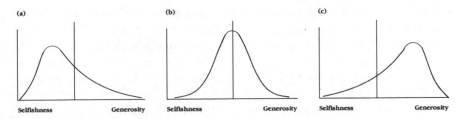

Figure 2.

cal degree of generosity. What I contribute could be twenty-five dollars or twenty-five cents depending on how generous or miserly others are being. The first contributor acts somewhat independently of others' value distributions, but even he or she may feel "compelled" to set a generous example for the rest, or in type (a) settings might be ashamed to show much compassion. People seldom act without thinking about what others are likely to do or think, which may be why altruism occurs and persists.[12]

Our decisions and actions are based *both* on what values we give most weight to *and* on others' apparent values. Since what they do is influenced also by others' values, not just by their own dispositions, we all act within a convoluted network of preferences, anticipations, and adjustments.[13] We never know how much of our behavior is really "ours" and what is due to the actions and statements of others. This became clear as we talked with Gal Oya farmers and IOs about why behavior there was changing.

Selfishness had been the prevailing orientation for the previous thirty years, but the value of generosity did exist within the farming community, though latent and dormant, seldom acted upon. When IOs came and encouraged people to think of the effects of their actions on others downstream, a few farmers fairly quickly supported efforts to save and share water, perhaps because their value disposition resembled (c), even though they acted before according to the selfish, left

12. If readers prefer, they can think of this as "enlightened self-interest," but I do not use this term because it enables us to avoid confronting altruism as a factor in social relations. As defined here and by others, it does not exclude, nor is it independent of, self-interest. Evidence that the latter is not as strong as "rational actor" theorists suppose has come recently from analysis of voting behavior in the United States, Britain, France, Germany, Italy, and Spain by Lewis-Beck (1988). Voters' choices are not significantly affected by their evaluation of their *own* economic situation but are correlated with their assessment of how the national economy is performing.

13. Munro (1928, 4) observed that "even as every molecule of physical matter is conditioned and directed by those with which it interacts, so the individual citizen is similarly motivated and controlled by the influence of those with whom he associates."

side of their distribution. Persons with disposition (b) were more likely than those with disposition (a) to join at first in such efforts. Once most agreed to collective action, the community's overall value preference shifted from (a) toward (c). This preference was accepted in open meetings. As the IO Sanda observed, people find it difficult to be selfish in public.

Gal Oya farmers did not become different people than they were before IOs arrived. But the salience and weighting of their different values was changed, partly by normative appeals and examples and partly by the structural innovations discussed above. The farmers were not selfish *or* generous. Like the rest of us, they were and are *both*, but their probabilities of being one or the other changed. Some part of the shape of one's value distribution (orientation) is probably determined by intrinsic disposition. Possibly it is even genetically inherited (Dawkins 1989). But it is in any case shaped by experience, persuasion, and reference group examples.

The first sharp challenge to my previous thinking was tall Punchibanda's short speech in January 1982. How was it possible, I wondered, that someone who had disregarded the needs of downstream farmers for twenty-nine years would suddenly not only start reducing his water offtake but even get up at night to make sure that water was being shared fairly? Rather than regard *him* as having been changed suddenly after the IOs came, it made more sense to see his *behavior* as different now, corresponding to a different shape of his probability distribution between selfish and generous values. The IOs helped create a climate of opinion where Punchibanda could more freely and confidently act according to his altruistic inclinations. Knowing him as I do now, I think his basic personal value orientation was skewed toward generosity (c). Yet by his own admission, for almost three decades he acted selfishly, like his neighbors, because that was the normative context in which he lived.

Reinforcing any tendency Punchibanda had for altruistic action was his role as a farmer-representative and the regular group meetings that made it possible for generosity to become effective. In a situation of anarchic water distribution, little benefit would have resulted from anyone's efforts to save water for tail-enders. If Punchibanda closed off his field to send water down the channel, some "undeserving" farmer in the middle could easily take and waste the water Punchibanda had saved. So why make an effort to be generous if it is not efficacious? Why forgo benefits or incur costs for the sake of others in such circumstances? Unless there was cooperation, a person who sought to be altruistic would look like a fool. But with farmer organizations in place, benefits from generous action could be achieved and ensured.

To explain the behavioral transformation in Gal Oya, a simple model came to mind, linking the distinction between selfishness and generosity as orientations to outcomes (whether one values only one's own benefit or others' benefits as well) with the difference between individualism and cooperation as orientations to action. The first continuum, from highly selfish to highly altruistic behavior, refers to the goals of action, while the second continuum, from individualistic to cooperative behavior, refers to the means of action. These two dimensions define an analytical space within which persons' value orientations can be located, shown in figure 3.

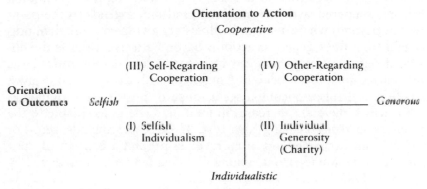

Figure 3. Alternative value orientations

What is meant by generosity or altruism? I use the terms interchangeably, though the latter has a more technical sound. Altruism, in keeping with the definitions of Axelrod (1984) and Margolis (1982), is a positive-sum relationship where one considers one's self better off if others' welfare is being enhanced and acts accordingly.[14] This does not require that one makes one's self worse off, though that could be voluntarily accepted if, for example, a small inconvenience or sacrifice

14. According to Margolis, "What defines altruistic behavior is that the actor could have done better for himself had he chosen to ignore the effect of his choice on others . . . altruism in the technical sense used here [means] that the individual's allocation of resources is influenced not only by the bundle of goods he obtains for himself but also by the effect of his choices on others or on his society, qualified only by the condition that the actor (not necessarily the recipients) regards this behavior as benign. An altruistic act need not have negative or zero value to the actor (1982, 15). For Axelrod, "Altruism is a good name to give to the phenomenon of one person's utility being positively affected by another person's welfare" (1984, 135).

could produce a greater benefit for someone whose well-being is valued. As suggested in figure 3, this orientation is *other-regarding* rather than *self-regarding* only.

The limiting condition of altruism, as conceived here, is that there be a *net addition* voluntarily made to total welfare—that one consider one's own loss less than another's gain and thus worth incurring for others' sake because their well-being is accorded some value. Punchibanda's getting up at night to patrol the field channel is an example of this. Such an action benefits more than one person because the person bearing the cost for it also derives satisfaction from it.[15]

A more restrictive definition would use the condition of Pareto optimality, which stipulates that the gains being made are achieved without anybody's suffering any loss. This would represent minimal altruism compared with the maximal kind that extends to the point (but not beyond) where some individuals are giving up more than others gain from these generous actions. Beyond altruism there is the domain of self-sacrifice, which can be negative-sum if substantial costs are incurred for others' benefit. Consistent with the definitions given in note 14, altruism encompasses a range of positive-sum outcomes within which there are increases in total net benefit acceptable to the person or persons bearing any cost. It does *not* include zero- or negative-sum exchange and often revolves around a concept of "fair share" that we found strong among Gal Oya farmers, as predicted by Margolis (1982).

An extended continuum of relationships can be sketched (fig. 4), where the horizontal axis of figure 3 represents the central portion (b and c) of a range of value orientations. Just as there can be self-sacrifice to the right—that is, beyond generosity—there can be destruction of others to the left, beyond selfishness. This latter is negative-sum, though it is probably positive for the selfish individual, since the costs are borne by others. Self-sacrifice can be positive-sum, to be sure, if others benefit and all the costs are absorbed by the "self-less" individual.

15. This aspect of altruism corresponds to Margolis's concept of *goods altruism*, where one performs generous acts to benefit others because one gets satisfaction from seeing their welfare improved; such satisfaction can result from others' generosity, not just one's own, toward the persons whose welfare is valued. *Participation altruism*, on the other hand, represents action done because it is regarded as good in itself—patriotic, moral, civic-minded, or whatever. It is done essentially for its own sake (like voting in elections where there is only an infinitesimal chance that one's vote will actually make a difference), because it affirms social values one wishes to perpetuate (Margolis 1982, 21–25). This latter form of altruism definitely comes within the definition proposed, but our discussion focuses more on goods altruism.

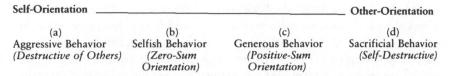

Figure 4.   Continuum of orientations toward self and others

The extremes are for various reasons not very common, though (a) is much more observable than (d). Within the range of selfishness (b), people are indifferent to each others' well-being, whereas when there is altruism or generosity (c), people attach some positive value to others' well-being. This does not mean they are indifferent to their own. Beyond altruism (d), others' benefits are valued so much that these are sought even to one's own detriment, while destructive selfishness (a) produces benefits for the individual by victimizing others.

When one acts selfishly, one regards others essentially as *strangers.* One enters into relationships as in the economic marketplace, looking only for one's own advantage, not caring whether exchanges benefit the other party. Such zero-sum interaction can produce positive-sum results, but one cares only about the net benefits that accrue to one's self. When one acts altruistically, conversely, one values others' well being. They are like *friends* with whom mutuality of interest and cooperation are expected to lead to positive-sum outcomes. To the left, beyond selfishness, others are regarded or treated as *enemies.* If they are not antagonists already, they are likely to start acting like foes.

Though some people are moved more often by self-serving motives and others tend to be altruistic, practically everyone has both inclinations. Few persons are saintly enough to be without any selfish instincts, and even confirmed egoists display occasional generous impulses. This both-and understanding of human nature suggests that everyone also has both individualistic and cooperative orientations to action, though most people are likely to incline more in one direction or the other. Not surprisingly, the same kind of alternative value dispositions sketched in figure 2 can be found for individualism and cooperation, with people's probability distributions either "normal" or skewed in one direction, depending on personality, experience, and present context.

In the absence of influences to the contrary, persons tend to act most often according to individualistic and selfish premises. This was certainly the prevailing situation when we started our work in Gal Oya in 1980. The tendency toward behavior associated with quadrant I in figure 3 can be attributed to the fact that people can be individualistic but not cooperative unilaterally. Working together is possible only if others also value and agree to collective action, and it appears to be easier to act selfishly than generously, as Punchibanda's example above suggests. Although altruism offers many positive-sum possibilities, achieving these is contingent on others' acting likewise. For effective and sustained generosity, there needs to be value consensus and cooperation. In the case of irrigation, a farmer at the head of a channel who reduces his offtake to help tail-enders may see someone in the middle take water he doesn't really need. Generosity is unlikely to persist if intended good results are not realized—if it is not efficacious.[16] It can also die out if not reciprocated. The combination of selfishness and individualism can thus produce a stable social equilibrium, even if the resulting level of social productivity is low.

One can have individualistic action for altruistic ends (quadrant II) and cooperative action for self-centered aims (quadrant III). Givers as well as receivers can benefit from charity or philanthropy, if only in terms of fame or self-respect, so quadrant II has positive-sum characteristics. Olson's concept of collective action that is undertaken for one's own advantage only, without regard for the benefits it may create for others, corresponds to quadrant III. In such a situation there is no assumption of interdependence of utility functions, to use economists' terminology.[17] It is usually more productive than quadrant I or even II, which is unpredictable and hard to sustain. But quadrant III behavior is vulnerable because it lacks the mutually reinforcing qualities of quadrant IV, which can make this something of an "equilibrium condition" like quadrant I.

When institutional organizers entered Gal Oya communities, they were proponents of cooperative action, suggesting that groups get together to tackle whatever problems they could not resolve by individual effort, and particularly endorsing shramadana campaigns of voluntary collective labor. They also encouraged sharing water equitably to help tail-enders who would otherwise not be able to cultivate

16. To follow Margolis's distinction, participation altruism may not be sufficiently valued by people who wish to practice goods altruism to justify continuing the effort (sacrifice) to be made.

17. Originally quadrant III was labeled "selfish cooperation, but Albert Hirschman regarded this description as unnecessarily pejorative. He correctly suggested that many positive outcomes can result from this, though his own work has shown the greater social value of altruistic cooperation. In figure 3 the term "self-regarding cooperation" is used, contrasted with "other-regarding cooperation" for quadrant IV, to distinguish between cooperative action that is unconcerned with others' well-being and that which takes others' welfare into account.

a crop and maintain themselves. The result was to shift farmer behavior from selfish individualism to other-regarding cooperation in a matter of weeks. The approach we proposed when training the first batch of IOs—promoting cooperation and other-regarding behavior—was not based on any explicit theory of social transformation. We were trying to deal with evident social and technical needs in the field. The strategy can now be diagramed as one of shifting value orientations from quadrant I to quadrant IV in figure 3. These two normative dimensions "frame" social reality and its alternatives in a fundamental way.

Explaining and Changing Orientations: Social Magnetic Fields

By their words and by their initiative, IOs created a climate in which it was congenial for farmers to act according to cooperative and altruistic values that they already had but that were not given expression in an environment hostile to such norms. In Gal Oya before 1981, a Hobbesian "state of nature" prevailed, according to officials and farmers. In the midst of a war of "all against all," acting according to other-regarding values was likely to be unrewarding or even self-destructive. Yet such values existed in that situation as possibilities, however remote. Though seldom expressed in action, their potential could be drawn upon, as the IOs and we soon discovered.

Moreover, acting in a more generous and cooperative way could produce benefits, as farmers came to realize. A fixed, even reduced volume of water could be made to serve more fields when shared systematically. Some costs were involved, but the rewards were substantial. Such outcomes are probably more quickly obvious in irrigation, and more obviously positive-sum, than in other areas. Both farmers and engineers could see crops growing in fields that had previously been barren in the dry season. Had we been conducting a public health campaign to reduce intestinal parasites, we could not have offered as much tangible positive reinforcement. But the understanding of norms and of their place in a theory of action that is suggested in figures 2 and 3, though it grows out of experience with irrigation, has general validity.

We were working in a situation where there were at least Pareto-optimal gains to be reaped. The water otherwise wasted that flowed through head-end fields into drainage channels could be sent to tail-end farmers, who gained by the redistribution of water without head-enders' losing output and income. True, the latter had to expend some labor to control the water that before they had let flow freely. Also, they had to attend meetings and contribute labor to maintenance and water distribution tasks.

If one regards people as motivated only by material rewards (or costs), the cooperation of head-enders is hard to explain. They could profit from purchasing fertilizer and other inputs collectively so that supplies would be cheaper and more assured. Or if the groups became channels for agricultural extension agents to give advice, because it was easier to reach farmers when they were organized, more knowledge of new technology could be gained. But these activities, undertaken by some field channel groups, were not widespread, at least not in the first stage of organizational development. So economic advantages as side payments benefiting head-enders cannot account for their cooperation in group formation. Some social "income" was derived from the shramadana campaigns, which were made festive occasions, but head-enders were willing to bear the miscellaneous time and effort costs of more intensive water management, which did not directly disadvantage them materially and possibly benefited them, partly because they derived some satisfaction from the improvement in their neighbors' situations, I believe based on discussions with farmers.

Certain "public goods" were created by the organizations. Head-enders and tail-enders alike benefited from having representatives who could deal with officials on behalf of all farmers on a variety of issues, not just regarding water (which would be adequate for head-enders in any case). Everyone was better off if there was any reduction in conflict over water. Quite apart from the hazard of possible violence and even murder, one never knows when one will need the cooperation and assistance of neighbors, such as for lobbying officials or dealing with flooding. Living in a situation embroiled in conflict creates unwelcome insecurity and tension. Whole families— wives, children, relatives—get caught up in recriminations over water distribution. So management and reduction of conflict represented a real if not measurable benefit for all, a kind of insurance policy keeping open the possibility of mutual assistance otherwise ruled out by feuding.

Status considerations appear also to have been important. Farmers who cooperated in collective efforts gained respect from their friends and neighbors. Conversely, if they declined to cooperate, they faced loss of respect and possible ostracism. One could enhance one's self-esteem by being an active contributor to this transformation of Gal Oya, and the philosophy of self-reliance underlying the program gave a basis for pride. That government officials showed respect to farmers, directly and through their representatives on the District Agriculture Committee, created a status reward of considerable value to Gal Oya farmers, who had been regarded by officials and sometimes by themselves as rejects from their original communities. (Recall Neeles's statement in chapter 4 about how farmers derived encouragement for their voluntary collective action from being listened to and treated respectfully by officials.)

Farmers' willingness to move from quadrant I to quadrant IV at IOs instigation would not have lasted had doing so gone against their perceived interests. But most farmers saw the new situation as much preferable both materially *and morally*. A few farmers lost some economic benefit, such as Waratenne, who had gotten Irrigation Department staff to give him special water access, which was eliminated when his channel was cleaned by fellow farmers (chapter 3). And any officials who had been taking bribes to alter water allocations lost some illicit income, as did those who profited from irregularities in awarding maintenance contracts.

In a climate of moral laxity, officials might as well exploit their positions, since they would probably be regarded as corrupt whether they were or not. Once a new climate was being established, those whose normative disposition was skewed in a selfish direction were in a weak position to object or resist because their advantages were illegal. A few influential farmers and colluding officials tried to sabotage the organizations, but once a higher standard was instituted, farmer-representatives could with confidence and support expose wrongful conduct.

Changes in behavior were accomplished by a combination of normative and structural changes, not just by roles and sanctions or by values and norms. Indeed, it was difficult to determine how much influence could be assigned to one or the other, because the effectiveness of each was enhanced by the other in a positive-sum manner, as with the combination of administrative and participatory means used to improve channel maintenance, reported in chapter 8. We had to resist the impulse to explain things in a reductionist way, attributing all improvement to just one set of factors or dividing up influence in a zero-sum way—50/50, 33/67, 90/10, 25/75.

One way to describe the shifts we observed is in terms of magnetic fields of force, which are polarized so that charged ions all align themselves in the same direction. It is said that "opposites attract," but this is true because north and south poles come together when magnets are *aligned in the same direction*. North-north or south-south poles repel each other because they are aligned in *opposite* directions.

One can describe the change in farmers' and officials' behavior in terms of magnets that have alternative poles—selfish (S) and generous (G). A farmer who puts his own interests ahead of others' (S>G, meaning S more than G) will find it easy to relate to a self-serving official

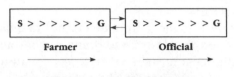

Similar Value Alignments Attract

Figure 5. Similar value alignments attract

whose value orientation is also S>G, as shown in figure 5.[18] They will each assess transactions in terms of what benefit it brings them personally, ignoring others' or the public's interests. If the farmer offers a bribe or if the official seeks one, accommodation can be easily reached, since they are on the same wavelength, to use a different electromagnetic metaphor. But if a farmer seeking purely personal advantages approaches a public-minded official—or if a group-regarding farmer comes to a self-centered official—we have the situation in figure 6, where there is no alignment between the two persons, and consequently some repulsion.

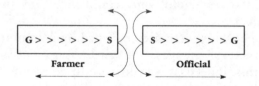

Opposite Value Alignments Repel

Figure 6. Opposite value alignments repel

We found that if either a farmer or an official in a relationship as depicted in figure 5 reverses his orientation firmly, this can swing the other person around toward the same new alignment. People find it

18. This formulation is consistent with Margolis's analysis (1988), which considers S and G in terms of their ratio, recognizing that both are usually present (probable) to some degree but one is more prevalent (more probable) than the other.

difficult to remain in the kind of disconnected relationship suggested in figure 6. Which person will turn to make a connection depends on who has the stronger value orientation. S (like I, individualism) may have a strong personal attraction, but G (like C, cooperation) is more tenable in public than S. This can be described in concrete terms.

If an official is basically self-oriented (S>G), any farmer with similar orientation can "do business" with him—for example, get an extra pipe inlet to his field installed in return for a bribe. A group-minded farmer (G>S), on the other hand, coming to ask for repair of a broken gate that affects the whole channel will be in a different position. He will not offer any special inducement because his request is legitimate and benefits others more than himself. In the absence of normative G>S "currents" such as created by the IOs and conducted through the structure of farmer organizations, since the official is a more powerful individual, an altruistic farmer will be either repelled or swung around to a self-seeking way of approaching the official.

But consider a farmer chosen to be a farmer-representative because he is more community-minded than his neighbors and "charged" by his organization in a group-serving direction. When he comes to the official, he projects more normative power. The official cannot do business as before and must change his way of acting, if not thinking, to G>S if he is to come into alignment with the representative. An S>G official dealing with a G>S farmer who seeks the interests of the group rather than, or in addition to, his own will be in an awkward position (fig. 6). At least publicly the official has to accept the farmer's requests as legitimate if backed by the unanimous consent of those the official is supposed to be serving.[19] If the official starts acting in a more generous way (G>S), he will be less approachable by self-seeking farmers (S>G); they will likewise have to become more other-oriented if they want to get cooperation from an official now aligned G>S.

The process of change in Gal Oya can be modeled by thinking of a jumble of normative alignments before the IOs' arrival, with most orientations pointed toward selfishness and individualism. Few pointed in the opposite direction. IOs came with their moral compasses firmly directed toward generosity (G) and cooperation (C). Those farmers with dispositions inclined this way (fig. 2c) found it easier to swing into open alignment with the IOs. As these farmers began to speak and

19. Here is where the rule of unanimity was very important for changing the moral climate in Gal Oya. Initiatives backed only by a majority would have left officials room to evade complying with its requests by aligning themselves with the minority. Unanimity among farmers left officials little room to maneuver.

act altruistically and to support cooperative action, others who had no strong disposition either way (fig. 2b) shifted that way too. Eventually even S>G farmers (fig. 2a) came around, at least publicly, and with everyone endorsing positive-sum values, these became reinforcing and self-fulfilling.

With moral compasses in the farming community reoriented, especially once the Government Agent and District Minister approved the program and the new deputy director of irrigation actively supported it, S>G officials found themselves drawn toward working in more generous and cooperative modes. They were not so much converted as realigned. References to moral compass may seem strange, but the idea of fields of moral force analogous to those produced by magnetic alignments corresponds to the observed dynamics in Gal Oya, suggesting that people's values got pointed in new directions.[20] It suggests by analogy how "social energy" can be generated by individuals and within groups.

It is important to view this dynamic from a both-and perspective. Otherwise one might emphasize cooperation and altruism to the neglect of individualism and self-centeredness. Exalting cooperation and altruism to the exclusion or neglect of their opposite poles would be as mistaken as the other way around. (The metaphor of electromagnetism is appropriate because the N and S orientations coexist; indeed, each depends on the other). Self-interest, enlightened or otherwise, has been a powerful force throughout history, and many constructive values can be attributed to individualism, as Bellah and his associates (1985) have showed. Moreover, various negative outcomes, ranging from mob psychology and tyranny to group-think and loss of identity, can be associated with an excessively strong collective orientation (Koestler 1967, 241–46). Harnessing the motive power of individuals' search for security, comfort, recognition, reputation, and so on, is more advantageous for society than trying fruitlessly to eliminate it. But we hear enough hosannas for individualism and selfishness these days that additional voices are not needed in that chorus.

Prevailing theories about "modernization and development" basically favor selfishness and individualism, the normative orientation of

20. These ideas arose from thinking about how IOs had come into chaotic communities, where value orientations were in disorder, like the iron filings sprinkled on a piece of paper in a science class. The IOs were like magnets placed under the paper, their lines of electromagnetic force aligning the iron filings according to their own polarization (S>G and C>I). Patterns can persist even after the magnet is withdrawn, though they cannot withstand shaking. Magnetic alignment of ions can be transferred to nonmagnetized metal if it remains in proximity to magnetic force for some time.

quadrant I in figure 3. It is seldom recognized and stated explicitly, but both mainstream economic theory and modernization theory in sociology rationalize the decline of quadrant IV. It is thought inevitable that people will give up their concern for one another as they "progress" beyond their "traditional" social relations.[21] People commonly advocate, though they do not as often wish to apply it to themselves, the general rule of competition, justifying it with neo-Darwinian if not neoclassical economic precepts. Adam Smith's metaphor in classical economic theory, the "hidden hand" is thought to legitimate individuals' narrow pursuit of their own interests, without regard to others' well being, presuming that respective individual successes will contribute to the benefit of all. It is forgotten that this is more ideology than science (Hirschman 1977).

To put the matter simply, mainstream theories of economic and social development point toward quadrant I as the most promising arena for fulfilling human potential and meeting human needs. One need not believe that quadrant IV offers the solution to all our problems to conclude that its orientations—altruism and cooperation—are currently underrated values. Their possible contributions are overshadowed by the emphasis on individualism and self-interest maximization in prevailing culture. Correcting the imbalance, which does not mean rejecting the opposite norms, is important. It involves recognizing and drawing on the energization that we know from personal experience can derive from altruistic and cooperative action (Kohn 1990).

Since individualism, like self-interest maximization, is a value that is widely held, it is a norm worth promoting in positive-sum ways. We should not have to choose between individualism and cooperation. Though analytically they may be opposed to one another, in practice they can be linked synergistically so that each value complements and enhances the other, which is also true for self-interest and altruism. In Gal Oya, individuals benefited from cooperation, and generous behavior enhanced their economic and social incomes.[22]

21. How quaint sounds the advice Governor Winthrop gave the Puritans as they were about to start a new life in America in 1630: "We must delight in each other, make others' conditions our own, rejoyce together, mourn together, labor and suffer together, always having before our eyes our community as members of the same body." This amounts to almost total interdependence of utility functions. Thomas Jefferson admonished Americans to "love your neighbor as yourself, and your country more than yourself" (Quoted in Bellah et al. 1985, 28, 31), but this did not become national doctrine or popular culture in the United States. Nor has it been advised for less developed countries.

22. Axelrod tells us, based on his computer simulations, that altruism is not needed for cooperation and that it is "rational" for egotistic individuals to cooperate (1984, 174). As

In reductionist philosophies and policies, impulses for altruism and cooperation are discounted as not part of "human nature," which is defined as individualistic and selfish. People's motivation is concomitantly seen as essentially materialistic, so that monetary incentives and disincentives get regarded as the main determinants of choice and action. This we did not see in Gal Oya, or in Nepal when a similar program was started there in 1987 to improve irrigation with farmer organization through catalysts. Social, psychological, and cultural values are also strong motivating forces.

Having made such observations, it was encouraging to read Peters and Waterman's book (1982) about how successful American corporations also find that "caring" about other people, not maximizing self-interest in the immediate or narrow sense, can be profitable. Whatever human nature is, it includes diverse and contrasting motivations. Which values get acted upon depends both on the individuals concerned—their own dispositions and goals—and on their contexts—what is thought possible, with what benefits, and what others are doing. This makes value preferences always relative, not absolute. They are products at least in part of their contexts, so there is no "bedrock" of human nature on which to build theories of motivation and behavior.[23]

Collective Action Reconsidered: The Limits of Individual and Material Motivation

Let us return to the question that opened this chapter, whether voluntary collective action can be a vehicle for improving people's well-being. The prevailing opinion in social science has been that free riding will in most circumstances deter such means for enhancing welfare. It is not enough to point out that in Gal Oya voluntary collective action could be mobilized fairly quickly and very impressively. We need to consider the theoretical underpinnings, if any, of this observation. A

suggested in connection with figure 3, however, there is a greater congruence and stability in quadrant IV than in quadrant III, where egoists engage in collective action only to advance their respective individual interests. Dawkins finds that cooperation and mutual assistance can flourish in a selfish world provided the "game" is iterated (1989, 224). There are, however, situations in which altruism is not advantageous (214) and some that induce zero-sum behavior, such as those created by divorce lawyers (221–22).

23. See contributions to the volume edited by Mansbridge (1990). She summarizes them as demonstrating "over and over that self-interest narrowly conceived fails to predict well in many areas of human life" (263).

survey of the economic literature on free riding has concluded "not that free-riding will never occur, but that it cannot be presumed always to occur" (McMillan 1979, 95). One of the empirical cases cited that contradicted the standard "dismal" view of the improbability of collective action was indeed an irrigation system in the Philippines (Kikuchi, Dozina, and Hayami 1978). Empirical efforts to measure how much people will take advantage of others in order to increase their own net benefit from a public good have shown surprisingly little deceit (Johansen 1977; Schneider and Pommerehne 1981).

Kimber (1981) points out a logical flaw in the argument of Olson, Hardin, and others of their school. They presume that rational actors can choose whether to free ride independent of others' choices on this subject. They present the alternatives confronting "rational actors" as either contributing toward collective creation of a benefit and getting a deserved share of it or not contributing to the benefit but getting some undeserved share anyway. This, however, is not a realistic or real choice, because it assumes that only the actor in question is rational and everyone else is not—that others will "irrationally" create benefits that the selfish rational actor can reap without cost. This is invalid reasoning. Unless one has evidence to the contrary, it is not reasonable to assume that everyone else is irrational. This means the real choices are to contribute to the creation of a benefit in which all will share or to not contribute but also not benefit, because all others made the same rational decision not to contribute. If an activity produces benefits greater than the cost of the individual's contribution, a rational chooser gains more by proceeding on the assumption that others—at least most or enough others—will similarly recognize the advantage of the first alternative and accordingly participate in collective action.[24]

Olson and Hardin have done a service by calling attention to the potentially debilitating effects of free ridership. In Gal Oya we had to be cognizant of this possibility when planning and promoting farmer organization. But groups were willing to tolerate some free riding if the benefits from collective action to those who were prepared to join in outweighed their own costs of contribution. Also, the costs as well as the benefits they reckoned were not simply material ones.

24. Kimber comments with appropriate irony on the notion that people will jointly contribute to create a public good *only* if assured of some exclusive personal benefits: "There is surely ... something whimsical in the idea that the Council for the Protection of Rural England is really organized, not to protect rural England, but to provide wine and cheese parties for its members" (1981, 196).

Some farmers—and not just the richer ones—were almost always willing to give up substantial amounts of time for the benefit of others, recognizing that they too gained from better water management. These farmers had significant opportunity costs for the time they gave up from income-earning activity, and moreover, they usually paid their travel costs and other expenses for attending meetings out of their own pockets. (Sometimes such expenses were reimbursed by collecting funds from all members.) As seen in chapter 9, they were also willing to forgo payment for attending meetings when they thought "fairness" was at stake.

Groups often chose tail-end farmers to be farmer-representatives, figuring that they had the most at stake in making sure the collective effort was successful and could best justify the expense of time and money. But this choice was also a surprising delegation of authority by head-end farmers that ensured attention to tail-end needs. A large part of farmer-representatives' reward was status from others and self-respect for having contributed to a more productive and legitimate system of local economic and social relations. Perhaps the farmers who became representatives had a higher demand than their neighbors for fairness, for no conflict, or for efficient use of resources.

Such observations called into question the presumption that people's behavior can be best or even adequately understood in terms of individual self-interest maximization, measured mostly or exclusively in terms of material benefits and costs.[25] Many persons may behave like "economic man" much or most of the time, but to assume in the name of analytical simplicity and theoretical rigor that everyone does so means forgoing opportunities that can be opened up by pursuit of collective interests and encouragement of more group-centered thinking, giving nonmaterial factors due consideration. These can mobilize ideas, energies, and talents now ignored or employed only for individual gain.

Tapping these resources depends not just on the normative environment but also on structural arrangements. Here is where institutions come in, to solve "the assurance problem" as Runge (1981, 1984) analyzes it, providing incentives and guarantees that all or enough persons will contribute to the creation of a public good, Kimber's first

---

25. March and Olsen (1984, 744) write: "Although self-interest undoubtedly permeates policy, action is often based more on discovering the normatively appropriate behavior than on calculating the return expected from alternative choices. As a result, political behavior, like other behavior, can be described in terms of duties, obligations, roles and rules."

alternative above. If it is expected that all will contribute their shares, and if therefore all (or enough) do, positive-sum public goods are created, greater in value than the resources that went into them.

These conclusions do not justify relying solely on altruistic norms. Structures of communication and authority for monitoring and decision making are important determinants of behavior. Incentives and not just ideas play a role in regulating collective action. In studying irrigation and other natural resource management in South Indian villages, Wade (1987) found that corporate institutions established voluntarily to regulate use of common property resources were found more frequently toward the tail-end of channels, where benefits were greater. His empirically based analysis leads to a conclusion similar to ours, that state control and private property arrangements imposed from outside are not the only viable alternatives for beneficial resource management.

Collective action based on appropriate normative and structural principles shaping behavior is feasible.[26] But to recognize and support such opportunities, we need to accord validity to cooperation and generosity as human values and motivations. By being more positive-sum in their dynamics, these are more productive orientations than their opposites, individualism and self-interest, as alternative principles of action.

Current economic theory presumes that these principles when expressed in zero-sum transactions will produce positive-sum results. But in irrigation, selfish individualism resulted in negative-sum outcomes—in substantial productive opportunities forgone. Once we analyze the social world in these terms, we see that this is not unique to the irrigation sector. The burden of proof should be on theorists, despite the elegance of their either-or neoclassical rationale, to show when and to what extent human welfare is enlarged by downplaying and even arguing against collective and nonmaterial factors.

---

26. Elinor Ostrom points out that "prisoner's dilemma" outcomes result from strange and unusual assumptions: participants are not able to construct their own rules on a cooperative basis, to link self-interest with commonly held values and goals: "Social scientists and philosophers have allowed themselves to be 'hung up' on simple, one-level paradoxical situations. Immense scholarly energy has been devoted to trying to prove that individual rationality in a perverse situation will somehow avoid an irrational outcome. Why should we expect perfectly rational individuals placed in highly irrational structures, with no opportunity to change them, to achieve collective rationality? What is a more irrational way to structure an enduring situation than that represented by the PD [prisoner's dilemma] game: no communication among participants, no previous ties among them, no anticipation of future interactions, and no capacity to promise, threaten, cajole, or retaliate" (1986, 10).

Once the normative map in figure 3 has been drawn, it is hard to believe that quadrant I is the best or only arena for societal or individual advancement. But such a formulation is itself reductionist and either-or. How can all four points on the normative compass be capitalized in positive-sum ways? The rehabilitation of altruism and cooperation does not mean the burial of individualism and self-interest. Rather, it should expand the social space in which human ideas, talents, and energies can be mobilized.

13

Social Energy as an Offset
to Equilibrium and Entropy

What about the impact of the grassroots projects, associations and movements themselves? A dense network of such movements, jointly with a large number of social activist organizations, is bound to change the traditional character of Latin American society in several ways, most of which are not yet well understood. But it seems safe to assert that, with such a network, social relations become more caring and less private.

Albert O. Hirschman, *Getting Ahead Collectively*

That previous barriers to collective action in Gal Oya could be removed by a combination of new structural arrangements and realigned normative orientations offers both insight and hope. But perhaps these changes could be explained in "social engineering" terms with electromagnetic forces that predate concepts of the new physics. Such formulations initially seemed satisfactory, given a disposition for parsimonious conclusions, for the sake of rigorous analysis and being able to communicate them easily. Much of what was happening could be accounted for within frames of reference generally accepted by social scientists, requiring only a shift from mechanistic to probabilistic ascriptions of cause and effect. Accepting the concept of uncertainty as an expansion upon Newtonian thinking seems easy, since social scientists are familiar with statistical concepts and the laws of probability. These are misleading, however, because this mode of reasoning assumes that certain fixed distributions underlie all observable phenomena and that a large enough or randomly enough drawn sample will reveal what is "true."

The larger problem of explanation was that the river was flowing uphill, to borrow a metaphor from William Calvin (1986). We were dealing with a situation where equilibrium was not an appropriate

357

model. Indeed, the physical and social system in Gal Oya had approached the nadir of disorganization, the definition of entropy, but it was now acquiring greater structure and productivity. This could be explained conventionally as raising the system to a higher level by increasing inputs, but then the physical analogy gets ambiguous. The work efficiency of inputs is usually reduced by something like friction. This means one needs to make investments greater than the improvements to be achieved. Or maybe somehow, through the leverage of farmer organization, the equivalent of mechanical advantage had been gained, so that inputs yielded outputs greater than themselves.

Such mechanistic conceptions had clear limitations. The IO program, with less than 5 percent of the total project investment, was transforming the situation in Gal Oya. We were witnessing a process of *energization* that itself needed explanation. Unappreciated and unexploited potentials were being tapped, involving both material resources and mental processes. We needed to explain how, in Popper's words, "non-physical things such as *purposes, deliberations, plans, decisions, theories, intentions* and *values* can play a part in bringing about physical changes in the physical world" (1972, 228–29). Outcomes were continually shaped by the investment of personal effort and by the effects of interpersonal relations, for better and for worse. The orderly worldview of classical physics, with its emphasis on equilibrium, seemed inappropriate, as did the view from thermodynamics that all is tending toward disorder. The disclaimer that Gal Oya was an open, not closed, system did not resolve the problem, because what closed social systems are there in which classical mechanistic rules operate? We were working in a very open system in which outcomes were being *made,* not simply observed.

Searching for Explanations

On one of the long drives from Colombo to Ampare, Sena, Lucky, and I reflected on the first two years of experience. How could we account for the changes we were seeing in the behavior of farmers and officials? We considered the influence of new organizations, of new roles and norms, discussed in the preceding chapter. These looked like sufficient "causes" because they were so proximate to the effects we wanted to explain. But how did these influences come about? They could be ascribed to the inventiveness, persistence, and inspiration of the institutional organizers. But to what could we attribute the qualities and performance of the IOs? Should the ARTI and Cornell staff get the credit for this? We had conceived the program and had recruited and trained the IOs, so the arrow of causality pointed our way. Yet this put us in a chain of infinitely regressing explanation, since our behavior and that of USAID and many others then had to be accounted for.

The role of the IO was understood from the outset, admittedly somewhat opaquely, as that of a *catalyst*.[1] IOs were to evoke potential within the farming community. But we saw farmers evoking potential within each other, so they too were catalysts, influencing each other's normative probability distributions, to use terms from the preceding chapter. Both farmers and IOs were eliciting new norms and behaviors from each other and also among engineers and other government staff, who in turn were energizing still other officials, and some of these were drawing out further potentials among farmers and IOs. We from ARTI and Cornell might be seen as supercatalysts who had started off this whole process, yet we were ourselves energized every time we visited Gal Oya. We were part of a web rather than a chain of causation.

We saw people bringing out the best in one another. This potential had been there all along, but it was not realized until the equilibrium of low expectations was broken and until normative alignments had been changed, as suggested by the figures in chapter 12 that depict people's orientations in probabilistic and electromagnetic terms. The IOs generated something like fields of social and moral force within Gal Oya, changing the direction of prevailing normative attraction. Reversing expectations was like switching north and south poles. This redirection deflected more and more people's normative compasses from selfishness to generosity and from individualism to cooperation. The change did not happen all at once but occurred in a succession of shifts in alignment where people's reoriented expectations influenced each other.

1. Many terms have been used for such a role—organizer, promoter, facilitator, animator, broker, intermediary, change agent, etc. (see Esman and Uphoff 1984, 253–58). The term *catalyst* has been objected to because, technically speaking, such an agent does not change as the ensuing reaction proceeds, and certainly community organizing is a transforming experience (Tilakaratna 1987, 44). But more important to us, the term implied that the catalyst does not itself create something but helps to manifest potential that *already exists* in the situation. Our plan was to withdraw the organizers at some point, expecting the process of change to continue at the direction of Gal Oya residents. The catalyst concept reduced the risk of creating attitudes of dependency. The IOs were to stimulate and support local initiative but not to become part of that local capacity.

This was itself too mechanistic a description, because all statements need to be understood in probabilistic terms. When one person's odds of acting generously were increased, so were the frequencies for others with whom the now more generous person interacted. Similarly, if even one person expressed a willingness to work cooperatively, others acted differently than if nobody had come forward with such a suggestion. Narangoda's speaking up for helping tail-enders affected his Sinhalese neighbors' readiness to save water for Tamils downstream. There was nothing deterministic about this, only a potential to be capitalized on.

How could we account for the broader set of influences that overarched individual choices and actions? This was not an either-or question. Interpersonal interactions, explainable with fairly proximate causal factors, appeared real and important, but we wanted some summary analytical framework to help make sense of all this, giving some deductive coherence to the empirical observations we were gaining confidence in.

Ideas

Structural explanations try to account for people's behavior in terms of their interests. These could be identified as arising from bureaucratic roles that conferred certain authority, status, and control over money; from economic class positions like that of landlord or moneylender; or from objective material advantages such as accrue from having one's field at the head of the channel. Why should engineers or local elites or head-enders be willing to give up some, if not all, of their advantages once IOs came and started talking about water sharing and group activity?

Part of any explanation should include the role of ideas, which made us take cognitive factors and analyses more seriously. Time and again we saw people respond positively to new ideas that were simply and clearly laid out, especially if they tapped into certain values people held, like efficient use of resources, respecting precedents, or acting fairly. Something as simple as the designation farmer-representative, instead of the widely used term farmer-leader, affected the behavior of thousands of people. Farmers acting as "representatives" rather than as "leaders" saw themselves, and were seen, in quite a different light. The latter, widely used designation had made fellow farmers into followers, not what a program promoting self-reliance needed. Officials could not make deals with farmer-representatives to be enforced from the top down. A process of negotiation and consensus building resulted from the introduction of a new word, conveying a powerful idea.

Ideas affected behavior by making new ways of acting conceivable and justifiable, often highlighting values already held or elevating interests previously given less weight. New relationships among ends and means acquired salience and motive power once described or reformulated. Just as conceptual confusion has an immobilizing effect on people, inhibiting them from taking action, their efforts could be accelerated by providing ideas that explicate relationships. Since people's interests and values are not neatly arranged in consistent hierarchies, ideas can evoke action by pointing out connections previously unseen. People do not often change behavior because they are persuaded outright to give up one thing in favor of something else, which is zero-sum. Usually they change when, thanks to new ideas, they see some positive-sum possibilities in novel ways of serving several values and/or interests at the same time.

That ideas could countervail and even alter material interests was suggested by the way engineers in Gal Oya accommodated to our proposal to give farmers control over water distribution at lower levels of the system. The Irrigation Department had responsibility for all the water delivered to farmers, deciding on amounts and schedules. In fact, the ID did not have enough personnel and information to exercise this authority consistently or effectively. This nominal control, however, meant that some technical staff could exploit their position to gain money or other benefits.

The idea of farmer participation in irrigation operations was at first unsettling for most engineers. Even those who would not lose financially would incur the psychic costs of reduced status and authority. In this situation, my engineer colleague Gil Levine proposed that the department redefine its responsibility, from one of "retailing" water to individual farmers to that of "wholesaling" water to groups of farmers. Rather than trying to deliver small amounts to each field, it could make bulk deliveries to the heads of distributary channels, letting farmer organizations divide and share the water. If they did not do this in a timely or equitable way, their water supply could be reduced, putting pressure on the groups to perform this task properly.

It took some time for this idea to sink in. It represented quite a shift in engineers' thinking about themselves, their profession, and their duties. Fortunately, the persons whose material interests were most affected were field-level technicians rather than senior engineers. The department acknowledged that its limited staff and funds might be better spent on rehabilitation and maintenance work than on water distribution. Once engineers had a rationale that preserved their professional self-respect, they made a change in their practices, agreeing to delegate authority for water distribution to farmer organizations.

It could be argued that there is more scope for illicit benefits from rehabilitation and maintenance work than from water distribution. But if interests explain behavior, this constellation of material incentives should have induced engineers to delegate operational responsibility to farmers earlier, without the prompting of our ideas. The concept of wholesaling water did not by itself cause the changed behavior, but it was an important contributing factor.

Farmers came up with one idea that was particularly powerful. One of the greatest dangers the farmer organizations faced was becoming politicized, since Sri Lanka has a long history of partisan competition and conflict. Someone suggested a homely saying that legitimated keeping party politics out of their organizations—"Water has no color"—a play on words inspired by the fact that each political party has its own color, green, blue, red, and so on. To reinforce the point, it was added that "putting color in water spoils it." With such a motto, farmers interested in improving water management, many of them partisans in other contexts and at other times, kept political considerations from entering into the selection of representatives, the allocation of water, and other decisions. Head-enders who might have used political connections to get an unfair share of water had to accept the principle of nonpartisan decision making once the compelling symbolism of water's having no color constrained and redirected their behavior, counter to their narrowly defined interests.

Engineers at first saw no place for IOs in their "technical" department and regarded expenditure for IOs' work as less valuable than money spent on channels, gates, and other equipment, thinking that farmers should simply do (or have to do) as they were told. When IOs were presented as specialists in farmer organization, they got more acceptance. And engineers became more sympathetic toward farmer organizations when we explained that they were "social infrastructure," requiring some substantial initial investment to create them, followed by reduced but continuing maintenance investments. If engineers remained unpersuaded, we described the organizations as "software" needed to make the hardware of physical structures productive, an explanation that carried more weight because computer use was spreading.

It is impossible to say just how much influence ideas like wholesaling or social infrastructure had on decisions and subsequent actions. That one cannot measure this influence does not justify ignoring it, however, because it is evidently important. Ideas by themselves cannot overcome material interests where these motives are strong, but ideas treated as soft but expandable resources gave us momentum and power. Ideas defy the laws of thermodynamics, which proclaim the conservation (really the limitation) of energy, establishing a zero-sum frame of reference that mandates an equal loss for every gain. Ideas can be given up to someone else without losing them (though one does lose control over them). This dynamic gives some hope that the destiny of entropy can be avoided or at least deferred (Boulding 1978), yet it requires that we understand the world in open system rather than closed system terms.

While material resources were limited in Gal Oya and were often the source of zero-sum reasoning and conflict, in the realm of ideas we were free to create positive-sum dynamics if we had the wit and persistence to use our minds to such effect. Ideas could not eliminate interests, but they could redefine, reorder, and even reverse them, since interests are, after all, ultimately more subjective than objective. Even if the planets of choice revolve around the sun of ego, their mass and consequently their orbits can be altered by revising people's evaluations of their interests and opportunities, so our solar systems of action are amenable to some alteration through exchange and change of ideas.[2]

Ideals

Ideals are a special kind of idea, expressing values of community welfare and common interest. IOs embodying and articulating such ideals found themselves able to enlist a great deal of support and cooperation by appealing to people's "better selves," represented by the right-hand side of their value distributions drawn in figure 2. The outcomes sought were not simply redistributive, with zero-sum outcome.

2. William Munro, using new physics concepts in his APSA presidential address, spoke of how human beings "respond to the stimulus of ideas, the electrons of the social universe," adding that "our entire process of civic education—in the schools and colleges, by the press and at the forum—consists of bombarding the human nucleus with ideas. Some get attached, but the majority do not. . . . The social atmosphere, like the physical universe, is filled with these invisible units of energy [ideas], moving at all rates of speed and penetrating power" (1928, 5).

Rather, they aimed at alternative arrangements where gains for some, mainly the disadvantaged, could be made at little though not necessarily zero cost to others. Tangible positive-sum outcomes were being fashioned. Improving the situation of the disadvantaged had very high legitimacy—nobody could speak against it. Thus, time and again we saw better-off persons willing to assist others less fortunate.

We may have benefited because this was a settlement scheme where all families had started out thirty years before with the same allocation of land. Land allotments had been subdivided over the years, however, often several times, and many parcels had been acquired by more successful farmers or by rich outsiders. Large de facto landholdings could be one hundred times bigger than the smallest ones. Still, it was accepted that everyone should have some access to water, since people's livelihood and survival was at stake. Selfish behavior could be countered by appealing to the principle that everyone deserves a chance to live. Similarly, acts of ethnic chauvinism could be restrained by invoking ideals found in the respective religions themselves, if someone was willing to remind transgressors of the teachings everyone respected.

What could explain the IOs' role in all this? They repeatedly worked well beyond the requirements we had set, motivated by what can best be described as idealism. Some farmers in turn and also some officials likewise gave expression to the bright side of their nature, which was not prepared to sacrifice others' well-being (or even to be indifferent to it) for their own sake. When Sena, Lucky, and I tried to explain our own involvement in the program, we had to acknowledge that our motivations were idealistic in our own ways. (We were, after all, traveling down to Gal Oya on a Sunday so as not to waste a working day on the road.)

Social scientists are not supposed to give expression to their normative beliefs, since our disciplines aspire to be value-free for the sake of objectivity. We could of course divide our time and ourselves in a zero-sum way between professional and personal activities, the latter being the acceptable sphere for value-driven public service. ARTI colleagues were in a similar situation where their professionalism supposedly excluded activism, but as public servants they were supposed to be contributing to development for their countrymen.

To try to understand all these energized efforts without some reference to ideals would be mistaken, since the most active IOs, farmers, and officials were the most idealistic in each group. Expressions of idealism were infectious, changing the way others defined and carried out their tasks. We were all catalysts in our own ways. The desire to see long-standing disadvantages eliminated was a motive force not to be underestimated or subsumed under reductionist notions that this activism was really only an expression of our egoism. We were not self-sacrificers or saints, but neither were we thinking only of ourselves. The influence of idealism was substantial even if by itself it was neither a necessary nor a sufficient cause. This realization was a useful insight coming from both-and thinking: self-interest and altruism can and do coexist. One should not ignore ideals because they are not by themselves determinant and their results are not always predictable.

Friendship

The program in Gal Oya derived its greatest thrust from friendship, that personalistic, idiosyncratic phenomenon we find everywhere "making the world go round." The core group at ARTI exhibited, at least in the first four years, strong bonds of mutual appreciation and support, mirroring the close connections already established among the Cornell faculty participating in this effort. The IO cadre established and maintained its high morale partly because organizers enjoyed each other's company and continually helped each other out. The farmer groups acquired more momentum where there was a spirit of friendship among members. The climate of mutual acceptance and assistance spread between groups. It started between ARTI and Cornell staff but then grew between them and the IOs, between IOs and farmers, between IOs and officials, and between farmers and officials.

References to friendship kept cropping up in discussions in the field, too often to be ignored.[3] The expressions "giving others the benefit of the doubt" and "bringing out the best in each other"—key characteristics of relations among friends—took on practical meaning with the emergence of cooperation and energization in Gal Oya. Among friends, technically speaking, one presumes positive-sum relationships and what economists call interdependence of utility functions. Probability distributions as sketched in figure 2 shift to the right, toward generosity and cooperation, when *friends* are involved.

3. Its importance in similar grass roots development efforts in India has been noted, for example, by Volken, Kumar, and Kaithathara (1982, 73), Field (1980, 139), and Clements (1986, 141). It is probably one of the most underreported factors in successful development work. At a workshop for the Irrigation Management Division in Sri Lanka in October 1989, the retiring state secretary for irrigation (and the division's first director), K. D. P. Perera, in a valedictory address described the IMD's task as "to bring farmers and officials together as friends," saying this was "the only way to get the benefits [from irrigation] everyone hopes for."

The importance of friendship was evident in good times but was even more dramatic in bad ones. The program went through some disappointing valleys of low morale and loss of effectiveness. These slumps were associated with breakdowns of friendship and trust, partly due to ethnic tensions after 1983, but more often due to personality conflicts and misunderstanding when some members of the institute no longer assisted each other in attaining goals and no longer got satisfaction from each other's success.[4]

Friendship is seldom taken seriously in social science, perhaps because it is particularistic, something Talcott Parsons disparaged in his scheme of analysis because it is not "modern" like its antithesis, universalism. The subject has rarely been treated with much theoretical interest or rigor.[5] Indeed, the most systematic discussion of friendship may still be Aristotle's two books on the subject in The Ethics, written 2,500 years ago (1953, 258–311).

The neglect is surprising considering how ubiquitous we all know friendship to be as a social force, seen for example in "old school ties" or in "connections" that arise from shared experience and mutual trust. Aristotle suggested that true friendship is held together not just by interests but also by shared ideals. There is a contemporary implication, however, that personal links, because they are particularistic, are somehow illegitimate. True, including some can mean excluding others. Yet most of the world's business, not just commercial but also administrative, gets expedited if not transacted on such bases. To ignore them only shows how unempirical social scientists have become in their search for abstract, universalistic explanations regarded as somehow more scientific than what their observations reveal.[6] Indeed,

---

4. A letter from Sena in March 1984 reported that the situation at ARTI was deteriorating: "The cohesiveness and the close infeeling is disappearing." Still, the program managed to sustain itself because the ideas and ideals undergirding it remained strong, and friendship at the field level was well entrenched. Sadly, in 1985 two key actors at ARTI even made physical threats against one another. Just because the program obtained positive results does not mean that social relations were always positive. My own role in the program shifted over time from that of a generator and consolidator of ideas to that of an arbitrator and go-between, trying to get some people to see each other in a favorable light. I was more successful in the first role than in the second.

5. Lillian Rubin, in one of the few extended recent studies of friendship, characterizes it as "the neglected relationship. . . . Until very recently, the subject of adult friendship has been largely absent from the literature of the social and psychological sciences, getting little more than passing reference in studies devoted to other issues" (1985, 9).

6. Social scientists have been interested in studying patron-client interaction and similar relationships (e.g., Schmidt et al. 1977; Eisenstadt and Roniger 1984), but mostly in terms of quadrant III kinds of behavior.

simply considering the origin of the word "social" suggests that the social sciences could well be understood as the sciences of friendship.[7]

Identifying ideas, ideals, and friendship as prime factors in and accelerators of the process of structural and normative change we observed in Gal Oya led to thinking about the theoretical connections among them. Though arrived at inductively, they are more than an ad hoc set, as analyzed below, having certain characteristics in common. We must also recognize that each of these factors can present a pathological negative-sum face. Some ideas, however attractive they may be at the time, subsequently prove to be wrong according to an informed consensus. We are now quite aware of the negative consequences of ideas associated with racial superiority or gender inequality. Ideals in extreme or perverted form can produce zealotry and destruction. Sadly, conflicts over ideals (or at least in the name of ideals) have been some of the most bitter and bloody. Further, cronyism and favoritism—friendship practiced at others' expense in a zero-sum manner—can produce unfortunate results. (According to Aristotle, these forms are not true friendship.)

Appreciating ideas, ideals, and friendship as potential sources of constructive social energy does not guarantee beneficent outcomes. The world we live in is probabilistic, and autonomous sources of invariant causation exist only in deductive models. Consequences are at least partly what we want and make them to be, since in the real world everything is interactive. But this gets into questions addressed in the concluding chapter. Here we are concerned with more proximate explanations.

Social Energy: Some Examples

About the same time we were reflecting on our experience in Gal Oya, Albert Hirschman was visiting grass roots development projects in Latin America that had received financial support from the Inter-American Foundation. When he tried to account for the dynamism and persistence of some of the more impressive projects he observed, he came up with the concept of social energy referred to previously, which animates individuals and groups to high levels of performance

7. The noun "society" and its adjective "social," like the French word *société*, stem from the Latin word for friend *(socius)*. The equivalents in German, *Gesellschaft* and *gesellschaftlich*, likewise come from *Gesell*, meaning companion or comrade.

for some collective purpose. Psychological factors appeared more crucial to Hirschman than material ones (1984, 56–57). Once the concept of social energy was formulated, other examples could be seen in the literature. The more closely I looked at development efforts in many countries, the more evident it seemed that our experience in Gal Oya was not unique, though each case is different in its own way.

One of the most impressive cases Hirschman observed was an association of five thousand *tricicleros* (tricycle drivers) who deliver goods on three-wheeled transport vehicles all around Santo Domingo, the capital of the Dominican Republic. To help these persons, some of the poorest of the poor, a loan fund was developed to enable them to buy their own vehicles, rather than pay 20 percent of their meager earnings each day as rent. To facilitate making the loans and to reduce the risk of default, hundreds of *grupos solidarios* were formed, consisting of five to seven tricicleros who would be responsible for each other's repayment of the loans that it was hoped would lead to their financial independence:

> At the level of social interaction, we were told, the arrangement tended to create close *friendship* (before the arrangement "I used to see him around, but I didn't known him," as one *triciclero* put it)—as well as probably some new conflicts. At the level of social organization, the effect was even more remarkable: once a number of *grupos solidarios* were formed, the *idea* arose to create a tie among the individual groups (each of which had an internal structure, with a President, Treasurer, Secretary, etc.) and an Association of Tricycle Riders (or rather of the *grupos solidarios*) came into being. . . .
>
> The Association soon developed some new activities of its own: it organized a rudimentary health insurance scheme and promoted contributions of members to funeral expenses for members and their immediate families, as an expression of *solidarity*. Plans are being made for a tricycle repair shop, with tools and spare parts to be available to members.
>
> Increasingly, moreover the Association began to act as an interest and pressure group, for example, by taking a stand against certain measures of the municipal traffic and police departments which often make the harsh life of the *tricicleros* even harder through various taxes, prohibitions, fines, and so on. As we were leaving, a protest against one such measure was being planned. "I can mobilize 500 tricycles to converge on any spot in the city and paralyze anything," said one of the Association's leaders, proud of their new-found strength.
>
> In this manner, a financial mechanism originally designed to do no more than protect a lending agency against default by individual borrowers is having powerful and largely unanticipated social, economic and human

effects, enhancing group solidarity and stimulating collective action. (Hirschman, 1984, 15–16; emphasis added)

As in Gal Oya, not only was the right thing done for the wrong reason, but large numbers of persons moved into quadrant IV in figure 3. Once people come together, many new improvements in their lives become possible, thanks to the inventiveness, commitment, and solidarity generated—three manifestations of ideas, ideals, and friendship.

Such microdynamics are not frequently documented in the social science literature. But evaluations for the International Fund for Agricultural Development of two major programs it has aided in South Asia (with loans of $30 million or more) report similar outcomes with the small groups fostered there. One of the most active groups under the Small Farmer Development Programme (SFDP) in Nepal was described as follows:

> Formed $2\frac{1}{2}$ years ago, the group has 19 members, including four landless. The maximum holding is 0.5 hectares. The group meets once a month and . . . there are four sub-committees dealing with group savings, group farming, purchase and sale of paddy, and loan utilization and repayment. . . . So far the group has not had any loan repayment problems. Nor has it experienced any serious conflict within the group. Its group savings based on a contribution of Rs. 15 per month [each] amount to Rs. 29,000. The group has instituted [a food bank] to be used in emergencies. It has a number of group activities such as potato cultivation, seed multiplication, and vegetable gardening. The members participated in road construction, donating a total of eight days' labour each last year. Fourteen members have undergone vasectomy; the others have not done so because they have only one child each.
>
> The group leader estimated that their incomes have been rising at the rate of 10 to 12 percent per annum and that differentiation among members has been reduced. Early this year the group made a loan of Rs. 1,000 to another group to set up a pharmacy in the village. This will also benefit their members with supplies of medicine [positive-sum altruism]. The group keeps complete and up-to-date records and has drawn up an impressive five-year investment and development plan comprising 13 items. (IFAD 1984a, 35; also see Ghai and Rahman 1981; Rahman 1984a)

The Grameen Bank in Bangladesh has recently received much favorable attention in international circles. It has assisted 800,000 of the urban and rural poor with small loans, averaging about $70. These have about a 98 percent repayment rate and boost incomes by 50 to

100 percent. They are channeled through borrowing groups of five members each, similar to the solidarity groups of tricicleros in the Dominican Republic. Like them and SFDP groups, Grameen Bank groups move beyond personal economic dealings into making social improvements like curbing dowry payments, stopping child marriages, and setting limits on what people are expected to spend on marriages and funerals. Such changes improve the status of women and generally benefit the poor, who go deeply into debt (and often lose their land) to meet customary obligations (IFAD 1984b; also see Fuglesang and Chandler 1986). Several other significant programs now operate in South Asia based on local organizations formed at the instigation of catalysts, using a methodology similar to that developed in Gal Oya.[8]

In some instances the impetus for change has come from within the community, but then usually from persons in somewhat differentiated roles who are in a position to shift normative distributions from quadrant I to quadrant IV. The Deedar cooperative in a Bangladesh village was started in 1960 by eight rickshaw pullers—among the poorest of the poor—who complained about their condition over cups of tea in a local tea stall. When they said they were too poor to escape their poverty, the stall owner challenged them to save each day the cost of one cup of tea, 8 annas, a trivial amount. He offered to save the same amount daily with them, giving up in an altruistic way the profit he would have made on their cups of tea. With their savings, members eventually were able to purchase their own rickshaws, like the tricicleros, and started on the road to financial independence and elevated social status. Thirty years later, the co-op has 1,200 members, with share capital of 2 million thaka and assets worth 6 million thaka. It operates

8. The Bangladesh Rural Advancement Committee is another good case from that country (Korten 1980, Streefland 1986). In Pakistan, the Aga Khan Rural Support Program is helping hundreds of thousands of households in that country's most remote northern districts improve their incomes and quality of life through newly formed village organizations promoted by AKRSP's social organizers (World Bank 1987), while the Orangi Pilot Project covering a periurban slum outside Karachi with 200,000 inhabitants has similarly promoted lane committees, of twenty to thirty households each, that construct self-help sewerage and waste disposal at a cost of only about $25 per household (Gafoor 1987). Orangi was the scene of violent conflict in 1987 between Afghan refugees and drug-running gangs, leading to the burning and destruction of large parts of the slum. The general manager of AKRSP said during a visit to Cornell in January 1988 that through the social infrastructure created by the project, most of the damage was repaired in a matter of months so that one could hardly know the earlier destruction; neighborhood associations also prevented further outbreaks of violence. Still more programs in Nepal, Sri Lanka, and India could be cited as mobilizing local energies and talents for self-managed development. Even so acerbic a reporter of human affairs as V. S. Naipaul found social energy catalyzed by the Shiv Sena political movement impressively improving life in Bombay slums (1977, 63–69). On solidarity groups generally, see Otero (1986).

its own brick factory, trucking service, tube wells, rice mill, retail store, and marketing center. Scholarships and free books are given to children of members to encourage educational advancement; the co-op operates its own model secondary school and a health insurance scheme. All this was done without outside financial assistance but with the remarkable leadership of the tea stall owner, who became the co-op's general manager (Ray 1983).

The catalyst for farmer organization in the Pithuwa irrigation system in Nepal, which is a model for self-managed improvement in that country, was a farmer who had previously been a community development worker and had knowledge of and confidence in organizational methods.[9] In both Pithuwa and Deedar, initiative to shift community norms and behavior came from someone who stood slightly apart from the rest and could therefore act more autonomously to promote cooperation and altruism.

An unusual example of catalytic efforts producing positive results from within a government is found in Thailand, where a provincial deputy governor undertook a quiet campaign of persuasion and motivation first of his provincial-level agriculture, education, health, and community development officials. They in turn reoriented their district subordinates, who selected and motivated subdistrict staff to create local development teams of public officials who helped to organize clusters of households (ten to fifteen each) in rural communities. Through discussion, the program reached consensus on a set of basic social needs that all people have—for food, schooling, employment, personal safety, religious activities, and so forth—defined in culturally valued ways.

Once a year each cluster is to assess its standing according to this list of basic needs, and its results are aggregated with those of other

9. Pithuwa farmers suffered water shortages in this government-constructed 600 hectare scheme because it was not properly operated and maintained until they formed groups for all sixteen distributary channels and an umbrella organization for the whole system (Laitos et al. 1986, 121–38). Before operation and maintenance responsibility was handed over to the organization, the government's expenditure of 90,000 rupees through "leaky" private contracting procedures netted only 20,000 rupees' worth of work done. When the engineer responsible for Pithuwa O&M was given only 40,000 rupees for work the next year, he decided to offer this money as a block grant to the organization to be used toward upkeep of the system. He later estimated that 100,000 rupees' worth of work was done on the system that year (personal communication, S. Poudyel). Hirschman would consider Pithuwa a case of the transfer of social energy, since the retired community development worker who started its organization grew up in the 3,000 hectare Chhattis Mauja system, which is about 100 miles away and which is managed entirely by water users through their own four-level indigenous organization. They contribute 60,000 man-days of labor annually for maintaining the main and secondary channels and still more for the rest of the system (Pradhan 1983).

clusters within the village. Village results are put together to make a subdistrict profile, and then district and provincial statistics are compiled. It was found that this process of gathering "bottom-up" statistics energized the civil servants involved. They became more aware of where there were concentrations of hunger, school dropouts, petty theft, and other problems. Also, they could see when year-to-year progress had been made. Those who neglected their agricultural, health, education, or community development responsibilities were shown up by the data, so both positive and negative incentives were mobilized.

As important or more so were the effects on the communities. Most of the needs being monitored were things that individuals, clusters, and villages could improve by self-help measures.[10] The simple process of taking an annual inventory of how well basic needs are being met can energize both the communities and the civil servants paid to assist them, with few additional material resources from the government. Organizing all households into clusters created a structural basis for getting people to think in more other-oriented ways and for planning and carrying out collective efforts. Normative shifts emerged alongside the new roles and procedures. Government staff who had been largely useless became more helpful as their value orientation shifted from S>G to G>S for a variety of motivations, reinforced by public demonstrations of altruism and cooperation. Communities that had existed mostly as geographic entities became more social ones—that is, communities of friends, with energies mobilized by ideas and ideals, some brought from outside and some generated from among themselves.

Most of the examples of self-help efforts described here come from Asia, but there are others from Latin America and some from Africa too.[11] They have many elements in common, of which social energy is

10. My favorite indicator, perhaps because my wife is a pediatrician, was how many of the babies born in a cluster during the past year weighed less than 3,000 grams at birth. This is a basic need for newborns, because adequate birthweight, reflecting good maternal prenatal nutrition, gives them better prospects for survival and good mental development. Public health nurses, though able to weigh all newborns, often did not do so because nobody paid attention to the statistics. Now villagers consider an underweight baby born in their cluster a black mark against them all. People started to take an interest in whether pregnant neighbors were eating well once they began to identify with each expected new member of their community. Richer households in a cluster now may give some assistance to expectant mothers in poor households, since a baby born in their cluster and weighing over 3,000 grams represents a "success" for all. Brief descriptions of the program are given in Heim et al. (1986, 35–39) and Uphoff (1986, 348–50). This approach has been accepted by the national government as a model for rural development.

11. Examples from Africa would include a self-help water supply program in Malawi

perhaps the most notable. Changing normative orientations away from individualism and self-centeredness releases remarkable energies within communities previously divided by conflict and often exploited by internal or outside forces. When I discuss the ideas presented here with persons involved in grass roots efforts elsewhere, they recognize the principles and techniques of motivation and group action and are quite satisfied with post-Newtonian concepts. [12]

Although social energy is complex, it can and should be analyzed, though this involves some reductionism and mechanistic imagery, describing "parts" as if they could achieve predictable outcomes if "assembled." The theorizing below about social energy, though oversimplified, makes a start at bringing this important factor in human affairs within the realm of investigation. In doing this, the way we think about social structures and dynamics will need to change, with either-or presumptions giving way to both-and insights. Possibilities for mobilizing underestimated and overlooked potentials in individual and collective endeavor should be identified and activated where these possibilities coexist with less promising probabilities.

---

based on community organization that has brought clean water to over a million villagers at a cost of about ten dollars per person plus contributed labor. This has a collective orientation, since all water is provided through community taps, with no private hookups (Liebenow 1981; Glennie 1983; see also Briscoe and de Ferranti 1988 on a similar program in Colombia). In Zimbabwe, a Savings Development Movement started by a priest and several associates grew to a quarter of a million members, almost all women, with over $2 million in deposits providing a base for improving agriculture and housing (Chimedza 1985; Bratton 1986, 1988). The "Six S" movement in nine countries of West Africa has mobilized social energy from half a million people through peasant federations in over two thousand communities to combat the hardships of the long dry season in the Sahel (LeComte 1986, 148–55; Harrison 1987, 279–84). Several remarkable examples of social energy mobilization in urban slums in Brazil and Peru are described by Durning (1989, 23–24). Around Lima, for example, people planted half a million trees, built 26 schools, 150 day-care centers, and 300 community kitchens, and trained hundreds of door-to-door health workers. Impressive cooperatives benefiting the rural poor in Guatemala and Mexico are documented in Gow et al. (1979, 153–70), Bunch (1982), and Rochin and Diaz Cisneros (1988).

12. For example, the director of the Bangladesh Rural Advancement Committee, F. A. Abed, and the general manager of the Aga Khan Rural Support Program in Pakistan, S. S. Khan, both found their experiences and explanations convergent with those offered here. The Ciamis project in Indonesia, which was able to mobilize vast amounts of social energy for soil conservation and agricultural development, constructed terraces at one-twelfth the cost of those built in a government program. Project savings were sufficient to improve 48,000 hectares of farmland for participating villagers. The analysis of this remarkable project by Craig and Poerbo (1988) draws on some of the same literature and many of the same ideas presented here. The head of ILO's program on participatory development, Anisur Rahman, writes (in Tilakaratna 1987, vii): "Any attempt at precise definitions of such phenomena as participation, self-reliance, democracy, creativity, self-awareness, etc. would reduce these notions to mechanical entities devoid of their living essence and dialectical movement (i.e., moving in association with one's opposite)." This is post-Newtonian thinking.

Material factors can present major constraints and also opportunities, but they are less determinant ultimately than the mental and moral factors to be explored.

Escaping Equilibrium and Avoiding Entropy through Positive-Sum Dynamics

Physics, with its prediction of ultimate entropy, outdoes economics as a dismal science, since economics at least points toward equilibrium as a normal condition and holds out the possibility of positive-sum growth. Within a closed system, the best that can be attained is a steady state, which is likely to run down over time. But in the social sciences, where closed systems are analytical artifacts, this expectation should not frame and limit inquiry and action. There are tendencies toward disorganization and loss of energy in the social universe no less than in the physical one, yet countervailing forces create complexity and new capacities. How else can the processes of biological and mental evolution be explained?[13] Societies as creations of ideas, ideals, and friendship, being open systems, are able to devise productive structures and catalyze new energies. These have indeed been gaining pace over time, though humans have at the same time created means to halt their forward movement through environmental degradation if not nuclear devastation.

In many ways the models of motion from Newton's mechanics are reasonable approximations for social science. There are many kinds of cycles observed, just as there are attractions and repulsions that give some predictability to human affairs. But these resemble the "strange attractors" of chaos theory more than the predictable orbits of celestial mechanics.[14] Any reduction of the social universe to the latter kinds of relationships, based on individualistic and materialistic presumptions about what is "real," leaves out the collective cooperative and mental possibilities that account for much of our forward if uneven movement through recorded history.

13. This is discussed in Jantsch (1980). Anthropologist Leslie White writes: "The second law of thermodynamics tells us that the cosmos as a whole is breaking down structurally and running down dynamically. Matter is becoming less organized and energy more uniformly diffused." With biological evolution, "the direction of cosmic process is reversed; matter becomes more highly organized and energy is more concentrated" (1958, 367). The same can result from social and cultural evolution.

14. When mapping what appear to be chaotic or random events, scientists find them often clustering around certain points (coordinates) in time and space, creating some pattern and stability within nonlinear processes (Gleick 1987, 121–53).

The big question for social scientists is how to become less dismal—how to construct explanations that account for and promote the constructive possibilities characterized here as emergences of social energy. Ideas, ideals, and friendship, as suggested already, share the characteristic of being positive-sum in their nature and dynamics. As such, they deserve more consideration as bases for social science work that can transcend equilibrium thinking and prevail over the downward pull of entropy.

Ideas

The first step is to take ideas more seriously as factors in human affairs. They are not simply reflections of material realities, even if they are, like Escher's hands, grounded to some extent in the realm of matter—starting with gray matter. Ideas shape material reality in many ways, just as they are influenced by the material realm. To follow the analysis of Eccles and Popper discussed in chapter 11, the realms of thinking and of things are highly interactive.[15]

Ideas, by not being diminished when given to someone else, defy the laws of thermodynamics. The value of certain ideas, if derived from a monopoly position (if they are "secrets"), can decline when they are shared, but their volume can expand indefinitely, since an idea can be disseminated endlessly without diminishing or losing the original. Many ideas in fact gain in value for being shared by others. What is language if not a set of ideas that is useful only to the extent that others attribute the same meanings to words?

The positive-sum potential of ideas is seen in mental creativity, which adds to the number of ways we can think and act by producing possibly infinite variations on known themes.[16] Not everything conceivable is attainable, but what is unthinkable cannot be purposefully done. The reality of self-fulfilling prophecies is widely recognized—

15. For a consideration of how mind could have evolved from matter by a Nobel laureate in biology who started out in theoretical physics, see Delbrück (1986). The dean emeritus of Princeton's School of Engineering and Applied Science and the manager of the Princeton Engineering Anomalies Research Laboratory report on experiments in which they found statistically significant influences of thoughts (consciousness) on physical events, replicated thousands of times in computer trials and other tests (Jahn and Dunne 1987). So the dualistic relation between mind and matter proposed by Descartes is questioned not only in quantum physics but now also in other sciences.

16. This is proposed in Hofstadter (1985, 232–59). A leading neurobiologist invokes chaos theory to explain the physiology of mental processes, suggesting that "chaos underlies the ability of the brain to respond flexibly to the outside world and to generate novel activity patterns, including those that are experienced as fresh ideas" (Freeman 1991, 78).

what people expect is more likely to happen. Most often this dynamic works in reverse—what is *not* regarded as possible thereby becomes impossible, at least through human agency. This principle should not be taken to a mentalist extreme of regarding all actions and outcomes as determined simply or only by subjective factors. Rather, it suggests how new ideas, or ones more widely shared, open up possibilities for action that did not exist before.

Robert Reich describes the contribution of ideas to positive-sum outcomes for American politics in terms similar to those we observed in Gal Oya, saying, "Citizens are motivated to act according to ideas about what is good for society. . . . such ideas determine how public problems are defined and understood. . . . government depends on such ideas for mobilizing public action" (1988, 122). He suggests that public participation in policy-making produces various benefits as problems and their solutions get redefined. Voluntary action can be generated, and individual preferences may be influenced, by considering publicly and together what is good for society (145). This was evident in Gal Oya. People may still choose to favor their own narrow or material interests. Public exchange of ideas does not guarantee that people will become more altruistic, but deliberation at least creates an opportunity for weighing and balancing personal and common interests, "to discover shared values about what is good for the community, and [to resolve] deeper conflicts among those understandings" (146): "The surfacing of conflicts, a less desired outcome of deliberation, is itself important and not necessarily destructive. Efforts to reconcile conflicting ideas, interests and perceptions should direct attention to the identification of overarching values [ideals] and a community of interest [friendship]. Such discoveries are not automatic but involve the exercise of leadership" (146).[17]

Exactly how ideas will be incorporated into social science theorizing is not yet clear, because they are intrinsically qualitative, having no weight, density, or mass and no standard units of measurement (Ostrom 1964, 85). By calling them "mental representations" as Gardner

17. Reich finds interest-group intermediation and expert calculation of net-benefit maximization less likely than deliberation to produce positive-sum outcomes in public policy-making: "The self-interested preferences of individuals as expressed through their market transactions do not reflect potential public ideas," he points out (1988, 146). Market approaches to public policy presume that quadrant I norms will produce social well-being. Mansbridge (1980) also supports deliberative processes, noting that people's perception of what their interests are can change as a result of both their own choices and everyone else's, particularly as they develop interest in the welfare of new friends (xii). This echoes the explanations proposed in chapter 12.

(1985) does, one can make ideas sound more solid and acceptable scientifically, but that resolves no problems of how to study and understand them. The best empirical solution may be to examine their effects as best we can. Not to consider them because their analysis must be inexact is short-sighted. That ideas are both ephemeral and eternal makes studying them difficult, but it also produces the positive-sum possibilities needed in the quest for development.[18]

Ideals

By definition, ideals are positive-sum, being ideas that have as their object some collective interest or common good. Even individualism and selfishness can be regarded as ideals if they are thought to contribute to some social good. Adam Smith argued two hundred years ago that, thanks to the "hidden hand," the greatest satisfaction of people's needs and wants will result from individuals' pursuing their self-interest through market transactions, because this would put scarce resources to desired uses most efficiently. The influence this principle has had as an idea should need no elaboration, but it also has had great force as an ideal, appealing especially, to be sure, to those in a good position to do well in a market economy given their resource endowments. Harnessing the motive power of self-interest can produce social benefits, but the greatest net benefits come from optimizing rather than from maximizing self-interested appeals.

Interest in collective motivation and outcomes should be restored by a both-and perspective, compensating for past neglect that resulted from either-or thinking emphasizing individualistic considerations. This does not mean that equal weight should always or necessarily be attached to both alternatives. Other-oriented thinking should be encouraged because it can create positive-sum payoffs from limited resources. At the same time, it is recognized that norms and institutions incorporating incentives of self-interest are likely to produce greater total benefits than those stressing only collective interests. Granted, individualistic approaches may benefit populations that have better resource endowments than the poor in developing countries. But where

18. Dawkins (1989, 199) has proposed that we understand ideas as *memes*, the intellectual equivalent of genes. Like genes, memes are transmitted intergenerationally with some mutations; those that survive produce more vigor in the species. A person is more likely to have some ideas still represented in the meme pool many generations hence than physical characteristics in the gene pool. The analogy is worth considering for the concreteness it gives to our understanding of ideas.

economic resources are particularly scarce, the consequences of this shortage may be ameliorated if people are prepared to derive some happiness from each other's well-being, thereby getting multiple satisfactions from the same limited material or social benefits.

Most ideals relate in some way to the norms of generosity and cooperation defining quadrant IV in figure 3. In the absence of such values, people will be drawn toward selfishness and individualistic action (quadrant I), with no interdependence of utility functions. They will be guided by self-interest maximization, though cooperation can evolve and persist even among egotists concerned only with their own welfare (Axelrod 1984) or under various specifiable conditions (Dawkins 1989). Persons strongly motivated by ideals may go beyond altruism to some self-sacrifice, for what is regarded as the greater good of their country, class, religion, or fellowman. Such commitment is the exception rather than the rule in any society. The example such persons set is significant, however, because it can move others who are inclined to be selfish and individualistic to act somewhat more generously and cooperatively out of inspiration or shame from others' evident positive-sum behavior.

Friendship

Friendship, as we have seen, can be defined as a positive-sum relation among people who value each other's well-being. At the micro level, this is a highly personalized phenomenon. The same principles and dynamics can operate at the macro level, in which case friendship is better understood as *solidarity*. I use the term friendship here because it is less abstract and has a noteworthy intellectual history going back to Aristotle, though solidarity can be nearly as concrete and has a very respectable social science pedigree deriving from Emile Durkheim's work (1933).

The concept of what is "social" has its origin in friendship, which Aristotle considered to be based on community and the bond that holds communities together (1953, 273, 258). Consistent with a both-and philosophy, Aristotle saw friendship as combining both love for others and self-love, and he spoke approvingly of self-love even as he exalted the loyalty and trust among friends that leads people to desire each other's good. True friends, he said, are drawn to one another in part because of their common attachment to ideals like goodness, truth, and beauty, and friends enjoy each other's company partly for the exchange of ideas. So in Aristotle's theory of friendship, one finds all three sources of positive-sum energization.[19]

"Society" depends on friendship inasmuch as others' welfare is bound up with our own. The popular concept of "enlightened self-interest" justifies taking others' needs into account because the possibility of postive-sum outcomes is recognized. Solidarity represents a generalized expression of friendship extended toward others we do not personally know, whether they are coreligionists, persons sharing the same class interest or the same ideology, like-minded hobbyists, or simply fellow human beings. Social scientists may find it more congenial to think and write in terms of solidarity, but the relationship derives basically from friendship as Aristotle discussed it long ago.

The theoretical argument that these three sources of social energy are related can be summarized as follows:

1. Ideas are positive-sum because *they can be given to others without being lost.* There is some loss of control over them, and there can be a loss of meaning and sharpness in the transmission. But passing an idea through someone else's mind, with its own experiences and insights, is likely to add to the richness of the idea rather than reduce it. And ideas shared with others usually acquire more significance and power. So the sharing of ideas usually has more pluses than minuses.

2. Ideals are positive-sum because *they direct thinking toward common interests* and away from purely selfish notions. They justify actions that serve interests beyond one's own by identifying one's own interests

19. "A friend wishes his friend's good for his friend's sake" and "his friend will wish for him the greatest goods possible for a human being," according to Aristotle (1953, 271); but at the same time in both-and terms he says, "everyone wishes his own good," and "all friendly feelings for others are extensions of a man's feelings for himself . . . a man is his own best friend. Therefore he ought to love himself best" (294, 300–302). Aristotle holds that "the self-love of most people is a bad thing" (301) because it is not tempered and ennobled by love for others. The motives of friendship with others for their own sake *and* for one's own benefit are hard to disentangle, since "good men are also useful to each other" (264). "Those who are friends for the sake of utility part as soon as the advantage ceases, because they were attracted not by each other but by the prospect of gain [quadrant III behavior]. . . . friendship in the primary and proper sense is between good men by virtue of their goodness, whereas the rest are friendships only by analogy" (264–65); "each party is good both absolutely and for his friend" (263). In explaining why persons with dissimilar personalities and traits may become friends, Aristotle cites Heraclitus: "Opposition unites," "from the different comes the fairest harmony," and "all things come from strife" (259). Bernstein (1983, xv, 164) comments that Habermas and Arendt among contemporary philosophers "presuppose and foster solidarity and friendship" in the tradition of Greek philosophers' concern with the primacy of friendship in ethics and politics.

with those of the community. The Golden Rule of "doing unto others as you would have others do unto you" is the most elementary of norms, but it can be elaborated in the more elegant philosophical theorizing.[20] Ideals, however insubstantial they may seem, can produce substantial results.

3. Friendship is positive-sum because *it leads people to mutually value each other's welfare,* expanding the satisfaction obtained from finite benefits to either person. At micro and macro levels, it multiplies the effort people will make to achieve common goals or to maintain the status quo in the face of adversity. Selfish motivation can accomplish the same, so it should not be disdained. But to reduce "human nature" to only selfish motivation and to exalt self-centered behavior ignores and even discourages the potentials for productivity and creativity that exist in quadrant IV even more than in quadrant III.

The power of ideas, ideals, and friendship (in operational terms, the interdependence of utility functions) was manifested in the initially inhospitable environment of Gal Oya, which appeared to be entropic or at least to operate at a very low-level equilibrium. Experience in that microcosm as well as in other situations reported above gives encouragement that transformative strategies can make quantum jumps toward improving the productivity, status, and security of disadvantaged people.

Pessimistic expectations based on equilibrium or entropic models are themselves part of the problem. It may be true that in the very long run we may not be able to escape the downward pull of entropy. But that perspective is for millennia. In the decades ahead we should challenge such assumptions, benefiting from whatever social energy can be mobilized. Ideas, ideals, and friendship can serve as booster rockets, propelling us like satellites around the earth, accelerating to elude gravitational forces. How long can we stay in orbit? Our thinking is what got us there in the first place, and it must be a continuing source of replenishable energy. Focused only on one's self, it is only intellec-

---

20. John Rawls, in *A Theory of Justice* (1971), tries to improve on classical utilitarian criteria of morality (those of Hume, Bentham, and J. S. Mill) by proposing an extension of traditional social-contract theories (Locke, Rousseau, Kant). He seeks to reconcile self-interest with principles that protect others' rights and well-being. Utilitarians, like Hobbes before them, aimed mostly at justifying quadrant III behavior. Going beyond such notions (moving toward quadrant IV), Rawls proposes that we see society as "a social union [where] the members of a community participate in each other's nature" and endorses the idea that human beings "as members of a social union seek the values of community" (565).

tual energy; the more it is directed toward others, the more it can produce social energy.[21]

The Energizing Effects of Expectations and Leadership

One of the best demonstrations of the connection between material and mental phenomena—indeed, of the inextricability of the objective and subjective realms—is the increasing attention being paid to expectations in the social and behavioral sciences. Though it has been a somewhat controversial subject in psychology, it is a recognized variable worthy of scientific research in that discipline. Moreover, it has been gaining scientific status with economists, who now examine "rational expectations" to improve upon their otherwise too mechanistic models. A whole range of objective and quantifiable variables—inflation, interest rates, wages, land prices, bond prices, foreign exchange rates, bulk ocean shipping—are more predictable when one considers what people think the future is likely to hold, or how they interpret the past.[22] Is the dollar value of a piece of real estate objective or subjective? The answer is, both. Does the supply of automobiles produced depend on material quantities of steel, labor, and other inputs? Yes, but equally on people's assessment of their needs, their psychological

21. Hirschman approves the proposal of the distinguished British economist Dennis Robinson that economists consider "love" within their theorizing, but he faults the idea that it should be regarded as a scarce resource, something to be economized. "We know instinctively that love or public spirit is not fixed or limited as may be the case for other factors of production. . . . these are resources whose supply may well increase rather than decrease through use; second, these resources do not remain intact if they stay unused; like the ability to speak a foreign language or to play the piano, these moral resources are likely to become depleted and to atrophy if not used. . . . once a social system, such as capitalism, convinces everyone that it can dispense with morality and public spirit, the universal pursuit of self-interest being all that is needed for satisfactory performance, the system will undermine its own viability, which is in fact premised on civic behavior and on the respect for certain norms to a far greater extent than capitalism's official ideology avows" (Hirschman 1985, 17). The expression "use it or lose it" appears to apply in the social realm. Social entropy is more attributable to the atrophy of values resulting from quadrant I expectations and justifications than to certain natural laws or dynamics.

22. When the *Social Science Index* was first published, only psychological articles were listed under the heading "Expectations." The heading "Expectations (Economic)" was introduced in 1980, with hundreds of articles thereafter so indexed. "Over the last several decades a number of economists [noted examples cited] have found that survey observations of expectational variables can be of assistance in the empirical modelling of economic behavior and econometric forecasting" (Lovell 1986, 110). The empirical reality of expectations gets somewhat masked, however, by economists's complex mathematical formulations that invoke heteroscedasticity, Lagrangian multipliers, Walrasian indeterminacy, and so forth.

desires, and also their beliefs about the future state of the economy—their positive or negative anticipations.

In the study of international relations, the so-called "realists" who sought to account for power in terms of its material bases and who emphasized national desires for domination are strongly challenged now by proponents of international regime theory. The latter emphasize the extent to which cooperation as well as shared norms structure international relations more than presumptions of conflict. Like economists, they are also highlighting the influence of expectations.[23] Having made the most of Cartesian or Newtonian models of decision making, some researchers of international relations are turning to cognitive analyses with a focus on perceptions, which in international dealings speak louder than so-called facts (Allison 1971). It makes a difference how "facts" are understood, all too often within different frames of reference (Lebow and Stein 1987). In the recent collapse of Soviet hegemony over Eastern Europe and, indeed, the challenges to Communist control within the Soviet Union, one sees what looked like formidable systems of physical power rapidly eroded by unseeable psychological currents. Ideas and volition (or lack thereof) can shape international relations more decisively than material forces, though the influences are best seen as complementary and often competing.

That expectations and perceptions are important is no news to psychologists. Some pathbreaking though controversial work has been done by Robert Rosenthal, who shook up his profession with experimental results showing that the effects of expectations showed up not just between people but between species. Randomly selected flatworms and rats performed better on "objective" tests when their trainers were told, as part of the experiment, that their subjects were more intelligent and should learn faster (Rosenthal and Halas 1962; Rosenthal and Lawson 1964). More significant for us is the Pygmalion effect found when teachers are told that some students were brighter than others—these students subsequently scored higher on tests (Rosenthal 1973).

These results have not always been replicated, but the number of times the effect has turned up has established it in the literature of psy-

23. Young emphasizes the role of institutions defined as "recognized patterns of practice around which expectations converge" (1980, 377), while Krasner defines *regimes* in international relations as sets of implicit or explicit principles, norms, rules, and decision-making procedures around which actors' expectations converge in a given area of international relations (1983, 1).

chology and social psychology.[24] There are of course "objective" limits to what expectations can produce; telling a teacher that an average student has an IQ of 150 might lead to more rapid learning for a time, but some deflation of expectations must be expected (Miller and Turnbull 1986). Yet there is no doubt that in interpersonal relations the consequences of expectations are widespread and significant. Whether teachers can turn typical students into geniuses is not as important as the frequency with which low expectations lead to underrealization of human potential.[25]

One of the most interesting findings in psychology is the well known Hawthorne effect, where it was found in a factory that assembly workers' productivity rose when they were told new lighting was being installed to improve visibility. Production continued to increase when the illumination was actually reduced, so long as workers thought the changes were being made for their benefit. More important than the actual level of lighting was the feeling that plant managers were concerned for the well-being of the workers. This relates to the observation in chapter 10 that people generally operate well below their potential, so there is the possibility of increasing output under various conditions with little or no added input.

Psychologists have tended to consider the Hawthorne effect an aberration, something to be avoided, the result of poor experimental design, even though field studies show Hawthorne effects to be more the rule rather than the exception (Sommer 1973). Rather than being discounted as disturbances of some otherwise "real" patterns in the Newtonian tradition, Hawthorne effects have an important place in post-Newtonian analyses. Peters and Waterman in their study of successful American businesses (1982) document repeatedly how

24. Students whom teachers considered smarter were given more feedback, a warmer socioemotional climate, more and more difficult material to study, and more time to respond. Curiously and sadly, when students they expected to do poorly did well, teachers tended to act resentful (Rosenthal, Baratz, and Hall 1974). Harris, Rosenthal, and Snodgrass (1979, 173) state: "That expectancy effects exist is no longer the issue. What is unknown at this point is exactly how they work. . . . The overall probability that there is no such thing as interpersonal expectancy is near zero." Miller and Turnbull (1986, 236) conclude: "There is relatively strong evidence that teacher expectancies influence student performance." They point to the possibility of self-disconfirming prophecies, where knowledge triggers preventive action, the converse of Robert Merton's idea of self-fulfilling prophecies.

25. In a symposium critiquing the review by Rosenthal and Rubin (1978), Elisha Babad comments that "while the more controversial Pygmalion studies showed that children may 'bloom' due to implanted positive expectations, we know that *most everyday influences of this type operate in the opposite direction—people being underestimated and performing below potential due to negative expectancies, preconceived notions and stereotypes*" (Rosenthal and Rubin 1978, 388; emphasis added).

significant material benefits result from personnel practices that show concern and respect for employees, that treat them as individuals, that encourage their creativity and initiative, that foster friendship and solidarity among workers and between workers and managers. The effects of such "soft" factors show up dramatically in "hard" corporate profits.[26]

This realization challenges the mechanistic images we have in our heads about people and organizations, that outputs are due to certain inputs which are processed to produce proportional results.[27] Inputs are certainly necessary, but they are not sufficient. Much more is required for the creation of results, since input-output relationships are neither as linear nor as predictable as presumed by machine models. Whether people will make appropriate and sufficient efforts to achieve objectives like greater factory production through conscientious labor input, more learning in pupils through dedicated teaching, or prevention of disease through maintaining public hygiene is a matter of probabilities. The range of alternative dispositions toward making greater and more innovative efforts is like that which was shown in figure 2. Many persons may tend toward inaction, as indicated with distribution (a). Individuals we regard as "active" people have distributions more like (c). To raise individual and collective performance, one wants to shift people's value dispositions from (a) or (b) toward (c), so that their probabilities for action are greater than those for inaction. These curves represent qualitative as well as quantitative dimensions of action, since how things are done can be as important as how much is done. This is why I referred above to conscientious labor and to dedicated teaching as well as to sanitation work that actually keeps things clean, not just going through the motions of cleaning.

When it comes to explaining the outcomes of individual and collective endeavor, the factor of *morale* is repeatedly referred to. The word in English for a dynamic that depresses outputs—demoralization—implies an opposite, *moralization*, that would make produc-

---

26. Peters and Austin (1985, xix) conclude paradoxically on the basis of further studies: "We have found that when it comes to achieving long-term success, soft is hard." Solidarity among employees has often been regarded as likely to reduce or constrain productivity. Examples of this are documented in the agricultural extension service in Kenya by Leonard (1977), but his conclusion is that "if properly involved in the decision-making processes governing their work, [extension] agents will use their informal ties to encourage one another to higher levels of productivity" (79). This was what we saw among institutional organizers in Gal Oya.

27. The "scientific management" school that gave rise to this machine model has been persuasively challenged by the "human relations" school (e.g., McGregor 1960). But mechanistic metaphors have lingered, especially in planning activities.

tion processes more efficient by returning more output for given inputs or expanding processes by attracting more and different inputs. The usual approach to boosting production is to offer incentives that appeal to individuals' self-interest, and there are good reasons to do this since we live in a both-and world. But appealing to sentiments and values that appreciate benefits not just for one's self but also for others can also expand production possibility frontiers. Expectations that we will benefit from an action can shift our dispositional curves to the right, toward (c) in figure 2, but so can the expectation that others will also benefit if the moral climate favors thinking altruistically.[28]

Offering material incentives may give more predictable results than appealing to moral values. But both approaches can mobilize effort, so if an either-or choice is forced, we forgo productive possibilities. Whether they are compatible depends to an important degree on how they are conceived and presented—whether people view them as mutually exclusive, either for the sake of "rigorous" analysis or out of ideological conviction. In this chapter we have been concerned with the social energy that arises when people value others' advancement alongside their own. This can transform productive systems, because persons acting socially and morally encourage one another to work cooperatively and altruistically rather than to act toward one another in zero- or negative-sum ways. Resources, effort, and imagination are elicited over and above what can be obtained by individual incentives. The expectation that others will act in a generous way makes such behavior more attractive to everyone.

How will this occur? The most general answer is leadership, speaking analytically as well as practically. The institutional organizers played this role initially but sought conscientious farmers to take it over as soon as possible. Some officials also played this role, acting beyond their formal roles of authority. A leader, we say, sets a good example, inspires confidence, maintains morale, and so on. But what does this mean?

A leader produces or at least articulates *ideas* that others can accept and respect. Good and clear ideas get people to expect that their con-

28. A remarkable study, *The Altruistic Personality*, by Samuel Oliner and Pearl Oliner (1988) deals with the extremes of altruism where the risk of ultimate self-sacrifice was very real, persons in Europe who risked their lives to save Jews from Nazi persecution and extermination. For such people, definite personality traits can be identified that account for the shape of their normative "distributions." But even then the standards of behavior set by others in their social circle had a definite influence on whether otherwise ordinary people would engage in acts of great heroism. For actions less self-sacrificial, the influence of *context* compared to that of *character* would probably be stronger, though no research has been done on this question like that undertaken by the Oliners.

tributions of resources and effort will help achieve certain worthwhile goals. The leader proposes or at least encourages ideas that produce consensus on goals, which has its own mobilizing effect. Leaders manipulate many other resources—funds, force, status, and the like—but it is often forgotten that they especially produce and exchange ideas, like "water has no color" (implying that politics should be kept out of irrigation management).

A leader can appeal to the more self-interested or to the more community-minded *ideals* of individuals. But in either case, the leader is tracing out a desired future, sometimes defending the status quo against one that is less desired but imminent. Values and value judgments are intrinsic to the exercise of leadership, since people are more willing to invest their efforts and hopes in enterprises that are not only comprehensible but that also promise a worthy outcome, one that people can justify to others, not just themselves. Leaders like Narangoda, Punchibanda, and Ratnayake in Gal Oya routinely supply justifications for action in an amalgam of ideas and ideals.

Further, leaders usually engage in the nurturing of *friendship,* either by nurturing personal friendship with and among supporters or by promoting some kind of group, community, or society that offers them identity, security, and other benefits. Some leaders are more concerned with accomplishing certain tasks than with maintaining bonds among group members.[29] But some kind of collective identification and solidarity is needed in any enterprise, and someone has to take the initiative to shape and preserve this. Almost by definition, this is expected of leaders.[30]

Analytically and operationally, leaders produce social energy by encouraging and promoting ideas, ideals, and friendship. They draw on otherwise unused or underutilized resources and talents, creating a positive-sum enterprise, at least for those who are part of it. It is important to have such resource mobilizing capabilities not just at the highest levels but really at all levels, as leaders at several levels reflect and replicate the ideas, ideals, and friendship expressed at the top.

29. One of the more instructive dichotomies in social science is the distinction between person-oriented, or supportive, leadership and task-oriented, or directive, leadership. This latter kind is sometimes called cognitive, in contrast to affective leadership. Leaders who are more concerned with interpersonal matters or simply better at dealing with them are involved more in friendship, while cognitively oriented leaders engage more with ideas (Stogdill 1974, 376–81; Yukl 1981, 147–69). Both types must deal with ideals, which serve as a bridge between ideas and friendship.

30. An excellent post-Newtonian analysis of leadership has been presented by Burns (1978); see also Bennis (1984), one of the main contributors to "the new business administration."

Dealings with other groups or states may be zero-sum or even negative-sum, but within the leader's group there must be positive-sum dynamics and perceptions that hold the enterprise together. We saw this in the Gal Oya case, where many persons—IOs, farmers, officials—helped those they associated with see mutually beneficial possibilities where before only redistribution of benefits was perceived as possible, or where people acted in ways that reduced the size of the pie for everyone.

Such generalized conceptions of social interaction and outcomes depend upon our ideas about the world and our assumptions about its nature and potentialities. This leads to some rethinking about how we can best think about the social universe. As I suggest in the concluding chapter, we can draw useful insights from the current intellectual revolution in the natural sciences, whose theories are gathered under the ominous label "chaos." A phenomenon like leadership, for example, is illuminated by the concept of "self-similarity" given by fractal geometry, which has contributed to the theories of chaos by highlighting the ubiquity and significance of repeating and replicating patterns. Studying the shapes of things as different as coastlines and snowflakes has revealed structures repeating similar forms at different scales. Leadership roles and functions, as noted above, are needed at several levels within any social enterprise, but when occurring on different scales they are similar, not identical. Rather than focus on individual units as discrete, chaos theory directs attention to "self-similar replication [of phenomena] across different levels of a system" (Hayles 1989, 313). Aggregate or atomistic analyses lose their effectiveness when called upon to deal with similar rather than identical versus different phenomena. In such a world, qualitative analysis becomes more important than straight quantitative studies, as structures assume greater salience in the construction of theory. Chaos concepts are thus more suited than classical ones for understanding and acting within an expanding and continually surprising universe, one that nevertheless exhibits some repeating principles of pattern and purpose.

14

Living and Learning
with Chaos

The social sciences have been accustomed to look for models in the most spectacular successes of the natural sciences. There is no harm in that, provided that it is not done in a spirit of slavish imitation. In economics, it has been common enough to admire Newtonian mechanics ... and to search for the economic equivalent of the laws of motion. But this is not the only model for a science, and it seems, indeed, not to be the right one for our purposes ... we can [now] see the role in science of qualitative structure and the power of qualitative as well as quantitative explanation.

Herbert A. Simon, "Rational Decision-Making in
Business Organizations"

The statement above by a Nobel Prize–winning economist, which I found while reviewing the literature on "rational expectations" theory, explicitly addressed some of the theoretical questions being explored in this book. That this article published twenty years ago in the leading journal of his discipline had had little impact on it or on the other disciplines he has contributed to—public administration, political science, and organization theory—was disappointing, but it appears to be another instance where ideas at variance with prevailing paradigms are not taken seriously (Kuhn 1962). By now, however, given the growing disenchantment with Newtonian modes of analysis and explanation over the past two decades, social scientists should be more open to such new modes of thinking.

The social sciences have long borrowed from the natural sciences, but as Simon says, a model based on "the equivalent of the laws of motion" no longer seems the best available. With a both-and orientation, one does not abandon previous theories and ways of thinking, but only their reductionist claims to a monopoly of explanatory power. The idea of social relativity means that the coexistence of divergent "truths" can be accepted so long as each can be validated within some intelligible *frame of reference,* some set of coherent concepts and premises, and most of all, some compelling purpose that holds these together.

Simon was remarkably prescient to identify the emerging alternative as involving analysis that is more qualitative and concerned with structure, rather than quantitative and preoccupied with aggregates. These are central features of chaos theory, the most important new scientific paradigm. Unfortunately, its designation seems to have been chosen more for its shock value than for descriptive accuracy, since within chaos there are universal structures despite disorder and recursive symmetries between levels (Hayles 1990). The order discerned, however, is nonlinear and unconventional. Simon correctly says there is no harm in looking into the realms of natural science for social science, so long as there is "no slavish imitation," no conflation of social science with physical science. In the preceding chapters, concepts from relativity and quantum theory have been discussed to shed light on social realities and dynamics. Here we will consider chaos theory, the third wave of scientific revolution in this century.[1]

It has been useful to imagine that we live in a world governed by Newtonian regularity and predictability. The order conjured up by celestial mechanics is seen as some kind of ideal, something we fall short of owing to our own deficiencies, much as we fail to draw perfectly straight lines and round circles or to construct flat planes without mechanical assistance. Having been taught Euclidian geometry, we tend to think of these forms as "true." But there can be other, nonlinear forms of geometry, like the fractal geometry developed by Benoit Mandelbrot, which "mirrors a universe that is rough, not rounded, scabrous, not smooth."[2]

1. On the three major "scientific revolutions" in twentieth-century physics, see the epigraph to chapter 10. Obviously, as a social scientist I cannot assess the validity of all ideas in this new field of theory, particularly its mathematical elements such as fractal geometry. I rely heavily on the widely read and acclaimed account of this new science by James Gleick (1987). His book shows how ideas develop and points out the crucial role of friendship in this development (e.g., 243–72).

2. "It is a geometry of the pitted, pocked and broken up, the twisted, tangled and intertwined," adds Gleick (1987, 94). Mandelbrot undertook to show "that such odd shapes carry meaning. The pits and tangles are more than blemishes distorting the shapes of Euclidian geometry. They are often the keys to the essence of a thing." The *Economist* reports that astronomers are also finding some relevance in this new science—as the solar system increasingly appears to be a shaky mechanism bearing the scars of catastrophes that have given it its shape (July 22, 1989, 85). According to the *The New York Times,*

Once linearity is no longer assumed to be the normal state of things, one can see that the irregular is much more common than the regular. An idealized sense of shape has led us to regard the former as an aberration or as accidental. Mandelbrot, however, taking the evidence of his senses seriously and disposed to both analysis and synthesis, went beyond description and discerned regularities and patterns within apparent chaos, finding order within disorder. His mathematical concepts have practical application in chemistry, metallurgy, seismology, even nuclear reactor safety, according to Gleick (1987, 83–118). They can also apply in the social sciences, as suggested at the end of chapter 13, since the fractal concept of self-replication throws light on the dynamics of leadership within organizations.

Chaos Theory

The study of chaos in the natural sciences can be described as that of deterministic disorder. It involves nonlinear, nonmechanistic relations, with wholes given as much attention as their parts. Chaos theory is concerned with emergent properties that can be explained but often not predicted (at least not with much precision) by the interaction of constituent elements and processes. Indeed, it is best understood as a science of *process* rather than one of stable states and conditions. In Heraclitean terms, it is a science of *becoming* more than of being. Simple processes can produce outcomes of great complexity, and small changes can push stable systems into qualitatively different kinds and levels of performance.

Harlan Cleveland, a distinguished political scientist with many years of experience in public affairs, writes of this "new science" this way:

> The word *chaos* has long been in my vocabulary, and yours. It defined what an educated person, educated in linear thinking by the logic (and the prestige) of the scientific method, couldn't hope to understand. *Chaos* described that inferno of unfathomable complexity lying just beyond that flat world of rational thinking and empirical evidence.
>
> But chaos now turns out to have its own patterns and probabilities. What had always seemed random—outbreaks of measles in New York

---

ecologists are moving away from their previous assumption that the "normal" condition of nature is a state of equilibrium (July 31, 1990, C1).

City, the fluctuations of Canada's lynx population, the ups and downs of cotton prices—now shows a complicated rhythm if our data base stretches the statistical string back far enough.

The same is true, apparently, of physical things and abstract numbers: They look random only when you look at them piece by piece, in the reductionist tradition of sciences that chop knowledge up into "manageable" chunks.[3]

Curiously, nonlinear processes when disturbed are more likely to return to their original or approximate starting points than are linear ones, which through cumulative divergence end up destroying equilibrium rather than restoring it. An understanding of chaotic dynamics can clarify, for example, why the generalization so often made in social science that "the rich get richer, and the poor get poorer" does not lead to an absolute concentration of wealth. If this proposition were literally true, the world would long ago have settled into an extreme and static division between haves and have-nots. Instead, there is a relatively stable pattern of income distribution across nations and over time, with the statistical mode always somewhere between the richest and the poorest. The dynamic of income concentration exists, but it operates within a system of chaotic—nonlinear—patterns of causation that check its expansion.

Paradoxically, the physical universe could not persist if it functioned according to laws of strict linearity. Such laws would lead to true chaos, to collapse into disorder. What is called the science of chaos explains dynamic equilibrium maintained through many recurrent patterns that are not exact replications.[4] Although these patterns lead to some change over time, they also produce a degree of concurrent stability. Physicists now recognize that our universe is too complicated to

3. This was written for the *Minneapolis Star Tribune*, November 6, 1988, 31A, and I serendipitously found a copy on a plane while returning to Cornell from a meeting in Chicago. The element of chance in my construction of this understanding of post-Newtonian social science was itself an example of "chaotic" influences.

4. Crone (1989) says that explaining change in social welfare policies and outcomes in Southeast Asian nations is "notoriously difficult." Linear explanations proposed by economists in terms of economic growth "cannot adequately explain the unevenness of social welfare change . . . [which] is very sensitive to initial conditions, and not very predictable or stable." The study of chaos, Crone suggests, offers "a fertile metaphor for political analysis. It sets out to analyze systems that are dynamic, non-linear, complex and unstable, such as meteorology or ecology, where standard scientific approaches capture only part of a system's dynamic. . . . Even within a system of phenomena that never repeats itself, and so cannot be predicted, there are similarities, patterns or clusters of outcomes when viewed at a macro level. For political analysis, this suggests that one look for the islands of recurrent behavior (or basins of attraction, where the system repeats itself over time, to use one chaos term), where usual approaches such as statistical analysis would show little generality"(2–4).

be replicated by linear rules and concepts. Trends can be identified in nature, but they can also vanish suddenly when crossing the border between calm and catastrophe. That this border is unmarked and fluctuating permits turbulence and coherence to coexist (Gleick 1987, 94, 56, 238). Should we expect social reality to be any less complicated and varying?

An understanding of chaos theory helps one appreciate the principle of relativity by stressing the importance of scale. Such supposedly fixed qualities as size and duration, even dimensions, depend on things like distance, magnification, and time horizon.[5] The new science also helps one understand that asymmetry is "normal." That biological evolution produced much symmetry (two eyes, two arms, two lungs, etc.) encouraged an idealization of matched pairs and shapes, but nonidealized observation of the natural universe suggests that asymmetry is, indeed, the more fundamental reality.[6]

Linear and mechanical systems, which exhibit strict proportionality and interchangeable parts, are usually artificial constructs, creations of the mind. With nonlinear and organic systems, which are more common and more "natural," one cannot assume that wholes are necessarily simply the sums of their parts or that one part can be freely substituted for another.[7] Social systems are neither linear nor mechanical, which is why Simon concluded that Newtonian models may not be appropriate for our purposes, and why qualitative approaches are likely to be more illuminating. Self-organizing systems that "wind up" rather than "wind down" are being increasingly studied by scientists,

---

5. See Gleick (1987, 103–10, 159–62), who suggests that an automobile's appearance varies according to whether it is viewed from a distance measured in meters or in microns or in kilometers. The first perspective, which we think of as "normal," shows a three-dimensional shape with a smooth surface. The second vantage point reduces the three dimensions to essentially two, though it reveals an immensely bumpy, irregular surface. Viewed from far above the earth, the car is a one-dimensional point with no apparent shape, mass, or surface features. An object's "real" characteristics thus depend on one's perspective and on the uses one would make of them.

6. Gleick (1988) quotes a researcher studying proteins at Duke Medical Center: "People expected they would all be stacked in neat little bundles. The models all tended to look kind of mechanical and regular, like polymer chemistry. The real structures are much more biological than that. Things are asymmetrical, and when they repeat they don't repeat exactly." This sounds likes social phenomena. Medical research is discovering that heart disease is signaled by regular rather than irregular heartbeats. Also, there is self-similarity of variation in heartbeats at different time scales (Goldberger, Rigney, and West 1990).

7. Biologists are discovering how genes, initially thought to operate in deterministic, mechanistic ways, affect growth within organisms in a much more complex, interactive manner (Beardsley 1991). Genetic elements perform more in ensemble than in solo fashion, affecting one another in various combinations.

who find them all the more interesting because they violate the assumptions of Newton's mechanics and classical thermodynamics.[8]

In nature, many outcomes are "sensitively dependent on initial conditions," so that small differences in cause can produce large variations in effects (Gleick 1987, 7–8, 144–53, 252–55). The same is true for social systems, where, for example, a shift of some share of income from consumption to savings can produce *either* economic growth or recession. Both possible outcomes coexist in the same cause. If people's views about the future are optimistic, their savings will be channeled into investments that keep output expanding. On the other hand, pessimistic forecasts will lead to a reduction in orders when consumer expenditure declines, causing a downward spiral of economic activity.

This effect is another example of how subjective interpretations can shape objective outcomes, with zero-sum transactions producing either positive-sum *or* negative-sum consequences. The system of dynamics is not linear, however, because economic booms peak and decline, just as recessions bottom out and are followed by recovery. Chaos models account for such patterns better than mechanistic ones.

Small changes in *rate* may have little effect on the stability of a system over some indeterminate range but may then suddenly destabilize it. The system may become irretrievably unstable if it contains no "strange attractors," as researchers of chaos call the loci of pattern appearing within otherwise random-seeming activity. More often, systems moving from one equilibrium eventually attain another (Gleick 1987, 23–31, 59–80).

Models developed by ecologists and epidemiologists to trace the dynamics of fish populations or the spread of disease can give clues to economists mapping fluctuations in the market price of commodities, for example. These contain both orderly and random influences that interact cumulatively, usually accelerating but sometimes canceling each other's effects. March and Olsen (1984, 745) suggest: "The policy path of two political systems [even] with identical underlying

8. See particularly Jantsch (1980), Prigogine (1980), and Prigogine and Stengers (1984), who marshal evidence from the natural sciences on concepts like "order through fluctuation" and "discontinuous effects of continuous causes." Conveney and Highfield (1990, 34) suggest: "The real world has little to do with the dreary state of equilibrium." They also present how, contrary to our usual understanding of the second law of thermodynamics, which holds that over time systems always run downhill to entropy, this law is actually compatible with the upward movement of evolution and energization. Entropy itself is probabilistic rather than certain.

political conditions will be radically different simply because of the way in which (possibly small) perturbations shift the focus of political pressure." Understanding this should restrict our reliance on mathematical models that presume linearity, since we live in a predominantly nonlinear world.[9]

What social scientists can derive from current rethinking in the natural sciences, at least at present, is not a full-blown theory or a set of methods but instead a more appropriate worldview. Fractal geometry may someday be applied to the study of voting patterns or the mapping of project implementation. But that is not necessary for the "chaos" perspective to be valuable. The new science cautions against mechanistic or reductionist modeling of social dynamics. Such analysis can and should be done, but it should be accompanied by many explicit qualifications and should be regarded as tactical exploration rather than as producing strategic conclusions.

"Chaos" as a designation should not be taken literally. The word suggests so much randomness that no theorizing or purposive effort would be meaningful. But since we are dealing with "deterministic disorder," where recurrent patterns can be attributed to strange attractors (whatever they are), there is no reason to retreat from all scientific endeavors. Systematic knowledge and effective action are possible if we learn to "go with the flow," trying to hit moving targets and influence evolving relationships. Hayles (1990) points out that chaos theory differs from deconstructionism in that it does not repudiate classical scientific interpretations but rather builds on them. This represents a both-and intellectual orientation.

The new science does not offer any explanations for the ultimate sources of dynamism and change in social relations. Where do ideas and idealism come from? How does an inclination toward cooperation or altruism originate? These questions reflect initial conditions to which social systems, large and small, are very sensitive. Normative orientations depend on individual and collective choices that are partly but not wholly shaped by external influences. They arise both from personal dispositions and from situational contexts. Just as market price patterns contain elements of both order and randomness,

---

9. A pioneer of chaos theory, Robert May, has concluded: "Not only in research, but also in the everyday world of politics and economics, we would all be better off if more people realized that simple nonlinear systems do not necessarily possess simple dynamical properties" (Gleick 1987, 80). In a seminal 1976 article in *Nature*, May suggested that breaking away from linear mathematical assumptions "would change the way people thought about everything from the theory of business cycles to the propagation of rumors."

prevailing norms are the result of both societal and individual factors. The two interact and shape one another so that outcomes are both caused and indeterminate. Personal purposes get transformed into public parameters, at the same time as social institutions and culturally shared values codetermine individual preferences. The micro becomes macro, and vice versa. Social energy from its individual sources becomes a collective force while, concurrently, people are energized or immobilized by what goes on around them.

If the social world were truly random and we confronted literal chaos, there would be no point in acting purposefully, in formulating ideas, in pursuing ideals, or in maintaining friendships. In fact, the possibilities for effective action are usually sufficient to justify planning and initiative, though with the knowledge that one can never be fully in control of the course of events, which are inherently dynamic and nonlinear. The bottoms of the rivers we try to cross have little bedrock, though there is usually enough solid matter to construct a crossing.

The uncertainty that attends all individual and collective action is not a matter just of lacking enough knowledge about all the actors and all their goals, strategies, and resources, as Laplace wanted us believe. Even in the natural realm where elements have no volition or autonomous purposes, we now know we should *expect* unpredictable system disequilibria and unexplained shifts in rates and outcomes of activity. Should this be any less true in social affairs, where people's interests and ideas come into play?[10]

In fact, living in a less deterministic world has some advantages, because it gives us more "degrees of freedom" than in a clockwork Newtonian universe. We need to plan our initiatives with the knowledge that disorder exists, but not without some pattern and structure. The importance attached to scale in the study of chaos is crucial, because it breaks from the reductionist presumptions of conventional science, natural and social, that the whole is simply made up of its parts. Instead, wholes should be seen in relation to their parts, which are themselves wholes to be viewed in relation to their own parts, and so on, practically indefinitely (Koestler 1967, 45–58).

10. Harlan Cleveland, cited above, says: "The chaos enthusiasts are discovering that the world of numbers and physical things is not so different from the untidy world of human affairs. We who study society have known all along, haven't we, that systems are never regular, never (despite 'scientific' polling) reliably predictable? Yet the chaotic confusions of politics and administration do yield to analysis, do prove to be what one chaos analysis calls 'randomness with its own underlying order.' "

Emergent properties at different levels, identified through observation and synthesis, are certainly as significant as the ever more elementary, yet often surprisingly complex, mechanisms that can be described through measurement and analysis. Unfortunately, emergent properties have received little attention in most disciplines because they appear less "scientific," deriving from intellectual activities as imprecise and unreplicable as speculation and synthesis.[11]

Recasting Social Science

Post-Newtonian social science treats emergent properties with more respect, taking cues from scientists who seek explanations in upward rather than simply downward examinations of hierarchical organization in nature (Gould 1985). We should consider how to increase the chances of achieving desired possibilities rather than to concentrate on estimating abstract probabilities based on actors' fixed preferences, choices, and means. A social science that is "action science" seeks to change the odds rather than simply reckon them.[12]

This requires a reorientation of thinking that can be compared to the game of chess. When we started working in Gal Oya, the ARTI-Cornell *task* was viewed in clear-cut terms. Over time we achieved a more ambiguous but realistic understanding that permitted us to move more effectively toward our project goals. The change in our way of thinking resembled what computer scientists now refer to as "fuzzy sets," categories that are without rigid boundaries or definitions.

The *board* was imagined initially as two-dimensional, with the squares at first seen as distinct blacks and whites. Gradually we came to see the game as having multiple dimensions, more like three-dimensional tic-tac-toe, and as changing, adding the fourth dimension

11. Mitch Feigenbaum, a major contributor to chaos theory, says: "There's a fundamental presumption in physics [and I would add, in social science] that the way you understand the world is that you keep isolating its ingredients until you understand the stuff that you think is truly fundamental. . . . The assumption is that there are a small number of principles that you can discern by looking at things in their pure state—this is the true analytic notion . . . [but] to understand you have to change gears. . . . It requires a different way of thinking about the problem . . . writing down partial differential equations is not to have done the [real] work on the problem" (Gleick 1987, 185–87).

12. In their analysis of "action science," Argyris, Putnam, and Smith (1981, 349–449) emphasize the importance of frames of reference, as discussed in chapter 11. They say one should expect mixed results and recognize the value of ambiguity and paradox. They too have found conceptual impetus and support in relativity theory and quantum physics (395).

of time. The board appeared more in alternating shades of gray that exhibited some, but rather imperfect, regularity.

The *teams* started out as us versus them—ARTI-Cornell against the Irrigation Department, which was resisting the introduction of farmer organization. After all, chess games are zero-sum, with a winner and a loser. But adopting an adversarial stance itself *reduced the prospects for success*. By nurturing common ideas and ideals and also friendship, we forged a single large team with us, the engineers and the farmers together, so that the game became positive-sum with gains seen as going to everyone.

All *players* were initially categorized as if they were kings, queens, rooks, bishops, knights, or pawns, with predictions of movements made accordingly. But we had to learn to treat this hierarchical classification only as a statement of probabilities, since sometimes persons classed as pawns could move several spaces at a time, and some bishops or rooks never moved more than one. Moreover, we found that everyone had some potential for what Johnson and Morse (1988, 2) call a "knight's move," which altered the probable course of events by moving in an unexpected, nonlinear way.

The *rules* we came to see as not clear-cut and as always changing according to who was in the game at that time, and what players thought the rules were or should be. Expected favorable outcomes could be undermined by certain players or by external events, and negative ones could be reversed by persistent efforts or by chance. It was hard to know who was ahead and who was behind, so the main thing was whether you were still in the game or had been sidelined. Given the free substitution of players on any team, the game was essentially continuous, and the teams were continually different.

Such an understanding of our work as a kind of "chaotic game" paradoxically gave us more influence, if not control, over its outcome. With a learning process approach, we did not expect to impose a linear logic on a nonlinear world. Blueprints would not succeed because the situation was inherently uncertain, and relations of cause and effect were probabilistic and contingent. Coming to think of people as quantum "probability distributions," as sketched in figure 2, rather than as self-contained Newtonian objects with given vectors and momentum was a major step forward.

A mechanistic, reductionistic way of thinking provides rationale for avoiding personal responsibility. As things stood in Gal Oya at the outset, it was unlikely we would succeed with our assignment, but this could be blamed on many things—procedures, personalities, prevailing values, economic resource constraints. Social scientists can become comfortably absorbed in identifying and analyzing obstacles beyond their control or anyone else's. But are these all beyond our influence? In social relations, we can go beyond recognizing emergent properties to actually creating them—in the interplay of interests and ideas, in the brokering of individual and collective perceptions, in fostering or drawing on friendship bonds, in weaving material and nonmaterial factors into new combinations, in capitalizing on potentials for imagination and energy that exist in persons and in groups.

A common denominator in promoting such transformations in Gal Oya was to avoid dichotomous and static thinking. There are, of course, significant differences in the social as well as the material world that deserve attention and classification. Careful and extensive analysis is essential for effective action. But instead of taking a binary either-or view of the world, a binocular both-and perspective can help us—at the same time distinctions are made—to try to transcend them.

Philosophers since Heracleitus have been intrigued by the complementary relation between apparent oppositions. In nature, many opposites are essential to producing valued effects, beginning with the union of male and female gametes. (Though these are treated as opposites, they actually differ only in a small but significant way.) Electricity is produced by the combination of positive and negative charges. Moving our limbs involves coordination of contrary sets of muscles, flexors and extensors, which are themselves controlled by a balance of nervous impulses that stimulate and inhibit. The functioning of the heart, lungs, and other organs is regulated by two counterpoised parts of the autonomic nervous system, sympathetic and parasympathetic. These various opposites can and should be understood as parts, being respectively real and important, but also as constituting wholes that are more than the sum of their parts.

Sometimes opposites are related in dialectical processes, whether Hegelian or Marxian, with thesis challenged by antithesis leading to some synthesis. But this is only one possibility. Opposites can eliminate each other, as when acid and basic chemicals neutralize each other. They can persist without fusion, like earth and sky. They can arrive at a compromise, as hot and cold weather fronts eventually equalize their temperatures and air pressures, being always superseded by new disequilibrium conditions.

Such examples make concrete the need to think about either-or alternatives in both-and terms. When working in Gal Oya, it was important to remain alert to differences, especially antagonistic ones, but not to regard them as decisive. Complementary differences warranted specific attention too, keeping in view both the distinctions and the larger whole they represented. Such a perspective helped us see more flux and less fixity in the configurations around us. Just as thermo-electric energy can be generated by putting two dissimilar metals to-gether to capture the current passing between their surfaces, so one can capitalize on the tension and energy between social oppositions. Ap-preciating this takes analysis and synthesis into the realm of philoso-phy and metaphysics. But it grew out of trying to understand and improve upon something as practical as introducing participatory ir-rigation management.

A Probabilistic World

Post-Newtonian social science starts by viewing the social world more in the way physicists now regard the world of quantum mechan-ics—as inherently and inescapably uncertain.[13] It is best perceived in terms of probabilities, and not always or necessarily with the bell-shaped curve of a normal distribution as its model. Probability distri-butions come in an infinite number of shapes and are themselves subject to change as they interact with the outcomes of other proba-bilities. Many disciplines, with the illumination of chaos theory, are moving away from the mechanistic worldview that previously shaped their theorizing. Even ecologists are less satisfied with their previous presumption that "nature" exists fundamentally in equilibrium (Levin 1989). Although many phenomena and relationships have extremely high probabilities, all are contingent. Causal relationships, or at least influences, can be identified or inferred. But consistent with a both-and worldview, varying degrees of determinism coexist with a persis-tent openness of outcomes.

Phenomena and relationships range along a continuum from near certainty to near impossibility, with few normal distributions because there are many more possibilities than probabilities. The tension be-tween probabilities and possibilities is ever present in the social uni-verse since, as noted already, the desirable possibilities are more numerous than are present probabilities. Especially with regard to

---

13. The implications of this view are explored in a volume edited by Becker (1991). See especially the essay "Quantum Theory and Political Theory" by Slaton (1991), contrasting this approach with the analysis of physics and politics in the Newtonian tradition.

development, the most probable outcomes are seldom the most desirable ones. It is fortunate that we live in a probabilistic world, because so long as the odds are not fixed, we can try, if we choose, to make what is desirable and possible more probable. We cannot simply remake the world as we would like it to be, but neither do we need to accept it as it is.[14]

Quantum physicists have a highly original concept of physical reality, conceiving of subatomic phenomena as existing not in themselves but as "quantum wave functions"—qwiffs for short (Wolf 1981, 170–75, 184–91). This concept is generalizable to the social universe, regarding all events and outcomes as probability distributions rather than as things. This extends to people, since their thoughts and actions are never completely fixed or certain. Before the fact (ex ante), particular events or outcomes exists as probabilities, somewhere between .9999 and .0001. Even highly probable things may not happen, and even those that are quite unlikely may nevertheless occur so long as they are possible. After the fact (ex post), an outcome's probability is either 1.0 or 0, since it occurred or did not occur. In quantum terms, the wave function has collapsed. But this condition is transient; in this Heraclitean world, there is no being, only becoming. As soon as one unknown outcome becomes known, like two-faced Janus it looks ahead into the unknown. Just as soon as any uncertainty is resolved about dynamic phenomena—whether persons, groups, or governments—a whole new set of uncertainties lies ahead. The passage of "time's arrow," as it is sometimes called (Coveney and Highfield 1990), continuously converts ex ante probabilities and possibilities into ex post events and outcomes. These are succeeded, even superseded, by still more ex ante probabilities and possibilities.

Any event or outcome, or any person, can be described before the fact as a distribution of probabilities. Indeed, its ontological status may be regarded as that of a frequency distribution. Though we may not know what it is, any event, outcome, or person will likely have some modal probability (such as 0.9, 0.1, or 0.5). Yet the tails of its distribution extend, perhaps imperceptibly, in two directions toward certainty and impossibility. A distribution with a high modal probability can approach certainty, of course, and one with a very low probability will tend toward impossibility.

---

14. In his book on the Kenyan bureaucracy, Leonard (1991) analyzes the biographies and strategies of "positive deviants," persons who "*use* their environments productively rather than be trapped by them" (11–12).

The significance of this view for social science and for development initiatives can be seen from another river metaphor, one that became clear after we had succeeded in getting the Irrigation Department to accept farmer participation. The ID's resistance to giving farmers a voice in irrigation management was like a logjam that ARTI and Cornell had to break up. How could this be done?

In a logjam, only rarely can one free the mass of logs by dislodging the foremost one. It is locked in by all the pressures from those behind or below it. One has most probability of success by finding strategically placed logs near the front that are movable and that will change the pressure on other logs. In the Gal Oya case, the director of irrigation was publicly unmovable, and his deputy director in Ampare was blocking our efforts behind the scenes. Fortunately, the chief irrigation engineer in Ampare was willing to make some moves toward working with farmers and to bring some of his junior engineers along to meetings with the new organizations (chap. 3). As improvements became evident, we got some favorable movement from other engineers within the Irrigation Department and some approval from other officials.

When Senthinathan became deputy director after September 1981, the pressures began shifting within the ID. Senthi was more supportive of our program, partly out of conviction but also because of farmers' accomplishments and because of the friendly approach of IOs. By January 1982 the director himself was willing to endorse the program (chap. 4). The logjam was partly broken, though that did not yet ensure a smooth flow downstream. Some smaller jams continued to form, for example, when farmers tried to get contracts to do maintenance work that technical assistants wanted for themselves (chap. 7). But our program was now in the mainstream, and the basic forces were favorable to its progress.

If each actor is viewed as a probability distribution, most of the engineers were skewed against farmer participation, though under certain conditions most could accept it. If ID staff thought their director was against farmer participation, of course, their orientations would be distinctly unfavorable. But when the chief irrigation engineer in Ampare was willing to be publicly cooperative, this change affected the attitudes and thinking of other engineers, and some of them got on board. He may have shifted his probability distribution because he saw a need to win farmer cooperation to carry out his assigned duties, or it may have been skewed in a favorable direction already because he believed in the participatory approach. What mattered were not his reasons so much as his actions. When the new deputy director started giving visible support, this in turn helped accelerate farmer initiative and responsibility, and that helped dislodge the director, to make a breakthrough against the bureaucratic barriers we faced.

Reflecting on how this logjam was broken made it more evident how interdependent are people's attitudes, values, and behavior. Our thoughts and actions are shaped only partly by our own rationality and decisions. They are influenced as much or more by other people, especially those we like and respect, who exercise authority over us, or whom we regard as more knowledgeable than ourselves. This contributes to the probabilistic nature of the social universe, because *which* persons will have an opportunity to influence others' behavior is to some extent "chaotic" (March and Olsen 1974, 745).

People like to think their choices are entirely their own. But the information and even the values we draw on in our decision making are derived mostly from others. This makes our respective probability distributions and normative orientations interdependent, linked in ways we do not fully comprehend.[15] As suggested in chapter 12, what others think and do affects what we think and do, and vice versa. This interdependence was evident throughout the unfolding of the farmer organization effort in Gal Oya. Although the general situation was chaotic, when IOs or farmers or certain officials articulated and acted on values that were more altruistic and cooperative than those prevailing, this orientation affected the frequency distributions of others' behavior. Whether people would regard their respective utility functions as interdependent was itself interdependent.

If the physical universe is probabilistic—and even the deductive realm of mathematics has this aspect (Chaitin 1988)—so must be the world of people individually and collectively. This calls for post-Newtonian thinking that moves from deterministic and mechanistic models of man to probabilistic assumptions in all of the social sciences.

A Phenomenological World

As someone whose orientation was more toward practice than theory, I have not steeped myself in philosophy, though I was always in-

15. Support for this view can be found in the field of demography, where Cleland and Wilson (1988) argue that the importance of culturally homogeneous populations to the rapid spread of population control programs implies that the fundamental forces reducing fertility operate at the societal level. Their analysis indicates that the best model for predicting fertility decline is not the microeconomic interpretation of individual or family decision making but that for diffusion of innovation.

trigued by its issues. The problems of explanation that we confronted after starting to work in Gal Oya prompted enough curiosity about broader questions of ontology and epistemology that I fortuitously acquired books like Bernstein (1983) and Baynes, Bohman, and McCarthy (1987) to read during travel time, though most of contemporary analytical philosophy was too abstract and self-referential to help with the puzzles of Gal Oya. A book manuscript on participatory development sent by a friend drew my attention to the relevance for Third World development practice of some of the ideas of Maurice Merleau-Ponty and Jürgen Habermas.[16] Previous reading on phenomenology has struck me as too technical to help with real-world problems, but now with experience in Gal Oya, the perspectives that phenomenologists proposed acquired more meaning.

There are long-standing, unresolved debates in philosophy over the nature of reality, commonly cast in terms of the relation between "subject" and "object," the perceiver and the perceived. Those who give priority to the latter consider themselves realists, while those who consider subjective processes the principal source of reality are generally classified as idealists. Phenomenology rejects the dualism this classification implies and seeks to resolve it with a systematic analysis and synthesis of how we acquire and use knowledge.

Descartes's classical resolution of the philosophic problem of understanding what is real ("I think, thereafter I am") reified the mind-body distinction, the ultimate in either-or formulations. It restricted "scientific" conclusions to what could be defended as absolutely certain, a criterion rejected in quantum physics. Descartes's reasoning proved nearly as influential for social science inquiry as Newton's contributions were for the natural sciences. Both were intellectual giants who contributed to broad advances in human comprehension. But to the extent that the world is now better understood as probabilistic and uncertain, their premises about what is real and how we can best learn about reality need to be superseded.

The philosophical school that best formulates post-Newtonian (or post-Cartesian) thought is broadly known as phenomenology. It offers a perspective increasingly appreciated in social science (Lincoln and Guba 1985). Even a cursory overview here, not using the technical

16. The author, A. Z. M. Obaidullah Khan, had helped Akhter Hameed Khan launch the Comilla experiment with small farmer cooperatives in East Pakistan (now Bangladesh) in the 1950s and later served as secretary and then minister of agriculture. After this he served as his country's ambassador to the United States and United Nations. His book *Creative Development* was published in 1990.

language that burdens all contemporary schools of philosophy, shows the relevance of phenomenology's central themes to the assumptions and directions I propose for social science. Its motivating concern is to overcome the split between physicalist objectivism and transcendental subjectivism (Husserl 1970). It regards the world as a field of possibilities; which of them are revealed and which are concealed depends on one's assumptions (Heidegger 1962).

A starting point is the acceptance that multiple realities can coexist (Schutz 1967), extending relativity theory into philosophy. In such a context, possibilities take on greater significance. The future is to be projected based on the possibilities that constitute it. Phenomenological philosophy, like quantum physics which developed contemporaneously, rejects the traditional opposition between subject and object. Like the new physics, it sees the process of gaining understanding as requiring some connection between the knowing subject and the object known, blurring the distinction between them.

Phenomenology tries to resolve the ancient controversy between Plato and Aristotle, referred to in chapter 11—whether essence is more real than existence or vice versa. According to Merleau-Ponty, the study of essences is fundamental to this philosophical method, but it "puts essence back into existence" (in Zaner and Ihde (1973, 72). Martin Heidegger, an equally prominent proponent of phenomenology, sees existence taking precedence over essence (1962, 67–68), so how are these views reconciled? By rejecting an extreme rationalist position that stresses reasoning at the expense of experience. Observation is taken seriously, as we found ourselves doing in Gal Oya, as no sharp distinction is made between appearance and reality (Arendt 1978, 19–30). With an open approach to concretely experienced phenomena, and proceeding with a minimum of conceptual presupposition, phenomenology is more a bottom-up than a top-down philosophy.

Universal concepts are accepted, but only if they derive from our experiencing of the world around us. While observation remains the fundamental methodology of phenomenological inquiry, neither perceiving nor sensing phenomena is sufficient. A critical role is assigned to the processes and products of thinking, since all facts and conclusions are matters of interpretation, which links phenomenology to the intellectual strategy of hermeneutics, discussed briefly in chapter 1.[17]

17. "Strictly speaking, there are no such things as facts, pure and simple. All facts are from the outset selected from a universal context by activities of our mind. They are, therefore, always interpreted facts, either facts looked at as detached from their context by an

Rather than get caught up in the debate over "rationality"—what is and what is not rational, and whether people are rational—phenomenology prefers to focus on *intentionality,* on people's purposes and the extent to which they can be achieved. Reasoning is important but is valued only as a means to human ends. In this and other ways, phenomenology resembles existential philosophy, discussed below. A major contributor to the phenomenological enterprise, who also stresses social action, writes of the duty of people to assume responsibility for their own choices and to take action for the effective transformation of their society (Merleau-Ponty 1962).

What Habermas calls purposive or communicative rationality (e.g., 1984) creates some common understandings though dialogue and action in the world. These reduce the distinction between subject and object and between practical and theoretical knowledge. Since no participant in this dialogue and action has a monopoly on the correct interpretation of surrounding realities, we all need to understand and incorporate others' interpretations into our own. This requires more explicit and systematic attention to frames of reference, discussed in chapter 11. It also challenges the theory of "rational decision making." In addition to philosophical objections, there are increasing empirical ones, as experimental evidence shows that different preferences arise from alternative descriptions of the same decision problem, contradicting the principle of invariant preferences that underlies the theory of rational choice (Tversky and Kahneman 1987).

The relevance of a post-Newtonian perspective was dramatized as I finished drafting this last chapter on a laptop computer in a hotel room in Colombo during a forty-eight-hour islandwide curfew in Sri Lanka, during which time perhaps a thousand people lost their lives (July 29–30, 1989). That conflict between a Sinhalese-majority government and a handful of secessionist Tamil guerrilla groups would bring this formerly peaceful country near collapse was beyond comprehension when we started work in Gal Oya. To have predicted this situation in 1980 would have sounded like lunacy. It was equally unlikely that the war in the north and east would give rise to a similarly violent and vicious war in the south between the government and young Sinhalese guerrillas. The JVP initially supported the Tamil secessionist demands but then reversed course, capitalizing on public

---

artificial abstraction or facts considered in their particular setting" (Alfred Schutz, in Zaner and Ihde 1973, 292). Heidegger (1962, 61–62) gives similar reasons why all knowledge is "interpretive." A hermeneutic approach is of course concerned with ideas; more unexpected is its link with friendship, pointed out by Bernstein (1983, xv).

opposition to an Indian peacekeeping force brought in to quell the war in the north. This government decision boosted the JVP's political viability and raised the possibility of a Pol Pot type of regime's coming to power. But the insurrection collapsed after its leaders were captured and killed by security forces in November 1989 (Gunaratna 1990). How could these events have been foreseen in linear or monotonic trends? With a "chaotic" conceptual framework, one could better comprehend such an outcome.[18]

The *meanings* that people attribute to events, institutions, policies, motives, and political appeals are as important as the phenomena themselves. In phenomenological terms, the meanings are extensions or manifestations of the phenomena, not something separate and separable. This conception may be perplexing if one thinks in either-or terms, but it is common sense from a both-and viewpoint. The meanings of things coexist and coincide with the things themselves. Distinctions should be made between thoughts and the things they refer to, but such differences are more analytical than concrete. At the same time differences are delineated, it is important to bear in mind the connectedness of phenomena, whether thoughts or things.

Frames of reference, which can be used to differentiate levels of existence or levels of analysis, help us recognize differences and similarities. The contexts that such frames provide are both empirical and normative. They confer meaning and value by establishing common bases for people's perceptions and evaluations. The more widely shared these frames are, the more consensus is possible on the meaning and utility of phenomena. Such sharing also tends to make generosity and cooperation more salient as values. This is why communication and interaction among people are ethical as well as practical functions in Habermas's view (1984) and why Peters and Waterman (1982) say that managers make meaning as well as profits.

In a phenomenological world there is no absolute reality, so we must construct a pragmatic reality by our collective thoughts and actions. The use of dichotomies such as subject and object is helpful instrumen-

18. On the significance of *context* for translating values, identities, and motivations into behavior, one can reflect on the fact that between 1800 and 1950 there were no recorded conflicts between Sinhalese and Tamils, despite clashes along other ethnic lines such as Sinhalese-Muslim and Buddhist-Catholic (Rogers 1987, 201). Ethnic identities as such did not lead to the killings and destruction in recent decades, as witnessed by the century and a half of peaceful Sinhalese-Tamil coexistence before administrative and political actions created insecurity, subversion, and conflict. Actions by foreign governments "perturbed" the situation within Sri Lanka, an open system, so that ethnic "essences" did not themselves determine outcomes.

tally and is necessary. But any frame of reference is a construction of the mind. It erects distinctions that are not ultimately real, since differences and similarities coexist everywhere, thrown together and pulled apart by the Heraclitean flow of events and outcomes. Understanding and becoming comfortable with this flux prepares one to take more effective initiatives, whether for promoting participatory development or for any other purpose.[19]

An Existential World

While the two worldviews discussed above help us to construct a relationship between probabilities and possibilities and between subjects and objects, a third perspective, existentialism, proposes a reconciliation between free will and determinism, an opposition that philosophers have long wrestled with. Thinking in either-or terms posed a dilemma when we approached our task in Gal Oya. With the odds against us, a deterministic model made success appear impossible. We were free to try to change the course of events, but was it foolish to try?

By coming to regard the situation as both determined (governed by causal relations) and open (amenable to attempts to change the presently most probable course of events), we could justify our efforts without having exaggerated expectations of what could be achieved. As experience showed, either perspective in the extreme was foolish. Existing relations, capabilities, orientations, and interests shaped outcomes quite forcefully, and any of the relationships, if altered, tended to revert to their previous form in the absence of continued effort and innovation. But consistent with a Heraclitean or chaotic understanding, they never returned to exactly their earlier form. Initiatives could make a difference, even if not a large or a permanent one. Since the systems were invariably changing and evolving, our actions could affect the course of events in small or sometimes significant ways.

Thinking of outcomes and even people's behavior in terms of probability distributions, like the quantum wave functions (qwiffs) discussed above, reconciled the contradiction between free will and

19. In a study of institutional development, Israel (1987, 29) suggests in post-Newtonian terms that a satisfactory unified theory of social organization is probably impossible "because social reality is always changing. . . . Theories have to be constantly adapted to new conditions and environments, and the theories themselves may influence, sometimes heavily, the social reality."

determinism if free will exists, in greater or lesser degrees, only before the fact. When an event occurs, the qwiff for it "collapses," and uncertainty at that moment is replaced by a definite, determined outcome. In the flow of time—that is, in real life—we at every instant confront a continuing array of possibilities, with probabilities ranging from highly likely to practically inconceivable. This means that we have some, even considerable, scope for initiative and action, but limits are set on outcomes by the configuration of probabilistic causal relations in our environment. This includes other people's values, knowledge, intentions and capabilities, which together with our own ideas and efforts determine events. From a both-and perspective, one can say that one's own actions matter little if at all in this vast network of relationships; yet at the same time they are crucially important. One should be realistic about the chances of having much influence but still recognize opportunities to affect the flow. Chaos theory supports ontologically the ethic and imperative of personal action associated with existential philosophy, because *small causes can have large effects.*

Once we recognize that the world should not be regarded as a vast mechanism in the tradition of Newton or Laplace, it is reasonable to conclude that individuals can and should act—to be sure, in a "learning process" mode—to affect the programs and contexts in which they operate.[20] Success may be limited, but in the large interstices between free will and determinism there is opportunity for contributing to positive outcomes. These results can set up the kind of cumulative dynamics observed in Gal Oya. Conversely, failure to act on behalf of altruistic and cooperative values and relationships will reinforce prevailing trends toward the selfish and individualistic poles of behavior considered in chapter 12. So while one can say that personal actions make little if any differences in a huge complex world where there is little direct causation anyway, nevertheless such actions make all the difference in the world. They can affect the course of life's river's flow, the more so when others are persuaded that efficacious action is possible. Those who believe they can affect outcomes, often regarded now as "the power elite," have their opportunities enlarged by others' lack of confidence or commitment that changes can and need to be made.

Reviewing a number of antipoverty programs throughout Africa, Asia, and Latin American (Uphoff 1988b), I found that many had built

20. This is eloquently argued by Robert Chambers in the conclusions to his books on rural development (1983) and irrigation management (1988). His plea for a "new professionalism" (1986) based on new thinking and changed values (putting "last" people, sectors, regions, technologies, etc., "first") makes a major contribution to development thinking.

on small-scale base organizations as we had done in Gal Oya, and most relied on organizers like our IOs, two structural features that have promise for mobilizing social energy. But a common factor was also individual initiative and leadership, at the top and throughout each program. Because such personal influences are idiosyncratic and value driven, they are generally discounted. Yet it is leadership that shifts or reinforces the normative orientations according to which people decide to act, or not to act, on behalf of others or themselves. Whether persons regard their interests and fates as interdependent is largely a consequence of some visible and respected individual's affirming this.

An existential perspective helps one to live and operate more effectively in a chaotic universe by stressing the need for individuals to make choices and take responsibility for their actions in an uncertain world. In the popular notion of existentialism, persons agonize over extreme either-or alternatives. But either-or analyses are best seen as tactical aids to thinking and choosing. Existential choices, to be optimizing, are best conceived and made in synthesizing positive-sum ways that derive from both-and reasoning.

This approach points to the desirability of harnessing the motive power of all four orientations described along the axes in figure 3. Development requires some ethic and practice of cooperation even as it opens opportunities for individual initiative and benefit. Both self-centered and other-regarding behavior can contribute to achieving higher levels of productivity. Development involves motivations in all four quadrants of figure 3, but prevailing theories and policies seem preoccupied with quadrant I—autonomous individual action—while depreciating quadrant IV—altruistic collective cooperation.

Quadrant I is consistent particularly with Newtonian kinds of thinking, being more reductionist and mechanistic in conception as well as materialistic and individualistic in assumptions. Much richness and variety of people's motives is lost in such a view of human nature. From an existential perspective, on the other hand, the complexity of situations and purposes is accepted as natural. Best solutions can be known only from practice and experience, though disciplined thought can help in finding more satisfactory directions. For social science to assist in this, it needs the kind of engagement with practice that. Volken and his colleagues suggested in the epigraph to part 2. As another analyst of development experience has recently argued: "We cannot change the world successfully unless we understand the way it works; but neither can we understand the world fully unless we are involved in some way with the processes that change it" (Edwards 1989, 125). Post-Newtonian social science is both experimental and hermeneutic.

There are still many more Gal Oyas awaiting the transformative power of social energy generated by ideas, ideals, and friendship. The analysis here is no justification for glossy "positive thinking" but rather a proposal for systematic positive-sum reconceptualizations. We should have no illusions that obstructive interests and institutions can be simply or swiftly swept aside by novel insights, creative categorizations, or satisfying explanations; by personally fulfilling values, persuasive criteria, or attractive futures; or by renewed appreciation of our common needs and destinies that make solidarity with others more than just an emotional bond or only a rational choice. Still, it is clear that our thinking and our choices have contributed substantially to the world of constraints around us, created despite intentions to the contrary.

If we cannot eliminate straightaway all material and mental obstacles to human fulfillment, we can begin by revising our scientific and day-to-day thinking along the lines that various disciplines are charting. We need not abandon or derogate all our present theories and concepts. That would be either-or thinking. Rather, we should dethrone those methodologies and assumptions that restrict positive-sum outcomes in the name of rigor, by equating the closed systems we create analytically though our minds with the multiple open and overlapping systems that exist all around us.

This general prescription may appear too broad, because its impetus was the desire and search for means to promote the well-being and self-management capability of people in disadvantaged communities in Sri Lanka. It can be said, however, not simply based on our experience in Gal Oya, but with knowledge of current experience in dozens of countries around the world, that achieving more participatory paths for development requires such an evolution in our formal and informal thinking. This mental reorientation is no substitute for altering material relationships. Rather, such reorientation is a requirement for these changes, since the two realms coexist.

Such a conclusion goes well beyond the domain of development problems in Asia, Africa, and Latin America. Gal Oya is something more than William Blake's allegorical grain of sand, in which truths extending to the macrocosm could be discerned. The analysis and synthesis here, prompted by the remarkable and anomalous outcomes of introducing farmer organization in Sri Lanka, draw on a broad literature encompassing diverse experience and several disciplines. The alternatives offered to conventional ways of promoting development and practicing social science thus do not rest on just a single case study.

Still, it would not be correct to conclude this far-reaching exposition without closing the circle, returning to where we started. For all of the rich ideas and principles derived from literature and experience elsewhere, the impetus and ingredients for these conclusions came from people living and working in Gal Oya—my ARTI and Cornell colleagues, the marvelous institutional organizers, and many fine engineers and other officials. Most of all, this new appreciation of what social science can and should contribute to improving human well-being comes from the many inspiring farmers there who showed where existing ideas and expectations could be wrong. They learned from each other as well as from us, and all of us can learn from them.

Diagram of Institutions and Roles
Involved in Gal Oya Project

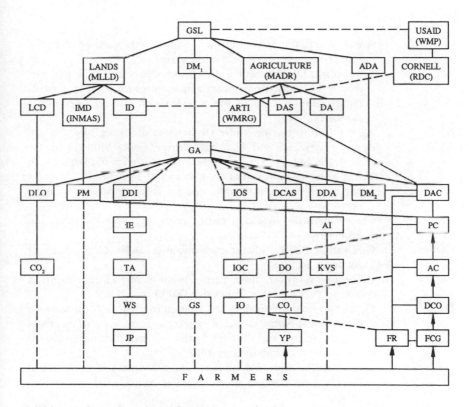

Solid lines indicate direct lines of supervision or funding.

Broken lines indicate cooperation or interaction. No decipherable diagram could include all the connections and lines of communication. ARTI and Cornell had informal lines of communication and cooperation with IMD and DAS, for example, and occasional contacts with LCD. The ARTI board of directors, chaired by the secretary of agriculture, had representatives of the Ministry of Lands and heads of DA, DAS, ID, IMD, and LCD as well as the Central Bank. The DA included all district heads of departments.

Arrows indicate bottom-up accountability; that for YP is weak.

413

Abbreviations and Acronyms

AC   Area Council, third tier of farmer organizations made up of all the farmer-representatives within each area.

ADA   Agricultural Development Authority, established in 1977 to coordinate the many departments and agencies involved in agricultural sector; reports to the president but has no operational authority.

AI   Agricultural instructor, under Department of Agriculture.

ARTI   Agrarian Research and Training Institute, under Ministry of Agriculture, but serves whole agrarian sector with ID, DAS, LCD, and so forth, represented on its board of directors.

ASC   Agrarian Service Centre, operated by Department of Agrarian Services.

CIE   Chief irrigation engineer, most senior in cadre of irrigation engineers.

CO_1 Cultivation officer, lowest staff member under Department of Agrarian Services.

CO_2 Colonization officer, under Land Commissioner's Department, to help settlers in new schemes like Gal Oya.

DAC   District Agriculture Committee, chaired by District Minister with Government Agent as executive secretary; made up of members of Parliament and district heads of departments, included farmer-representatives after 1982.

DAS   Department of Agrarian Services, under Ministry of Agricultural Development and Research; handles agricultural inputs and credit.

DC   Distributary channel (D-channel), secondary level of irrigation distribution system.

DCAS District commissioner of agrarian services, under Department of Agrarian Services.

DCO   Distributary channel organization, second tier of farmer organization, made up of field channel group representatives (farmer-representatives).

414

DDA	Deputy director of agriculture, head of Agriculture Department at district level.
DDI	Deputy director of irrigation, head of Irrigation Department's administration for a range, usually one or two districts.
DI	Director of irrigation (Irrigation Department).
DLO	District land officer, under Land Commissioner's Department and Government Agent.
DM$_1$	District minister, appointed by president with cabinet rank.
DM$_2$	District Manager for Agricultural Development Authority.
DO	Divisional officer, of agrarian services, under Department of Agrarian Services.
FC	Field channel, the tertiary level of irrigation distribution system.
FCG	Field channel groups; base level units for farmer organization.
FO	Farmer organization, referring to field channel groups or to whole system of organization under ARTI program.
FR	Farmer-representative, chosen by consensus by farmers to represent their field channel group.
GA	Government Agent, head of district administration.
GS	*Grama sevaka*, village-level administrative officer.
GSL	Government of Sri Lanka.
ID	Irrigation Department, established 1900, under Ministry of Lands and Land Development.
IDO	Institutional Development Officer, permanent cadre that grew out of IO program.
IE	Irrigation engineer, senior technical officer in Irrigation Department.
IIMI	International Irrigation Management Institute, based in Sri Lanka.
IMD	Irrigation Management Division, established 1983, under the Ministry of Lands, but located in Irrigation Department facilities; a parallel agency to Irrigation Department, to be interdisciplinary; responsible for INMAS program
IMPSA	Irrigation Management Policy Support Activity, funded by USAID to develop consensus, strategy, roles, and procedures to implement policy of participatory irrigation management.
INMAS	Integrated Management System for Major Irrigation Schemes, established by Irrigation Management Division in 1984 to implement Irrigation Systems Management Project and introduce farmer organization in all major irrigation schemes.
IO	Institutional organizer.
IOS	Institutional organizer supervisor.
ISMP	Irrigation Systems Management Project, designed by USAID to succeed Gal Oya Management Project and apply its lessons elsewhere.
JP	*Jala palaka*, literally water manager, lowest-level employee of Irrigation Department.
JVP	Janatha Vimukthi Peramuna (People's Liberation Front), insurrec-

tionary movement of mostly Sinhalese youths, which almost toppled the Sri Lankan government in 1971 and which revived in the late 1970s, becoming a threat to overthrow the government in 1987–89. The rise and fall of the JVP is chronicled in Gunaratna (1990).

KVS   Agricultural extension agent for Agriculture Department.

LCD   Land Commissioner's Department, headed by land commissioner responsible for administering and adjudicating land matters, especially in settlement schemes like Gal Oya; under Ministry of Lands and Land Development.

LTTE   Liberation Tigers of Tamil Eelam, major secessionist group among more than three dozen; referred to commonly as "Tigers." Best account of LTTE is Gunaratna (1987).

MADR   Ministry of Agricultural Development and Research, usually referred to as Ministry of Agriculture.

MLLD   Ministry of Lands and Land Development, otherwise known as Ministry of Lands.

MO   Maintenance overseer, also known as work supervisor (WS), supervised by a technical assistant.

MP   Member of Parliament.

NIA   National Irrigation Administration, in the Philippines.

O&M   Operation and maintenance.

PC   Project Committee, established under INMAS in 1986; made up of officials and majority of farmer-representatives; chaired by project manager at first, now by farmer.

PLC   Project-Level Committee, started by IOs in 1985; replaced by Project Committee under INMAS.

PM   Project Manager, assigned by Irrigation Management Division to a specific irrigation project to implement INMAS program; liaison with farmers and staff of all departments; also designation for prime minister of Sri Lanka.

PMB   Paddy Marketing Board, established to maintain good income for farmers by buying paddy (rice) at guaranteed price.

R&TO   Research and training officer, ARTI.

RDC   Rural Development Committee, interdisciplinary group of faculty at Cornell University, with special interest in participatory development and irrigation management.

RDS   Rural Development Society, voluntary village-level organization sponsored by Ministry of Rural Development.

SLFP   Sri Lanka Freedom Party, one of the major parties, governed most recently 1970–77.

TA   Technical assistant, junior technical officer in Irrigation Department; works under direction of irrigation engineer.

UNP   United National Party, one of the major parties, governing since 1977.

USAID   United States Agency for International Development; designed and financed Gal Oya Water Management Project and Irrigation Systems Management Project.

WMP   Water Management Project, designed by USAID and known as the Gal Oya Project because it focused on improving water management in the Gal Oya irrigation system; implemented by Irrigation Department with funding from USAID and government of Sri Lanka.

WMRG   Water Management Research Group, initially responsible for ARTI's work in the Gal Oya Water Management Project; later merged into Irrigation and Agrarian Relations Division of ARTI.

WS   Work supervisor, other name for maintenance overseer (MO).

YP   *Yaya palaka*, literally tract manager, elected by farmers in a certain area (tract) as liaison with Agrarian Service Centre, under Department of Agrarian Services.

Participants in The Gal Oya Experience

Most Active Participants

Ariyaratne (D. M. Ariyaratne): Government Agent for Ampare district, 1979–86; director of the Irrigation Management Division after 1989.

Bhawani: Started as research assistant for Jeff Brewer; hired as research and training officer by ARTI in 1983 to help with Tamil IO program; married Manoharan in 1985.

Dayaratne (P. Dayaratne): District minister for Ampare district, 1978–89; appointed minister of lands, irrigation, and Mahaweli development in 1989.

Dissanayake: Active farmer-representative on M16; member of District Agriculture Committee.

Dixon (Dixon Nilaweera): Participant in Cornell Rural Development Committee seminar in 1974; Government Agent for Batticaloa, 1977–82, then additional secretary of agriculture, and from 1990, secretary of agriculture.

Doug (Douglas Merrey): Anthropologist with consulting firm advising Irrigation Department on Gal Oya project, 1981–83; from 1985, on staff of International Irrigation Management Institute in Sri Lanka.

Ed (Edward Vander Velde): First resident consultant for Cornell with ARTI, 1981–82, on leave from Aquinas College; later on staff of International Irrigation Management Institute in Pakistan.

Godaliyadde: Chief irrigation engineer responsible for water distribution and maintenance in Gal Oya, 1985–87; then did Ph.D. work in agricultural engineering at Cornell.

Godfrey (Godfrey de Silva): Engineer who experimented with farmer participation in Minipe irrigation scheme, 1978–80; later became chairman of Mahaweli Engineering and Construction Authority; appointed director of irrigation in early 1989 and state secretary for irrigation in late 1989; director of Irrigation Management Policy Support Activity from 1990.

Hammond (D. Hammond Murray-Rust): Cornell Ph.D. student in agricultural engineering; did research in Gal Oya in 1979 and 1981–82, returned as consultant 1983 and 1984; now on staff of International Irrigation Management Institute in Sri Lanka.

Herb (Herbert Blank): Second USAID manager for Gal Oya project, 1983–86.

Jayantha (Jayantha Perera): Sociologist, colleague during my 1978–79 sabbatical year at ARTI; later deputy director of ARTI responsible for project activity.

Jeff (Jeffrey Brewer): Anthropologist; third Cornell resident consultant at ARTI, 1983–84.

Joe (Joseph Alwis): Director of water resource development in Ministry of Lands, 1978–85; director of Irrigation Management Division, 1985–86; director of ARTI, 1986–88; additional secretary of agriculture, 1988–89.

Kalumahattea: Rich farmer and merchant; initially opposed IO program, became active farmer-representative and District Agriculture Committee member; later opposed program again.

Kari (Kariyawasam): Engineer, assistant USAID project manager, 1980–81, then part-time Cornell consultant with ARTI to help with technical issues.

K. D. P. (K. D. P. Perera): Senior deputy director of irrigation; then director of Irrigation Management Division, 1984–86, and director of irrigation, 1986–89; state secretary for irrigation, 1989–90.

Ken (Kenneth Lyvers): First USAID manager for Gal Oya project, 1980–83; previous experience with water user associations in Pakistan.

Kum (Kumarasamy): Deputy director, then senior deputy director of irrigation for water management, 1981–86.

Lucky (Lakshman Wickramasinghe): Sociologist member of ARTI group and briefly its head; joined UNICEF in 1983.

Mano (Manoharadas): First chief irrigation engineer with Gal Oya project; cooperative with IO program, left in 1981; later chief irrigation engineer in Batticaloa district.

Mano (Manoharan): Second chief irrigation engineer with project; cooperative with IO program; married Bhawani.

Mark (Mark Svendsen): Second resident consultant for Cornell with ARTI, 1982–83; agricultural engineer writing Ph.D. dissertation after research on irrigation management in Philippines; then with International Irrigation Management Institute and International Food Policy Research Institute.

Mune (S. Munasinghe): Senior colonization officer, seconded to ARTI to supervise IO program in Ampare; concurrently district land officer.

Nalini (Nalini Kasynathan): Lecturer in geography at University of Peradeniya; Cornell training consultant to assist ARTI with Tamil IO program, 1983–86.

Nanda (S. M. B. K. Nandaratna): Initially research assistant for Jeff Brewer; supervised IO program and became ARTI research and training officer in 1986.

Narangoda: Farmer-representative from Gonagolla who gave unusual leadership at all levels of farmer organization.

Neeles: Active farmer-representative on UB16.

Norman (Norman Uphoff): Chairman of Cornell Rural Development Committee; spent sabbatical year with ARTI in 1978–79; helped plan IO program and worked with it from 1980.

Ponrajah (A. J. P. Ponrajah): Director of Irrigation Department, 1981–84.

Punchibanda: Name of two farmer-representatives who gave unusual leadership in Gonagolla: both members of District Agriculture Committee.

Raj (Rajendra Ariyabandu): Agronomist research and training officer helping with program in 1983–84.

Rane (Ranasinghe Perera): One of the first IOs, then hired by ARTI as research and training officer in 1982; assigned to work with IO program in 1983–84, before earning master's degree in United States.

Ratnayake: Name of two farmer-representatives, both elderly, who provided unusual leadership, one in Gonagolla, the other in Weeragoda.

Razaak (M. G. M. Razaak): Sociologist, became research and training officer in 1986.

Sena (Piyasena Ganewatte): Sociologist, hired as training consultant by Cornell to assist ARTI 1980–85; later consultant for USAID-funded Irrigation Systems-Management Project introducing farmer organizations in Polonnaruwa district.

Senthi (Senthinathan): Deputy director of irrigation in charge of Gal Oya project 1981–85, then again from 1986; very supportive of farmer organizations.

Subasinghe (T. B. Subasinghe): Director of ARTI, 1977–86.

Thurai (Thurairajaratnam): Deputy director of irrigation in charge of Gal Oya project 1980–81; not supportive of IO program, but later endorsed farmer participation.

Tilak (Tilakaratne): Irrigation Management Division project manager for Gal Oya from 1986.

Vida (Vidanapathirana): Economist, headed ARTI group, 1983–84, left for job as United Nations volunteer in Sudan; later joined Irrigation Management Policy Support Activity as agricultural economist in 1990.

Waratenne: Initially *yaya palaka* for UB9 who opposed program; later farmer-representative and District Agriculture Committee member.

Warren Leatham: Head of technical assistance team for Gal Oya project, 1982–86; later head of team for Irrigation Systems Management Project after 1989; in between was head of USAID Nepal Irrigation Management Project, which Barker, Levine, and Uphoff assisted with.

Wijay (C. M. Wijayaratna): Economist, first head of ARTI's Water Management Research Group, 1980–83; did Ph.D. work at Cornell in agricultural economics 1983–86, then resumed as head of group at ARTI; later joined International Irrigation Management Institute staff in Philippines.

Most Active Institutional Organizers
(batch number in parentheses)

Abu Bakr (3): Tamil-speaking IO who stayed with program longest.

Anula (1): Name of two excellent IOs. One worked in Uhana until 1984, married to Jayalath; other worked in Uhana and then Weeragoda through 1983, married to Gunasena.

Ari (Ariyadasa) (1): Active IO, but first to leave; became vice-principal in school near Uhana after marrying Ratna.

Ariyapala (1): Second coordinator of program; worked near Uhana.

Batagoda (3): Coordinator for downstream Gonagolla area.

Chuti (1): Worked at head of Uhana area; unusual leadership; became teacher in 1984, remained in area.

D. S. (D. S. Ranasinghe) (1): Third, excellent coordinator of program; became institutional development officer.

Dissa (Dissanayake) (2): Became coordinator for Weeragoda and eventually institutional development officer.

Gunasena (1): Coordinator of UB9 group; "converted" Waratenne.

Hemachandra (1): Worked in Uhana, having been one of my research assistants during 1978–79; developed simple doctrine of IO strategy.

Jaufer (2): Tamil-speaking IO working in Mandur area.

Jayalath (1): Worked in Uhana, later chosen as assistant project manager for irrigation scheme in north.

Jina (Jinadasa) (1): Initially not taken into program, later was excellent coordinator for Gonagolla area; now institutional development officer.

K. B. (K. B Dissanayake) (1): First coordinator of program; "converted" Kalumahattea; became teacher.

Kule (Kularatna Perera) (2): Coordinator for Paragahakela area at head of Left Bank system; now institutional development officer with Irrigation Systems Management Project in Polonnaruwa.

Nilame (4): Nickname for second Dissanayake; got tail of UB2 organized after six others failed; now institutional development officer with Irrigation Systems Management Project.

P. B. (P. B. Ratnayake) (1): Stalwart IO throughout program; now institutional development officer with Irrigation Systems Management Project.

Prema (Premachandra) (1): Coordinator for Weeragoda; left teaching job to return as institutional development officer.

Ratna (1). Excellent IO, continued as IO after marrying Ariyadasa.

Ratnasiri (1): Second training coordinator, later hired by International Irrigation Management Institute as research assistant.

Sarath Silva (3): Worked in tail of Gonagolla, eventually only IO left in Gal Oya; appointed institutional development officer in Gal Oya.

Sarath Wijesiri (2): worked on M5, then coordinator for Weeragoda appointed district manager of Agricultural Development Authority.

Siri (Siriwardena) (2) First training coordinator; became teacher.

Sita (1): Excellent IO, worked with Chuti at head of Uhana.

Subramaniam (2): Trilingual IO; put out of program on false charge of passing counterfeit bill.

Sugi (Sugatadasa) (2): Tamil-speaking IO who eventually left to become university philosophy lecturer.

Sumanapala (1): IO in Gonagolla area.

Sumane (Sumanadasa) (2): IO who brought Sinhalese and Tamil farmers together on M5.4.4

Wimalasena (2): Enthusiastic IO, working in Weeragoda.

Less Central or Less Frequent Participants

Ben Bagadion: Deputy administrator of National Irrigation Administration in Philippines, where he gave leadership for participatory irrigation management; visited Sri Lanka for our project in 1980 and 1981.

Carlos Isles: Trained organizers for National Irrigation Administration in the Philippines, helped plan IO training during 1980.

Chris Panabokke: Former director of agriculture; friendly critic of program while on ARTI board of directors; now with International Irrigation Management Institute.

Dave Korten: Ford Foundation consultant in Philippines assisting National Irrigation Administration on participatory irrigation management; made visits for our project in 1980 and 1981.

Gamini Dissanayake: Minister of lands and Mahaweli development, 1977–89.

George Wickramasinghe: Commissioner of agrarian services when project started, then an additional secretary of agriculture.

Gil Levine: Professor of agricultural engineering at Cornell, vice-chairman of Rural Development Committee, made a number of visits to work with Irrigation Department engineers and ARTI, particularly on technical issues.

Hassan: Irrigation engineer, responsible for construction in Gal Oya project, 1982–85; farmers considered him a foe of program.

Ismael: Acting deputy director of irrigation, Ampare, 1985–86; not strong supporter of program.

Kapilaratne: Commissioner of agrarian services, 1983–88; had been temporary registrar at ARTI; later state secretary for agriculture.

Kuruppu: District manager for Agricultural Development Authority, 1980–84; later project manager at Kaudulla under INMAS.

Mahinda: ARTI research and training officer, in charge of IO program in 1984.

Milan Rodrigo: Sri Lankan student at Cornell in communication and education, did thesis research in Gal Oya.

Nancy St. Julien: Cornell Ph.D. student in regional planning, assisted ARTI group on organizational issues in 1984 and 1985.

Nanda Abeywickrama: Secretary, Ministry of Lands, 1977–88; now Director of Operations for International Irrigation Management Institute.

Oscar Zolezzi: Cornell Ph.D. student in agricultural engineering from Mexico, did thesis research in Gal Oya in 1985–86.

Rabindran: Irrigation engineer in Gal Oya at start of project; left in 1981, later chief irrigation officer in Batticaloa.

Randy Barker: Professor of agricultural economics at Cornell; made a number of visits to work with ARTI on monitoring and evaluation.

R. B. Senakaarachchi: Statistician, member of ARTI group in first year; had grown up in Gal Oya, left program to earn master's degree in United Kingdom.

Shyamala Abeyratna: Rural sociologist on ARTI staff member doing master's and Ph.D. work at Cornell, did thesis research in Gal Oya in 1982.

Terence Abeyesekere: Agricultural economist on ARTI staff, did Ph.D. work at Cornell, 1980–84, head of ARTI water management group in 1984.

Walt Coward: Professor of rural sociology at Cornell; visited ARTI twice to help on farmer organization issues.

Wijayadasa: Irrigation engineer, responsible for operation and maintenance in Gal Oya project from 1987.

References

Abeyratne, Shyamala. 1982. *The impact of second-generation settlers on land and water resources in Gal Oya, Sri Lanka*. Ithaca: Rural Development Committee, Cornell University.

Adams, Richard N. 1975. *Energy and structure: A theory of social power*. Austin: University of Texas Press.

Allison, Graham. 1971. *Essence of decision: Explaining the Cuban missile crisis*: Boston: Little, Brown.

Almond, Gabriel A. 1970. *Political development: Essays in heuristic theory*. Boston: Little, Brown.

Almond, Gabriel A. (with Stephen Genco). 1990. Clouds, clocks, and the study of politics. In *A discipline divided: Schools and sects in political science*, ed. G. A. Almond, 32–65. Newbury Park, Calif.: Sage.

Anderson, Philip W., Kenneth J. Arrow, and David Pines, eds. 1988. *The economy as an evolving complex system*. Reading, Mass.: Addison-Wesley.

Apter, David E. 1965. *The politics of modernization*. Chicago: University of Chicago Press.

Arendt, Hannah. 1978. *The life of the mind*. New York: Harcourt Brace.

Argyris, Chris, and Donald Schön. 1978. *Organization learning*. Reading, Mass.: Addison-Wesley.

Argyris, Chris, Robert Putnam, and Dianna McLain Smith. 1981. *Action science: Concepts, methods and skills for research and intervention*. San Francisco: Jossey-Bass.

Aristotle. 1953. *The ethics of Aristotle: The Nichomachean ethics*. Harmondsworth, U.K.: Penguin Books.

Arthur, W. Brian. 1990. Positive feedbacks in the economy. *Scientific American* 262 (2): 92–99.

Axelrod, Robert. 1984. *The evolution of cooperation*. New York: Basic Books.

Bagadion, Benjamin U., and Frances F. Korten. 1985. Developing irrigators' organizations: A learning process approach. In *Putting people first: Sociological variables in rural development*, ed. M. Cernea, 52–90. New York: Oxford University Press for the World Bank.

423

Barnes, Jonathan. 1979. *The Presocratic philosophers.* Vol. 1. *Thales to Zeno.* London: Routledge and Kegan Paul.

Baynes, Kenneth, James Bohman, and Thomas McCarthy, eds. 1987. *After philosophy: End or transformation?* Cambridge: MIT Press.

Beardsley, Peter. 1991. Smart genes. *Scientific American* 265 (2): 86–95.

Becker, Theodore L., ed. 1991. *New physics, new politics: Thought experiments in quantum politics, political science and constitutional physics.* New York: Praeger.

Bell, Roderick, David V. Edwards, and R. H. Wagner, eds. 1969. *Political power: A reader in theory and research.* New York: Free Press.

Bellah, Robert, et al. 1985. *Secrets of the heart: Individualism and commitment in American life.* New York: Harper and Row.

Bennis, Warren. 1984. Transformative power and leadership. In *Leadership and organizational culture,* ed. Thomas Sergiovanni and John Corbally, 64–71. Urbana: University of Illinois Press.

Berger, Peter L., and Thomas Luckmann. 1966. *The social construction of reality: A treatise in the sociology of knowledge.* New York: Anchor Books.

Bernstein, Richard J. 1983. *Beyond objectivism and relativism: Science, hermeneutics, and praxis.* Philadelphia: University of Pennsylvania Press.

Best, Michael H. 1990. *The new competition: Institutions of industrial restructuring.* Cambridge, Mass.: Polity Press.

Boff, Kenneth R., et al., eds. 1986. *Handbook of perception and human performance.* 2 vols. New York: John Wiley.

Bohm, David, and F. David Peat. 1987. *Science, order and creativity.* New York: Bantam Books.

Boulding, Kenneth. 1978. *Ecodynamics: A theory of societal evolution.* Beverly Hills, Calif.: Sage.

——. 1989. *Three faces of power.* Newbury Park, Calif.: Sage.

Bratton, Michael. 1986. Farmer organization and food production in Zimbabwe. *World Development* 14 (3): 367–84.

——. 1988. Poverty, organization and policy: Towards a voice for Africa's rural poor. Paper presented at Colloquium on the Changing Nature of Third World Poverty, Michigan State University, March 13–15.

Briggs, John P., and F. David Peat. 1984. *Looking glass universe: The emerging science of wholeness.* New York: Simon and Schuster.

Briscoe, John, and David de Ferranti. 1988. *Water for rural communities: Helping people help themselves.* Washington, D.C.: World Bank.

Brohier, R. L. 1933. *Ancient irrigation works in Ceylon.* 3 vols. Colombo: Government Printer.

Bruner, Jerome. 1986. *Actual minds, possible worlds.* Cambridge: Harvard University Press.

Bunch, Roland. 1982. *Two ears of corn: A guide to people-centered agricultural development.* Oklahoma City: World Neighbors.

Burns, James MacGregor. 1978. *Leadership.* New York: Harper and Row.

Calvin, William H. 1986. *The river that flows uphill: A journey from the big bang to the big brain.* New York: Macmillan.

Cernea, Michael. 1984. The social organization of water users: Experience with water users associations in bank projects. Draft Staff Working Paper. Washington, D.C.: World Bank.

Chaitin, Gregory J. 1988. Randomness in arithmetic. *Scientific American* 259 (1): 80–85.

Chambers, Robert. 1983. *Rural development: Putting the last first*. London: Longman.

——. 1986. *Normal professionalism, new paradigms and development*. Discussion Paper 227. Sussex: Institute of Development Studies.

——. 1988. *Managing canal irrigation: Practical analysis from South Asia*. New Delhi: Oxford and IBH Publishing.

Chanmugam, J. C. 1976. Gal Oya odyssey. *Journal of the Ceylon Survey* 34 (2): 48–54.

Chimedza, Ruvimbo. 1985. Savings clubs: The mobilization of rural finances in Zimbabwe. In *Rural development and women: Lessons from the field,* ed. S. Muntemba, 1: 161–74. Geneva: International Labor Organization.

Cleland, John, and Christopher Wilson. 1988. Demand theories of the fertility transition: An iconoclastic view. *Population Studies* 41 (5): 5–30.

Clements, Paul. 1986. A conceptual framework for analyzing, managing and evaluating village development projects. *Sociologia Ruralis* 26 (2): 128–45.

Coleman, James A. 1969. *Science, order and creativity.* New York: Bantam Books.

Coveney, Peter, and Roger Highfield. 1990. *The arrow of time: A voyage through science to solve time's greatest mystery.* New York: Fawcett Columbine.

Craig, J. G., and H. Poerbo. 1988. Food security and rural development: The case of Ciamis, Indonesia. *Journal of Rural Cooperation* 16 (1–2): 111–24.

Crone, Donald. 1989. Social welfare in Southeast Asia: The "chaos" of politics. Paper for meeting of Association of Asian Studies, Washington, D.C., March.

Daniel, Sheryl B. 1983. The tool box approach of the Tamil to the issues of moral responsibility and human destiny. In Keyes and Daniel (1983, 27–62).

Davies, P. C., ed. 1989. *The new physics.* Cambridge: Cambridge University Press.

Dawkins, Richard. 1989. *The selfish gene.* 2d ed. Oxford: Oxford University Press.

Delbrück, Max. 1986. *Mind from matter? An essay on evolutionary epistemology.* London: Blackwell Scientific Publications.

Derthick, Martha, and Paul Quirk. 1985. *The politics of deregulation.* Washington, D.C.: Brookings Institution.

de Silva, I. M. 1958. The Gal Oya scheme. Paper for Center on Principles and Policies of Land Settlement for Asia and the Far East, November 21, 1958. Colombo: Gal Oya Development Board. Mimeographed.

de Silva, K. M. 1981. *A history of Sri Lanka.* New Delhi: Oxford University Press.

de Silva, N. G. R. 1985. Involvement of farmers in water management: Alternative approach at Minipe, Sri Lanka. In FAO (1985, 133–48).

d'Espagnat, Bernard. 1979. The quantum theory and reality. *Scientific American* 241 (5): 158–81.

Drescher, Anne, Josef Esser, and Wolfgang Fach. 1986. *Die politische Ökonomie der Liebe.* Frankfurt: Suhrkamp.

Dunn, Edgar S., Jr. 1971. *Economic and social development: A process of social learning.* Baltimore: Johns Hopkins University Press.

Durkheim, Emile. 1933. *The division of labor in society.* Glencoe, Ill.: Free Press. (Original French version, 1902.)

Durning, Alan B. 1989. *Action at the grassroots: Fighting poverty and environmental decline.* Worldwatch Paper 88. Washington, D.C.: Worldwatch Institute.

Dyer, Wayne W. 1989. *You'll see it when you believe it.* New York: Morrow.

Eccles, John. 1970. *Facing reality.* Berlin: Springer-Verlag.

Eccles, John, and Daniel N. Robinson. 1984. *The wonder of being human: Our brain and our mind.* New York: Free Press.

Edwards, Michael. 1989. The irrelevance of development studies. *Third World Quarterly* 11 (1): 116–35.

Eisenhardt, Kathleen M. 1989. Building theories from case study research. *Academy of Management Review* 14 (4): 532–50.

Eisenstadt, S. N., and L. Roniger. 1984. *Patrons, clients and friends: Interpersonal relations and the structure of trust in society.* Cambridge: Cambridge University Press.

Esman, Milton J., and Norman Uphoff. 1984. *Local organizations: Intermediaries in rural development.* Ithaca: Cornell University Press.

Etzioni, Amitai. 1961. *A comparative analysis of complex organizations.* New York: Free Press.

FAO (Food and Agriculture Organization). 1975. *Water management for irrigated agriculture (Gal Oya irrigation scheme), Sri Lanka: Project findings and recommendations.* Rome: Food and Agriculture Organization of the United Nations.

———. 1985. *Participatory experiences in irrigation water management,* Proceedings of the Expert Consultation on Irrigation Water Management, Yogyakarta and Bali, Indonesia, July 16–22, 1984. Rome: Food and Agriculture Organization of the United Nations.

Festinger, Leon. 1957. *A theory of cognitive dissonance.* Evanston, Ill.: Row, Peterson.

Field, John O. 1980. Development at the grassroots: The organizational imperative. *Fletcher Forum* 4(2): 145–65.

Foster, George M. 1965. Peasant society and the image of limited good. *American Anthropologist* 67 (2): 293–315.

Frank, Robert H. 1988. *Passions within reason: The strategic role of emotions.* New York: W. W. Norton.

Freeman, Walter J. 1991. The physiology of perception. *Scientific American* 264 (2): 78–85.

French, John, and Bertram Raven. 1959. The bases of social power. In *Studies in social power,* ed. D. Cartwright, 150–65. Ann Arbor: University of Michigan Press.

Fuglesang, Andreas, and Dale Chandler. 1986. *Participation as process: What we can learn from Grameen Bank, Bangladesh.* Oslo: Norwegian Agency for International Development.

Gafoor, Abdul. 1987. A social engineering experiment in Pakistan: A study of Orangi. *Regional Development Dialogue* 8 (2): 108–18.

Galbraith, John K. 1983. *The anatomy of power.* Boston: Houghton Mifflin.

Gardner, Howard. 1985. *The mind's new science: A history of the cognitive revolution.* New York: Basic Books.

Geertz, Clifford. 1973. Thick description: Toward an interpretive theory of culture. In *The interpretation of cultures: Selected essays by Clifford Geertz,* 3–30. New York: Basic Books.

Ghai, Dharam, and M. A. Rahman. 1981. The small farmers' groups in Nepal. *Development* (Society for International Development, Rome) 1 (2): 23–28.

Gillis, Jerry, and Canute Jamtgaard. 1982. "Overgrazing" reconsidered: A review of case studies. *Sociologia Ruralis* 21 (2): 129–41.

Gleick, James. 1987. *Chaos: Making a new science.* New York: Viking.

——. 1988. Secret of proteins is hidden in their folded shapes. *New York Times,* June 14, C10.

Glennie, Colin. 1983. *Village water supply in the decade: Lessons from field experience.* New York: John Wiley.

Goldberger, Ari L., David R. Rigney, and Bruce J. West. 1990. Chaos and fractals in human physiology. *Scientific American* 262 (2): 43–49.

Goldstein, Judith. 1989. The impact of ideas on trade policy. *International Organization* 43 (1): 31–71.

Goodenough, Ward. 1970. *Description and comparison in cultural anthropology.* Chicago: Aldine.

Goodman, Nelson. 1984. *Of mind and other matters.* Cambridge: Harvard University Press.

Gould, Stephen Jay 1985. *The flamingo's smile: Reflections in natural history.* New York: W. W. Norton.

Government of Ceylon. 1971. *Report of the Gal Oya Project Evaluation Committee.* Sessional Paper I-70. Colombo: Government Printer.

Gow, David G., et al. 1979. *Local organizations and rural development: A comparative reappraisal.* Washington, D.C.: Development Alternatives.

Gribbin, John. 1984. *In search of Schrödinger's cat: Quantum physics and reality.* New York: Bantam Books.

Guba, Egon G. 1985. The context of emergent paradigm research. In Lincoln (1985, 79–104).

Gunaratna, Rohan. 1987. *War and peace in Sri Lanka.* Colombo: Institute of Fundamental Studies.

——. 1990. *Sri Lanka—A lost revolution? The inside story of the JVP.* Colombo: Institute of Fundamental Studies.

Gunawardana, R. A. L. H. 1971. Irrigation and hydraulic society in early medieval Ceylon. *Past and Present* 53 (1): 3–27.

Habermas, Jürgen. 1984. *The theory of communicative action: Reason and the rationalization of society.* Trans. Thomas McCarthy. Boston: Beacon Press.

——. 1987. Philosophy as stand-in and interpreter. In Baynes, Bohman, and McCarthy (1987, 296–315).

Hall, Peter A., ed. 1989. *The political power of economic ideas: Keynesianism across nations.* Princeton: Princeton University Press.

Hampton, Jean. 1987. Free-rider problems in the production of collective goods. *Economics and Philosophy* 3 (3): 245–73.

Hardin, Garrett. 1968. The tragedy of the commons. *Science* 168:1243–48.

Harris, Monica, Robert Rosenthal, and Sara Snodgrass. 1979. The effects of teacher expectations, gender and behavior on pupil academic performance and self-concept. *Journal of Educational Research* 79 (3): 173–79.

Harrison, Paul. 1987. *The greening of Africa: Breaking through in the battle for land and food.* New York: Penguin Books.

Hayles, N. Katherine. 1989. Chaos as orderly disorder: Shifting ground in contemporary literature and science. *New Literary History* 20 (2): 307–22.

——. 1990. *Chaos bound: Orderly disorder in contemporary literature and science.* Ithaca: Cornell University Press.

Headland, Thomas N., Kenneth L. Pike, and Marvin Harris. 1990. *Emics and etics: The insider/outsider debate.* Newbury Park, Calif.: Sage.

Heginbotham, Stanley J. 1977. The study of South Asian conceptual systems. *Items* (Social Science Research Council, New York) 21 (3): 34–36.

Heidegger, Martin. 1962. *Being and time* Trans. J. Macquarrie and E. Robinson. New York: Harper and Row.

Heim, F. G., et al. 1986. *How to develop the small farming sector: The case of Thailand.* Feldafing: Deutsche Stiftung für Internationale Entwicklung.

Hirschman, Albert O. 1958. *The strategy of economic development.* New Haven: Yale University Press.

——. 1967. *Development projects observed.* Washington, D.C.: Brookings Institution.

——. 1970. *Exit, voice and loyalty: Responses to decline in firms, organizations and states.* Cambridge: Harvard University Press.

——. 1971. *A bias for hope: Essays on development and Latin America.* New Haven: Yale University Press.

——. 1977. *The passions and the interests: Political arguments for capitalism before its triumph.* Princeton: Princeton University Press.

——. 1982. *Shifting involvements: Private interest and public action.* Princeton: Princeton University Press.

——. 1984. *Getting ahead collectively: Grassroots experiences in Latin America.* New York: Pergamon Press.

——. 1985. Against parsimony: Three easy ways of complicating some categories of economic discourse. *Economics and Philosophy* 1 (1): 7–21.

Hochman, Harold M., and James D. Rodgers. 1969. Pareto optimal distribution. *American Economic Review* 59 (3): 542–57.

Hofstadter, Douglas. 1979. *Gödel, Escher, Bach: An eternal golden braid, a metaphorical fugue on minds and machines in the spirit of Lewis Carroll.* New York: Vintage Books.

——. 1985. *Metamagical themas: Questing for the essence of mind and pattern.* New York: Basic Books.

Husserl, Edmund. 1970. *The crisis of European sciences and transcendental phenomenology.* Evanston, Ill.: Northwestern University Press. (Original German version, 1936.)

Hussey, Edward. 1972. *The Presocratics.* London: Duckworth.

IFAD (International Fund for Agricultural Development). 1984a. Kingdom of Nepal: Small farmer development programme—mid-term evaluation report. Rome: International Fund for Agricultural Development.

———. 1984b. People's Republic of Bangladesh: Small farmer agricultural credit project—mid-term evaluation report on Grameen Bank. Rome: International Fund for Agricultural Development.

IIMI (International Irrigation Management Institute). 1990. *Workshop on resource mobilization for sustainable management of major irrigation schemes: Workshop papers.* February, 22–24, 1990, Kandy. Colombo: International Irrigation Management Institute.

Ilchman, Warren, and Norman Uphoff. 1969. *The political economy of change.* Berkeley: University of California Press.

Indrapala, K., ed. 1971. *The collapse of the Rajarata civilization in Ceylon and the drift to the south west.* Peradeniya: University of Ceylon.

Ingram, Helen, and Lawrence Scaff. 1987. Politics, policy and public choice: A critique and a proposal. *Polity* 19 (4): 613–36.

Israel, Arturo. 1987. *Institutional development: Incentives to performance.* Baltimore: Johns Hopkins University Press.

ISTI (International Science and Technology Institute). 1985. *Final evaluation of Sri Lanka water management project (no. 383-0057).* Washington, D.C.: International Science and Technology Institute.

Jahn, Robert G., and Brenda J. Dunne. 1987. *Margins of reality: The role of consciousness in the physical world.* New York: Harcourt Brace Jovanovich.

Janowitz, Morris. 1962. *The role of the military in underdeveloped countries.* Princeton: Princeton University Press.

Jantsch, Erich. 1980. *The self-organizing universe: Scientific and human implications of the emerging paradigm of evolution.* New York: Pergamon.

Jervis, Robert, Richard Ned Lebow, and Janice Gross Stein. 1985. *Psychology and deterrence.* Baltimore: Johns Hopkins University Press.

Johansen, L. 1977. The theory of public goods: Misplaced emphasis. *Journal of Public Economics* 7 (1): 147–52.

Johnson, Kersten, and Richard Morse. 1988. Interpreting participatory development: Histories and projections. Honolulu: Resource Systems Institute, East-West Center. Photocopied.

Johnston, Bruce F., and William C. Clark. 1982. *Redesigning rural development: A strategic perspective.* Baltimore: Johns Hopkins University Press.

Jones, Roger. 1982. *Physics as metaphor.* Minneapolis: University of Minnesota Press.

Keyes, Charles F., and E. Valentine Daniel, eds. 1983. *Karma: An anthropological inquiry.* Berkeley: University of California Press.

Khan, A. Z. M. Obaidullah. 1990. *Creative development: An unfinished saga of human aspirations in South Asia.* Dhaka: University Press.

Kikuchi, M., G. Dozina, and Yujiro Hayami. 1978. Economics of community work programs: A communal irrigation project in the Philippines. *Economic Development and Cultural Change* 26 (2): 211–25.

Kimber, Richard 1981. Collective action and the fallacy of the liberal fallacy. *World Politics* 33 (2): 178–96.

Koestler, Arthur. 1964. *The act of creation.* New York: Macmillan.
———. 1967. *The ghost in the machine.* New York: Macmillan.
Kohn, Alfie. 1990. *The brighter side of human nature: Altruism and empathy in everyday life.* New York: Basic Books.
Korten, David C. 1980. Community organization and rural development: A learning process approach. *Public Administration Review* 40 (5): 480–511.
Korten, Frances F. 1982. *Building national capacity to develop water users' associations: Experience from the Philippines.* Staff Working Paper 528. Washington, D.C.: World Bank.
Korten, Frances F., and Robert Siy, eds. 1988. *Transforming a bureaucracy: The experience of the Philippine National Irrigation Administration.* West Hartford, Conn.: Kumarian Press.
Krasner, Stephen, ed. 1983. *International regimes.* Ithaca: Cornell University Press.
Kuhn, Thomas. 1962. *The structure of scientific revolutions.* Chicago: University of Chicago Press.
———. 1977. *The essential tension: Selected studies in scientific tradition and change.* Chicago: University of Chicago Press.
Kuttner, Robert. 1985. The poverty of economics. *Atlantic Monthly* 255 (2): 74–84.
Laitos, M. Robert, et al. 1986. *Rapid appraisal of Nepal irrigation systems.* WMS Report 43. Fort Collins, Colo.: Water Management Synthesis Project, Colorado State University.
Laitos, M. Robert, Kanda Paranakian, and Alan Early. 1987. *A short history of the farmer irrigation participation project in Thailand.* Fort Collins, Colo.: Water Management Synthesis Project, Colorado State University.
Lakoff, George, and Mark Johnson. 1980. *Metaphors we live by.* Chicago: University of Chicago Press.
LaPalombara, Joseph, and Myron Weiner, eds. 1966. *Political parties and political development.* Princeton: Princeton University Press.
Lassen, Cheryl. 1980. *Reaching the assetless poor: Projects and strategies for their self-reliant development.* Ithaca: Rural Development Committee, Cornell University.
Lebow, Richard Ned, and Janice Gross Stein. 1987. Beyond deterrence. *Journal of Social Issues* 43 (4) 5–72.
LeComte, Bernard J. 1986. *Project aid: Limitations and alternatives.* Paris: Organization for Economic Cooperation and Development.
Leibenstein, Harvey. 1976. *Beyond economic man: A new foundation for microeconomics.* Cambridge: Harvard University Press.
Leonard, David K. 1977. *Reaching the peasant farmer: Organization theory and practice in Kenya.* Chicago: University of Chicago Press.
———. 1991. *African successes: Four public managers of Kenyan rural development.* Berkeley: University of California Press.
Lerner, Daniel. 1958. *The passing of traditional society: Modernizing the Middle East.* New York: Free Press.

Levin, Simon A. 1989. Challenges in the development of a theory of community and ecosystem structure and function. In *Perspectives in ecological theory*, ed. Jonathan Roughgarden, Robert M. May, and Simon A. Levin, 242–55. Princeton: Princeton University Press.

Lewis-Beck, Michael. 1988. *Economics and elections: The major Western democracies*. Ann Arbor: University of Michigan Press.

Liebenow, J. Gus. 1981. Malawi: Clean water for the rural poor. *American Universities Field Staff Reports, Africa*, no. 40.

Lincoln, Yvonna S., ed. 1985. *Organizational theory and inquiry: The paradigm revolution*. Beverly Hills, Calif.: Sage.

Lincoln, Yvonna S., and Egon S. Guba. 1985. *Naturalistic inquiry*. Beverly Hills, Calif.: Sage.

Lovell, Michael C. 1986. Tests of the rational expectations hypothesis. *American Economic Review* 76 (1): 110–24.

Luce, R. Duncan, and Howard Raiffa. 1957. *Games and decisions: Introduction and critical survey*. New York: John Wiley.

Maass, Arthur. 1983. *Congress and the common good*. New York: Basic Books.

McCloskey, Donald. 1983. The rhetoric of economics. *Journal of Economic Literature* 21 (3): 481–517.

———. 1985. *The rhetoric of economics*. Madison: University of Wisconsin Press.

McGregor, Douglas. 1960. *The human side of enterprise*. New York: McGraw-Hill.

McMillan, John. 1979. The free-rider problem: A survey. *Economic Record* 55 (1): 95–107.

Mansbridge, Jane J. 1980. *Beyond adversary democracy.* Chicago: University of Chicago Press.

———, ed. 1990. *Beyond self interest*. Chicago: University of Chicago Press.

March, James, and Johan Olsen. 1984. The new institutionalism: Organizational factors in political science. *American Political Science Review* 78 (3): 734–50.

Margolis, Howard. 1982. *Selfishness, altruism, and rationality: A theory of social choice*. Chicago: University of Chicago Press.

———. 1988. Dual-utilities and rational choice. Paper presented to American Political Science Association meetings, Washington, D.C.

Mayer, Albert, Kim McMarriott, and Richard L. Park, eds. 1958. *Pilot project India*. Berkeley: University of California Press.

Merleau-Ponty, Maurice. 1962. *Phenomenology of perception*. London: Routledge and Kegan Paul. (Original French version, 1945.)

Merrey, Douglas J., and D. Hammond Murray-Rust. 1987. People's participation in the Gal Oya rehabilitation project as viewed by agency personnel. Paper for Workshop on People's Participation in Irrigation Management, June 28–July 7. Hyderabad: Administrative Staff College.

Metcalf, Barbara Daly, ed. 1984. *Moral conduct and authority: The place of Adab in South Asian Islam*. Berkeley: University of California Press.

Milgram, Stanley. 1974. *Obedience to authority: An experimental viewpoint*. New York: Harper and Row.

Miller, Barbara Stoler, trans. 1986. *The Bhagavad-Gita: Krishna's counsel in time of war.* New York: Bantam Books.

Miller, Dale T., and William Turnbull. 1986. Expectancies and interpersonal processes. *Annual Review of Psychology* 37: 233–56.

Moore, M. P. 1981a. Report on visit to Gal Oya irrigation rehabilitation project and institutional organizer programme, 6–9 December 1981. Sussex: Institute of Development Studies. Photocopied.

——. 1981b. The sociology of irrigation management in Sri Lanka. *Water Supply and Management* 5 (2): 117–33.

Morgenstern, Oskar. 1968. Game theory. In *International encyclopedia of the social sciences,* 6: 62–69. New York: Macmillan and Free Press.

Morss, Elliott, John Hatch, Donald Mickelwait, and Charles Sweet. 1976. *Strategies for small farmer development.* 2 vols. Boulder, Colo.: Westview Press.

Munro, William Bennett. 1928. Physics and politics: An old analogy revised. *American Political Science Review* 22 (1):1-11.

Murray-Rust, D. Hammond. 1983. Irrigation and water management in Sri Lanka: An evaluation of technical and policy factors affecting operation of the main channel system. Ph.D. diss., Department of Agricultural Engineering, Cornell University.

Murray-Rust, D. Hammond, and Russ Cramer. 1979. An evaluation of water management and water-users' organizations in three selected irrigation schemes in Sri Lanka. Colombo: U.S. Agency for International Development. Photocopied.

Murray-Rust, D. Hammond, and M. P. Moore. 1984. *Formal and informal water management systems: Cultivation meetings and water delivery in two Sri Lankan irrigation systems.* Cornell Studies in Irrigation 2. Ithaca: Irrigation Studies Group, Cornell University.

Myrdal, Gunnar. 1957. *Economic theory and under-developed regions.* London: Duckworth.

Naipaul, V. S. 1977. *India: A wounded civilization.* Harmondsworth, U.K.: Penguin Books.

Netting, R. M. 1976. What Alpine peasants have in common: Observations on communal tenure in a Swiss village. *Human Ecology* 4 (2): 135–46.

Neustadt, Richard. 1960. *Presidential power: The politics of leadership.* New York: John Wiley.

Nietzsche, Friedrich. 1955. *Friedrich Nietzsches Werke.* Munich: Carl Hanser.

O'Donnell, Guillermo, and Philippe C. Schmitter. 1986. *Transitions from authoritarian rule: Tentative conclusions about uncertain democracies.* Baltimore: Johns Hopkins University Press.

O'Flaherty, Wendy, ed. 1980. *Karma and rebirth in classical Indian tradition.* Berkeley: University of California Press.

Oliner, Samuel, and Pearl Oliner. 1988. *The altruistic personality: Rescuers of Jews in Nazi Europe.* New York: Free Press.

Olson, Mancur. 1965. *The logic of collective action: Public goods and the theory of groups.* Cambridge: Harvard University Press.

Ordeshook, Peter C. 1986. *Game theory and political theory: An introduction.* Cambridge: Cambridge University Press.

Ostrom, Elinor. 1986. How inexorable is the "tragedy of the commons"? Institutional arrangements for changing the structure of social dilemmas. Distinguished Faculty Lecture, Indiana University, Workshop in Political Theory and Policy Analysis. Photocopied.

——. 1990. *Governing the commons: The evolution of institutions for collective action.* Cambridge: Cambridge University Press.

Ostrom, Vincent. 1964. Culture, science and politics. In *The making of decisions: A reader in administrative behavior,* ed. W. Gore and J. Dyson, 85–92. New York: Free Press.

Otero, Maria. 1986. *The solidarity group concept: Its characteristics and significance for urban informal sector activities.* New York: Private Agencies Collaborating Together.

Ouchi, William. 1981. *Theory Z: How American businesses can meet the Japanese challenge.* Reading, Mass.: Addison-Wesley.

Oye, Kenneth. 1992. *Economic discrimination and political exchange: World political economy in the 1930s and 1980s.* Princeton: Princeton University Press.

Pascale, Richard, and Tony Athos. 1981. *The art of Japanese management.* New York: Simon and Schuster.

Pattee, Howard H. 1978. The complementarity principle in biological and social structures. *Journal of Social and Biological Structures* 1 (2) 191–200.

Pearce, Joseph Chilton. 1971. *The crack in the cosmic egg: Challenging constructs of mind and reality.* New York: Washington Square Press.

Penrose, Roger. 1989. *The emperor's new mind: Concerning computers, minds, and the laws of physics.* New York: Oxford University Press.

Perera, Jayantha. 1985. *New dimensions of social stratification in rural Lanka.* Colombo: Lake House.

Peters, Thomas J., and Robert H. Waterman, Jr. 1982. *In search of excellence: Lessons from America's best-run companies.* New York: Warner Books.

Peters, Tom, and Nancy Austin. 1985. *A passion for excellence: The leadership difference.* New York: Random House.

Pike, Kenneth L. 1967. *Language in relation to a unified theory of the structure of human behavior.* The Hague: Mouton.

Popper, Karl R. 1972. *Objective knowledge: An evolutionary approach.* Oxford: Clarendon Press.

Popper, Karl R., and John C. Eccles. 1977. *The self and the brain.* Berlin: Springer-Verlag.

Pradhan, Prachanda P. 1983. Chhattis Mauja. In *Water management in Nepal,* 218–42. Kathmandu: Agricultural Projects Service Centre.

Prigogine, Ilya. 1980. *From being to becoming: Time and complexity in the physical sciences.* San Francisco: W. H. Freeman.

Prigogine, Ilya, and Isabelle Stengers. 1984. *Order out of chaos: Man's new dialogue with nature.* New York: Bantam.

Pye, Lucien. 1962. *Politics, personality and nation building.* New Haven: Yale University Press.

Rabibhadana, Akin. 1983. The transformation of Tambon Yokkrabat, Changwat Samut Sakorn. *Thai Journal of Development Administration* 22 (1): 73–104.

Rabinow, Paul, and William M. Sullivan, eds. 1987. *Interpretive social science: A second look*. Berkeley: University of California Press.

Rae, Alastair. 1986. *Quantum physics: Illusion or reality?* Cambridge: Cambridge University Press.

Rahman, M. A. 1984a. The Small Farmer Development Programme of Nepal. In Rahman (1984b, 121–51).

———. 1984b. *Grass-roots participation and self-reliance: Experiences in South and South East Asia*. New Delhi: Oxford University Press and IBH Publishing.

Rappoport, Anatol. 1969. *Two-person game theory: The essential ideas*. Ann Arbor: University of Michigan Press.

Rawls, John. 1971. *A theory of justice*. Cambridge: Harvard University Press.

Ray, Jayanta Kumar. 1983. *Organising villagers for self-reliance: A study of Deedar in Bangladesh*. Comilla: Bangladesh Academy for Rural Development.

Reich, Robert B. 1988. *The power of public ideas*. Cambridge, Mass.: Ballinger.

Rochin, Refugio, and Heliodoro Diaz Cisneros. 1988. Peasant cooperatives: A Mexican model. *Rural Sociologist* 8 (3): 219–25.

Rogers, John D. 1987. *Crime, justice and society in colonial Sri Lanka*. London: School of Oriental and African Studies, University of London.

Rosenthal, Robert. 1973. The Pygmalion effect. *Psychology Today* 7 (4) 56–63.

Rosenthal, Robert, Stephen Baratz, and Clay Hall. 1974. Teacher behavior, teacher expectations, and gains in pupils' rated creativity. *Journal of Genetic Psychology* 124 (1): 115–21.

Rosenthal, Robert, and Edward Halas. 1962. Experimenter effect on the study of invertebrate behavior. *Psychological Reports* 11 (1): 251–56.

Rosenthal, Robert, and Reed Lawson. 1964. A longitudinal study of the effects of experimenter bias on the operant learning for laboratory rats. *Journal of Psychiatric Research* 2: (2) 61–72.

Rosenthal, Robert, and Donald B. Rubin. 1978. Interpersonal expectancy effects: The first 345 studies. *Behavioral and Brain Sciences* 1 (3): 377–435.

Rubin, Lillian B. 1985. *Just friends: The role of friendship in our lives*. New York: Harper and Row.

Runge, C. Ford. 1981. Common property externalities: Isolation, assurance and resource depletion in a traditional grazing context. *American Journal of Agricultural Economics* 63 (4): 595–606.

———. 1984. Institutions and the free rider: The assurance problem in collective action. *Journal of Politics* 46 (1): 154–81.

Sandford, Stephen. 1983. *Management of pastoral development in the Third World*. London: John Wiley.

Schmidt, Steffan, Laura Guasti, Carl H. Landé, and James C. Scott, eds. 1977. *Friends, followers and factions: A reader in political clientelism*. Berkeley: University of California Press.

Schneider, Friedrich, and Werner Pommerehne. 1981. Free riding and collective action: An experiment in public microeconomics. *Quarterly Journal of Economics* 96 (4): 689–904.

Schoggen, Phil. 1989. *Behavioral settings: A revision and extension of Roger G. Barker's "Ecological Psychology."* Stanford, Calif.: Stanford University Press.

Schutz, Alfred. 1967. *The phenomenology of the social world.* Evanston, Ill.: Northwestern University Press. (Original German version, 1932.)

Sen, A. K. 1977. Rational fools: A critique of the behavioral foundations of economic theory. *Philosophy and Public Affairs* 6 (2): 317–44. Also in Mansbridge (1990, 25–43).

Sheth, D. L. 1987. Alternative development as political practice. *Alternatives* 12 (2): 155–71.

Simon, Herbert A. 1945. *Administrative behavior.* New York: Free Press.

———. 1969. Rational decision-making in business organizations. *American Economic Review* 69 (4): 493–513.

Skinner, B. F. 1971. *Beyond freedom and dignity.* New York: Vintage Books.

Slaton, Christa Daryl. 1991. Quantum theory and political theory. In Becker (1991).

Smith, Peter. 1987. Understanding the other man's point of view. In *Irrigation design for management,* 157–69. Papers from Asian Regional Symposium, International Irrigation Management Institute, Kandy. Wallingford, U.K. Overseas Development Unit of Hydraulics Research.

Sommer, Robert. 1973. The Hawthorne dogma. *Psychological Bulletin* 70 (6): 592–95.

Stein, Arthur. 1982. When misperceptions matter. *World Politics* 34 (4): 505–26.

Steiner, Roy. 1990. An analysis of water distribution strategies using an irrigation land management simulation model. Ph.D. diss., Department of Agricultural Engineering, Cornell University.

Stogdill, R. M. 1974. *Handbook of leadership: A survey of theory and research.* New York: Free Press.

Stone, I. F. 1988. *The trial of Socrates.* Boston: Little, Brown.

Streefland, Pieter. 1986. *Different ways to support the rural poor: Effects of two development approaches in Bangladesh.* Amsterdam: Royal Tropical Institute.

Streeten Paul. 1987. International cooperation and global justice. *Journal für Entwicklungspolitik* 3:5–15.

———. 1989. Global institutions for an interdependent world. *World Development* 17 (9): 1349–60. (Presented at conference on the Role of Institutions in Economic Development, Cornell University, November 14–15, 1986.)

Sweet, Charles, and Peter Weisel. 1979. Process versus blueprint models for designing rural development projects. In *International development administration: Implementation analysis for development projects,* ed. G. Honadle and R. Klaus, 127–45. New York: Praeger.

Tendler, Judith. 1976. *Inter-country evaluation of small farmer organizations in Ecuador and Honduras.* Program Evaluation Study. Washington, D.C.: U.S. Agency for International Development.

Tilakaratna, S. 1987. *The animator in participatory rural development: Concept and practice.* Geneva: World Employment Programme, International Labour Organization.

Tversky, Amos, and Daniel Kahneman. 1987. Rational choice and the framing of decisions. In *Rational choice: The contrast between economics and psychology,* ed. Robin H. Hogarth and Melvin W. Reder, 67–94. Chicago: University of Chicago Press.

Uphoff, Norman, ed. 1982–83. *Rural development and local organization in Asia.* 3 vols. New Delhi: Macmillan.

——. 1985. People's participation in water management: Gal Oya, Sri Lanka. In *Public participation in development planning,* ed. J. C. Garcia-Zamor, 131–78. Boulder, Colo.: Westview Press.

——. 1986. *Local institutional development: An analytical sourcebook, with cases.* West Hartford, Conn.: Kumarian Press.

——. 1987. Activating community capacity for water management: Experience from Gal Oya, Sri Lanka. In *Community management: Asian experience and perspectives,* ed. D. C. Korten, 201–19. West Hartford, Conn.: Kumarian Press.

——. 1988a. Participatory evaluation of farmers organizations' capacity for development tasks. *Agricultural Administration and Extension* 30 (1): 43–64.

——. 1988b. Assisted self-reliance: Working with, rather than for, the poor. In *Strengthening the poor: What have we learned?* ed. John C. Lewis, 47–59. New Brunswick, N.J.: Transaction Books.

——. 1989. Drawing on social energy in project implementation: A learning process experience in Sri Lanka. *Sri Lanka Journal of Development Administration* 6 (2): 103–37. Paper for meeting of the American Society for Public Administration, Boston, March, 1987.

——. 1990. Distinguishing power, authority and legitimacy: Taking Max Weber at his word, using resource-exchange analysis. *Polity* 22 (2): 295–322.

Uphoff, Norman, and Milton Esman. 1974. *Local organization for rural development in Asia: Analysis of Asian experience.* Ithaca: Rural Development Committee, Cornell University.

Uphoff, Norman, and Warren Ilchman. 1972. *The political economy of development: Theoretical and empirical contributions.* Berkeley: University of California Press.

Uphoff, Norman, and R. D. Wanigaratne. 1982. Local organization and rural development in Sri Lanka. In Uphoff (1982–83, 1:479–549).

Uphoff, Norman, M. L. Wickramasinghe, and C. M. Wijayaratna. 1990. "Optimum" participation in irrigation management: Issues and evidence from Sri Lanka. *Human Organization* 49 (1): 26–40.

Volken, Henry, Ajoy Kumar, and Sara Kaithathara. 1982. *Learning from the rural poor: Shared experiences of the mobile orientation and training team.* New Delhi: Indian Social Institute.

von Neumann, John, and Oskar Morgenstern. 1953. *Theory of games and economic behavior.* Princeton: Princeton University Press.

Wade, Robert. 1987. The management of common property resources: Finding a cooperative solution. *Research Observer* (World Bank) 2 (2): 219–34.

——. 1988. *Village republics: Economic conditions for collective action in South India.* Cambridge: Cambridge University Press.

Wang, Paul. 1983. *Advances in fuzzy sets, possibility theory, and applications.* New York: Plenum.

Weber, Max. 1947. *The theory of economic and social organization.* New York: Free Press.

——. 1949. *The methodology of the social sciences.* Glencoe, Ill.: Free Press.

White, Leslie. 1958. *The science of culture: A study of man and civilization.* New York: Grove Press.

Whyte, William Foote, Davydd Greenwood, and Peter Lazes. 1989. Participatory action research: Through practice to science in social research. *American Behavioral Scientist* 32 (5): 513–51.

Wickham, Thomas, and A. B. Valera. 1978. Practices and accountability for better water management. In *Irrigation policy and management in Southeast Asia,* 61–75. Manila: International Rice Research Institute.

Wijayaratna, C. M. 1985. Involvement of farmers in water management in Gal Oya, Sri Lanka. In FAO (1985, 117–32).

Williams, Colin C. 1988. *Examining the nature of domestic labour.* Aldershot, U.K.: Avebury.

Wolf, Fred Alan. 1981. *Taking the quantum leap: The new physics for nonscientists.* San Francisco: Harper and Row.

World Bank. 1987. *Aga Khan Rural Support Program in Pakistan: An interim evaluation.* Washington, D.C.: Operations Evaluation Department, World Bank.

Yin, Robert. 1989. *Case study research: Design and methods.* Newbury Park, Calif.: Sage.

Young, Oran R. 1980. International regimes: Problems of concept formation. *World Politics* 32 (3): 331–56.

Yukl, Gary. 1981. *Leadership in organizations.* Englewood Cliffs, N.J.: Prentice-Hall.

Zadeh, Lofti. 1987. *Fuzzy sets and applications: Selected papers.* New York: John Wiley.

Zaner, Richard, and Don Ihde, eds. 1973. *Phenomenology and existentialism.* New York: G. P. Putnam.

Zolezzi, Oscar. 1986. *Gal Oya water management project: Preliminary report on a special study carried out in Ampare during Yala 1984.* Colombo: Agrarian Research and Training Institute and Cornell Irrigation Studies Group.

Zukav, Gary. 1979. *The dancing Wu Li masters: An overview of the new physics.* New York: Bantam Books.

Author Index

Abeyratne, Shyamala, 27, 28
Adams, Richard, 291
Allison, Graham, 382
Almond, Gabriel, 20, 312
Anderson, Philip, 24
Apter, David, 312
Arendt, Hannah, 379, 404
Argyris, Chris, 11, 396
Aristotle, 366
Arrow, Kenneth, 74
Arthur, W. Brian, 24
Athos, Tony, 15
Austin, Nancy, 384
Axelrod, Robert, 17, 341, 351–52, 378

Bagadion, Benjamin, 22
Baratz, Stephen, 382
Barnes, Jonathan, 283
Baynes, Kenneth, 403
Beardsley, Peter, 392
Becker, Theodore, 399
Bell, Roderick, 291, 350–51
Bellah, Robert, 350, 351
Bennis, Warren, 386
Berger, Peter, 315
Bernstein, Richard, 22, 23, 377, 403, 405
Best, Michael, 15
Boff, Kenneth, 316
Bohm, David, 14, 280, 320
Boulding, Kenneth, 17, 324, 363
Bratton, Michael, 373
Briggs, John, 14
Briscoe, John, 373
Brohier, R. L., 27
Bruner, Jerome, 19, 317
Bunch, Roland, 373
Burns, James, 386

Calvin, William, 357
Chaitin, Gregory, 402
Chambers, Robert, 408
Chandler, Dale, 370
Chanmugam, J. C., 26
Chimedza, Ruvimbo, 373
Clark, William C., 13
Cleland, John, 402
Clements, Paul, 273, 365
Cleveland, Harlan, 390–91, 395
Coveney, Peter, 14, 314, 393, 400
Craig, J. G., 273, 373
Cramer, Russ, 37
Crone, Donald, 391

Daniel, E. V., 281
Daniel, Sheryl, 281
Davies, P. C., 14
Dawkins, Richard, 328, 340, 352, 377
de Ferranti, David, 373
Dellbrück, Max, 375
Derthick, Martha, 19
de Silva, I. M., 26
de Silva, K. M., 27
de Silva, N. G. R., 42
d'Espagnat, Bernard, 280
Diaz Cisneros, H., 373
Dozina, G., 353
Drescher, Anne, 17
Drucker, Peter, 378
Dunn, Edgar, 11
Dunne, Brenda, 375
Durkheim, Emile, 378
Durning, Alan, 373
Dyer, Wayne, 316

Eccles, John, 315–17, 375
Edwards, Michael, 410
Eisenhardt, Kathleen, 22

Eisenstadt, S. N., 366
Esman, Milton, 33, 103, 359
Etzioni, Amitai, 324

FAO, 5, 9, 29
Festinger, Leon, 336
Field, John, 365
Foster, George, 285
Frank, Robert, 17
Freeman, Walter, 375
Fuglesang, Andreas, 370

Gafoor, Abdul, 370
Galbraith, John K., 324
Gardner, Howard, 18, 317, 376–77
Geertz, Clifford, 23
Ghai, Dharam, 369
Gilles, Jerry, 329
Gleick, James, 14, 290, 292, 294, 302,
    310, 374, 389, 390, 392, 393, 394, 396
Glennie, Colin, 373
Goldberger, Ari, 342
Goldstein, Judith, 19
Goodenough, Ward, 320
Goodman, Nelson, 19
Gould, J. Stephen, 302, 313, 396
Government of Ceylon, 5
Gow, David, 373
Greenwood, Davydd, 313
Gribben, John, 14
Guba, Egon, 12, 19, 403
Gunaratna, Rohan, 239, 260, 406, 416
Gunawardana, R. A. L. H., 27

Habermas, Jürgen, 315, 379, 403,
    405, 406
Halas, Edward, 382
Hall, Clay, 382
Hall, Peter, 19
Hampton, Jean, 328
Hardin, Garrett, 327–30, 353
Harris, Monica, 383
Harrison, Paul, 373
Hayami, Yujiro, 353
Hayles, Katherine, 387, 389, 394
Headland, Thomas, 320
Heginbotham, Stanley, 281
Heidegger, Martin, 404, 405
Heim, F. G., 372
Highfield, Roger, 14, 314, 393, 400
Hirschman, Albert O., 1, 16, 32, 292,
    301, 324, 329, 334, 344, 357, 368, 381
Hochman, Harold, 287
Hofstadter, Douglas, 302, 319, 375
Husserl, Edmund, 404
Hussey, Edward, 321

IFAD, 369, 370
Ihde, Don, 404, 405
IIMI, 268
Ilchman, Warren, 291, 311
Indrapala, K., 27
Ingram, Helen, 288
Israel, Arturo, 293, 407
ISTI, 227

Jahn, Robert, 375
Jamtgaard, Canute, 329
Janowitz, Morris, 312
Jantsch, Erich, 14, 292, 322, 374, 393
Jervis, Robert, 19
Johansen, L., 353
Johnson, Kersten, 397
Johnson, Mark, 279
Johnston, B. F., 13
Jones, Roger, 279

Kahneman, Daniel, 405
Keyes, C. F., 281
Khan, A. Z. M. Obaidullah, 403
Kikuchi, M., 353
Kimber, Richard, 353
Koestler, Arthur, 314, 350, 395
Kohn, Alfie, 351
Korten, David, 11, 370
Korten, Frances, 22
Krasner, Stephen, 382
Kuhn, Thomas, 4, 23, 388
Kuttner, Robert, 19

Laitos, M. Robert, 22
Lakoff, George, 279
LaPalombara, Joseph, 312
Lassen, Cheryl 37
Lawson, Reed, 382
Lazes, Peter, 313
Lebow, Ned, 19, 382
LeComte, Bernard, 373
Liebenstein, Harvey, 293
Leonard, David, 384, 400
Lerner, Daniel, 312
Levin, Simon, 399
Lewis-Beck, Michael, 339
Liebenow, J. Gus, 373
Lincoln, Yvonna, 12, 403
Lovell, Michael, 381
Luce, R. Duncan, 284
Luckman, Thomas, 315

Maass, Arthur, 337–38
McCloskey, Donald, 23, 280
McGregor, Douglas, 384
McMillan, John, 353

Mansbridge, Jane, 17, 352, 376
March, James, 354, 393–94, 402
Margolis, Howard, 17, 326, 328, 341, 342
Mayer, Albert, 9
Merleau-Ponty, Maurice, 403, 404, 405
Merrey, Douglas, 8
Metcalfe, Barbara, 281
Milgram, Stanley, 338
Miller, Barbara Stoler, 281
Miller, Dale, 383
Moore, M. P., 29, 32, 38, 76
Morgenstern, Oskar, 284, 285
Morse, Richard, 397
Morss, Elliott, 11
Munro, W. B., 291, 324, 339, 363
Murray-Rust, D. Hammond, 5, 8, 29, 37, 76
Myrdal, Gunnar, 296

Naipaul, V. S., 370
Netting, R. M., 329
Neustadt, Richard, 337
Nietzsche, Friedrich, 304

O'Donnell, Guillermo, 324
O'Flaherty, Wendy, 281
Oliner, Pearl, 385
Oliner, Samuel, 385
Olsen, Johan, 354, 393–94, 402
Olson, Mancur, 327–30, 337, 344, 353
Ordeshook, Peter, 284
Ostrom, Elinor, 328, 329, 334, 355
Ostrom, Vincent, 317, 376
Otero, Maria, 370
Ouchi, William, 15
Oye, Kenneth, 20

Pascale, Richard, 15
Pattee, Howard, 322
Pearce, Joseph, 18
Peat, F. David, 14, 280, 320
Penrose, Roger, 14
Perera, Jayantha, 39
Peters, Tom, 15, 301, 352, 383, 384, 406
Pike, Kenneth, 320
Poerbo, Hasan, 273, 373
Pommerehne, Werner, 353
Popper, Karl, 23, 312, 315, 358, 375
Pradhan, Prachanda, 371
Prigogine, Ilya, 14, 290, 292, 322, 393
Pye, Lucien, 312

Quirk, Paul 19

Rabibhadana, Akin, 336
Rabinow, Paul, 22

Rae, Alastair, 14, 280
Rahman, M. A., 369
Raiffa, Howard, 284
Rappaport, Anatol, 284
Rawls, John, 380
Ray, J. K., 371
Reich, Robert, 19, 376
Robinson, Daniel, 315–17
Rochin, Refugio, 373
Rogers, James, 287
Rogers, John, 406
Roniger, L., 366
Rosenthal, Robert, 382, 383
Rubin, Donald, 383
Rubin, Lillian, 366
Runge, C. Ford, 354

Sandford, Stephen, 329
Scaff, Lawrence, 288
Schmidt, Steffan, 366
Schmitter, Philippe, 324
Schneider, Friedrich, 353
Schoggen, Phil, 338
Schön, Donald, 11
Schutz, Alfred, 404, 405
Sen, A. K., 17
Sheth, D. L., 331
Simon, Herbert, 337, 388
Siy, Robert, 22
Skinner, B. F., 290
Slaton, Christa, 399
Smith, Peter, 299
Snodgrass, Sarah, 383
Stein, Janice, 19, 382
Steiner, Roy, 332
Stengers, Isabelle, 14, 292, 322, 393
Stogdill, R. M., 386
Stone, I. F., 21
Streefland, Pieter, 370
Streeten, Paul, 17
Sullivan, William, 22
Sweet, Charles, 11

Tilakaratne, S., 273, 359
Turnbull, William, 383
Tversky, Amos, 405

Uphoff, Norman, 33, 103, 125, 227, 291, 306, 311, 359, 372, 408

Valera, A. B., 332
Volken, Henry, 275, 365, 409
von Neuman, John, 284

Wade, Robert, 355
Wang, Paul, 322

Wanigaratne, R. D., 306
Waterman, Robert, 15, 301, 352, 383, 406
Weber, Max, 23, 324, 325
Weisel, Peter, 11
White, Leslie, 374
Whyte, William F., 313
Wickham, Thomas, 332
Wickramasinghe, M. L., 125
Wiener, Myron, 312
Wijayaratna, C. M., 125, 237
Williams, Colin, 283

Wilson, Christopher, 402
Wolf, Fred Alan, 14, 277, 338, 400
World Bank, 370

Yin, Robert, 21
Young, Oran, 382
Yukl, Gary, 386

Zadeh, Lofti, 322
Zaner, Richard, 404, 405
Zolezzi, Oscar, 10
Zukav, Gary, 14, 279

Subject Index

Agrarian Research and Training Institute, ix, 3, 4, 11, 32–41, 43, 45–46, 48, 50, 71, 73, 104–5, 110, 112, 119, 128, 134–38, 140, 141, 168–72, 173, 197, 200–202, 205, 207, 208, 230, 232, 239, 244, 245, 256, 257, 258, 259, 269, 272, 280, 292, 306, 359, 365, 366, 396–97, 401, 411

Agrarian Service Centres, 77, 81, 137, 153, 164, 245, 255

Agricultural Development Authority, 154, 166, 202, 206, 210

Altruism, 17, 25, 66, 69, 76, 104, 326, 338–45, 348–51, 356, 365, 376, 378, 385, 406, 408

Alwis, Joe, 38, 49, 64, 66, 198, 206, 236, 258, 300

Analysis, 283–84, 295–96, 313, 316, 319, 323, 396

Area Councils, 79, 99, 130, 138, 164, 174, 203, 205, 210, 216, 223, 229, 230, 234, 241, 244, 249, 253–54, 270

Aristotle, 311, 318, 366, 367, 378–79, 404

Bagadion, Benjamin, 40

Barker, Randy, 34, 36, 39, 41, 141

Behavioral analysis, 330–32

Behaviorism, 290

Bhawani, 135, 141, 143, 144, 156, 171, 173, 194, 197, 201, 249

Bicycles, 54, 65, 66, 70, 90, 113, 114, 115, 118, 121–22, 123, 124, 128, 152, 159, 196, 235, 240, 244, 255, 256

Binocular vs. binary thinking, 283, 398

Bohr, Niels, 309, 322

Both-and thinking, 25, 282–84, 295–96, 301, 308, 310, 311, 315, 318, 325, 327, 343, 350, 373, 388, 406, 408, 409

Bottom-up approach, 80–81, 103

Breakage, of irrigation structures, 10, 29, 36, 83, 118–19, 135–36, 162, 180, 209

Brewer, Jeff, 112, 113, 116, 118, 135, 141–42, 171

Cartesian thinking, 25, 382, 403

Caste relations, 67, 122, 138

Catalysts, 37, 200, 359, 371

Causation, 14, 20, 24, 52, 358–60, 362, 365, 391, 395, 399, 407

Celestial mechanics, 13, 289, 290–91, 388

Channel numbering, 55

Chaos theory, 14, 24, 277, 294, 387, 389–96, 397, 399, 402, 406, 407

Chuti, 54–55, 89, 90, 92, 123–24, 210, 233

CIA, 126–27, 250

Civil defense, 174, 178–79, 180, 191

Closed systems, 291, 294–96, 358, 374

Cognitive analysis, 18–19, 330–32, 360–63

Cognitive dissonance, 336–37

Collective action, 17, 81, 126, 327–30, 332, 344, 352–55

Complementarity, 24, 322, 324, 399

Contingence, 51, 299, 321, 325, 399

Convention, farmer, 153, 171–72, 173, 175–79, 183, 195–96, 250–51

Cooperation, 15, 17, 21, 25, 45, 67, 69, 76, 88, 96, 104, 108, 127, 134, 142, 267, 326, 332–33, 341–45, 346, 356, 365, 406, 408

Cornell University, 32–33, 40, 165, 213, 268, 359, 365, 396–97, 401, 411

Corruption, 47, 64, 100, 122, 138, 188, 212, 220, 243, 246, 247–48, 347–48, 361–62

443

Coward, Walter, 33, 35, 36
Credit, 137, 148, 185, 213, 217, 267, 271
Crop insurance, 70, 148, 214, 246–47,
    249, 251, 253
Crop protection, 100, 185, 204

Dependency, 63, 85, 359
Deputy director of irrigation:
    Senthinathan, 9, 73, 79, 92, 118–19,
    133–34, 136, 140, 178, 209, 232, 233,
    237, 240, 249, 260, 263, 266, 267,
    401; Thurairajaratnam, 36, 48–50,
    71, 140
Descartes. See Cartesian thinking
de Silva, Godfrey, 42, 262, 265
Detachment, 13–14, 22
Determinism, 281, 297, 311, 319, 323,
    324, 394, 395, 407–9
Dialectical thinking, 265, 290, 320,
    373, 398
Dichotomies, 13, 273, 302, 323–24, 386,
    398
Director of Irrigation (Ponrajah), 42–43,
    73, 90–94, 105, 139–40, 401
Dissanayake (FR), 190–91, 211, 254
Dissanayake (IO), 118, 216, 245, 249, 257
Distributary channel organizations
    (DCOs), 150, 159–64, 175, 177, 178,
    182, 192–93, 203, 216–17, 229, 241,
    244, 254, 266, 267, 268, 270
District Agriculture Committee (DAC),
    106, 109, 127, 135, 138, 146–47, 154,
    164, 234, 247, 254, 347
District Minister (P. Dayaratne), 9, 39, 47,
    107, 109, 167, 171, 200, 202, 206, 209,
    232, 234, 261, 269
D. S. Ranasinghe, 143, 152–53, 157–58,
    194, 201, 202, 212–13, 229, 232, 259

Einstein, Albert, 14, 290, 302, 304, 305–
    9, 319
Either-or thinking, 4, 25, 273, 282–83,
    295–96, 298, 314, 317, 321, 327, 373,
    385, 403, 409, 410
Elephants, 6, 27, 152, 169, 304
Emergent properties, 313–14, 396, 398
Enemies, 343
Energy, 104, 290, 292–93, 303, 305, 316,
    319, 358, 363, 364, 380, 399
Engagement, 313, 315, 409
Entropy, 14, 291, 303, 358, 363, 374–75,
    380, 393
Equilibrium, 8, 19, 24, 291, 292, 303,
    344, 357, 374–75, 380, 391, 393, 399
Escher, M. C., 319–20, 325, 375
Essentialism, 310–14, 404

Evaluation of program, 222, 227–29, 255
Existentialism, 311–14, 407–9
Expectations, 381–85

Farmer-official relations, 36–39, 76, 79,
    107, 130–31, 145, 222, 241, 242, 252,
    347–50
Farmer organizations, 15, 30, 32–35, 37–
    38, 106, 109, 119, 127, 139, 151, 164,
    170, 173, 213, 241, 250–51, 269; con-
    trol over maintenance budgets, 203,
    242–43, 263; legal recognition of, 128,
    150, 161, 178, 229, 248
Farmer participation, 32, 45, 47–49, 73,
    91, 236, 248, 261, 361, 401
Farmer-representative, role of, 78, 99,
    109, 129, 167, 199–200, 213, 335–36,
    354, 360–61
Field channel groups, 78, 150, 229, 243–
    44, 247, 250, 253, 257, 263, 334, 336
Fitzgerald, F. Scott, 301
Food and Agriculture Organization
    (FAO), 28–30, 205, 224, 246
Fractal geometry, 310, 387, 389
Frames of reference, 306–10, 314, 317–
    18, 324, 363, 389, 396, 406
Free riding, 149, 326, 328–29, 334, 353
Free will, 281, 311, 319, 323, 407–9. See
    also Determinism
Friendship, 104, 108, 109, 134, 343,
    365–67, 368, 374, 378–79, 380, 384,
    386, 405
Fuzzy sets, 321–22, 396

Gal Oya irrigation system: initial condi-
    tions, 5–6, 28–30, 36–38; map of, 7;
    new project, 30–33, 29–32, 37; origi-
    nal project, 26–28
Game theory, 17, 284–85, 328
Ganewatte, Sena, 41, 43, 46, 54, 59, 66,
    73, 75, 80, 91–92, 94, 98, 102, 112,
    113, 117, 123, 131–33, 136, 143, 146,
    149, 156, 157, 165, 171, 172, 173, 177,
    178, 186, 197, 201, 202, 206, 258, 260,
    267, 300, 364, 366
Generosity. See Altruism
Godaliyadde, 183, 190–91, 197
Gonagolla, 64, 75, 97, 101, 123, 130–31,
    137, 147, 153, 163–66, 170, 204–5,
    220–23, 241–48, 263–64
Government Agent: D. M. Ariyaratne, 6,
    8–9, 39, 44, 46–47, 72, 105–6, 109,
    127, 162, 178–79, 203, 232, 262; Nan-
    dasena, 209, 210, 228–29, 234
Grameen Bank (Bangladesh), 369–70
Gunasena, 56–57, 86, 107

Hawthorne effect, 383–84
Heisenberg, Werner, 14, 293, 302, 324
Hemachandra, 80–81, 92, 94–95, 106–7
Heracleitus, 3, 20–21, 284, 294, 304, 311, 312, 320, 321, 325, 379, 390, 398, 400, 407
Hermeneutics, 21–23, 404–5, 410
Hierarchical systems, 313–14, 395–96
Hirschman, Albert, 16–17, 32, 52, 367–69
Holistic thinking, 284, 296, 315
Holons, 314
Human nature, 352, 380, 409
Human potential, 12, 104, 184, 292–93, 373–74

Idealism, 168, 336, 364
Idealist thinking, 12, 14, 20–21, 311, 317, 403
Ideals, 163, 363–65, 367, 374, 377–80, 386
Ideas, 18–19, 25, 51, 52, 104, 196, 311, 317, 330–31, 356, 360–63, 367, 374–77, 379, 385–86
IFAD, 369
Ilchman, Warren, 298
Individualistic thinking, 24, 122, 288, 302–3, 325, 327, 341–45, 349, 350, 352–55, 377–78, 408
INMAS, 207, 211, 238, 252
Institutional organizers, 6, 10–11, 37–44, 48, 74, 92–93, 167, 213, 219, 236–37, 257, 411; dropouts among, 10–11, 103–4, 114, 129, 136–38, 140, 143, 152, 154, 168, 172, 173, 187, 195–96, 202, 208, 240; permanent cadre of, 105, 133, 136–37, 140, 149, 169, 198, 200, 207, 236, 237, 257, 259; recruitment and training of, 39–44, 133, 134–35, 141, 168, 172, 189, 200, 206, 230, 255, 266; shortcomings of, 51, 170, 253, 257; supervision of, 11, 40, 46–47, 114, 169, 240; tests of program, 110–11, 126–27, 138–39. See also Tamil program
Inter-American Foundation, 16, 367
Interdependence of utility functions, 287–89, 336, 341, 344, 365, 378, 380
Interest rates, 146, 185, 203, 213, 268, 271. See also Credit
International Irrigation Management Institute (IIMI), 31, 198, 238
International Labor Organization, 273, 373
IOs. See Institutional organizers

Irrigation Department, 5, 9, 32, 36, 42–43, 70, 127, 137, 156, 201–2, 221, 232, 247, 255, 263, 361–62, 397, 401–2; resistance from, 45–50, 236; reversal of attitude, 139–40, 236, 400
Irrigation Management Division, 140, 170, 198, 207, 235, 236, 238, 255, 257, 261, 262, 267
Irrigation Management Policy Support Activity, 265, 268–69
Irrigation Systems Management Project, 237, 258, 260

Jala palakas, 60–62, 64, 67, 182, 189, 242
Jayalath, 108, 118, 129, 153, 194
Jinadasa, 153–54, 157–58, 174, 175, 205, 220, 240–41, 245, 253, 257, 259
Job Bank, 38–39, 41, 107
JVP, 57, 158–59, 259, 260, 262, 263, 264, 265, 405–6

Kalumahattea, 80, 84–85, 108–9, 177, 179, 200, 271
Kariyawasam, H. C., 42, 74, 136
K. B. Dissanayake, 84–85, 118
Keerthi, 187–88, 190, 194, 245
Korten, David, 11, 35, 36, 168
Kularatna Perera, 226, 257, 268
Kumarasamy, 59, 140, 141, 198
Kuruppu, 154, 166, 242

Land tenure, 52, 65, 76–77, 90, 122, 124, 217–18, 268, 335, 364
Laplace, Pierre, 277, 302, 395, 408
Leadership, 95, 257, 264, 266, 269, 371, 376, 385–86, 387, 408
Learning process, 11–12, 35, 69, 77, 168, 267, 397, 408
Levine, Gil, 33, 35, 278, 361
Liberation Tigers of Tamil Eelam. See "Tigers"
Linear thinking, 24, 390, 391, 392, 393, 394, 397
Love, 17, 380
Lyvers, Ken, 42, 71, 74, 92, 139

Magnetic analogies, 347–50, 359
Maha (rainy) season, 9, 114, 146, 170
Mahaweli project, 137, 205
Mahinda, 171, 201
Maintenance, of irrigation systems, 227–29, 234, 242, 247, 268, 333. See also Shramadana
Malaria, 26, 67, 125
Mandelbrot, Benoit, 389–90

Manoharadas, 49, 50–51, 59–62, 71, 139, 401
Manoharan, 119, 152, 156, 174
Marxist thinking, 21, 290, 318
Materialistic thinking, 12, 14, 15, 18, 24, 302–3, 311, 325, 346, 352–53, 361, 385, 410
Matter, 305, 316, 319, 324
Meanings, 389, 406–7. See also Hermeneutics
Mechanistic thinking, 18, 24, 277, 289–94, 297, 302–3, 327, 358, 360, 373, 381, 384, 394, 397, 399, 409
Merrey, Doug, 74, 80, 91, 98, 100–101, 102, 265
M5 distributary area, 10, 119–23, 124–25, 138, 183–90
Mind, 315–17, 324
Minister of Lands (Gamini Dissanayake), 38, 42, 46, 73, 91, 133, 140, 171, 175, 189, 207, 257
Ministry of Lands, 10, 136, 140, 232
Moore, Mick, 32, 74
Morale, 365, 366, 384–85
Motorcycles, 143, 196, 201, 235
Munasinghe, S., 28, 40, 43, 46–47, 54, 62, 66, 68, 70, 73, 80, 92, 106, 107, 109, 110, 114, 117, 118, 123, 129, 133, 141, 143, 145, 149, 157, 167, 174–75, 195, 202, 208, 210, 212, 214, 226, 228–30, 240, 245, 247, 253, 256–57, 259, 266, 300
Murray-Rust, D. Hammond, 40, 47–48, 50–51, 112, 113, 120, 127, 133, 143, 145, 151, 164–65

Nalini Kasynathan, 135, 141, 143, 144, 156, 168–69, 171, 202, 206, 234, 236
Nandaratna, 143, 171, 173, 176, 178, 191, 192, 193, 197, 200, 201, 202, 208, 209, 214, 227, 237, 255
Narangoda, 106, 147, 163–66, 170, 176–79, 197, 204, 205, 211–14, 220, 222, 241, 242, 244, 247, 248, 263, 264, 266, 292, 360, 386
National Irrigation Administration (Philippines), 11, 22, 35, 37, 40, 44, 300
Neeles, 95–97, 103, 118, 181–82, 211, 212, 347
Negative-sum outcomes, 285, 287, 303, 324, 333, 342, 355, 385, 387, 393
Nepal, 22, 233, 237, 260, 369, 371
New business administration, 15, 386
New physics, 13–14, 279, 280, 291, 363
Newtonian thinking, 13, 14, 19, 20, 24, 277, 279, 280, 284, 290–91, 302–3, 304, 312, 327, 357, 374, 382, 383, 388, 389, 392, 393, 395, 397, 399, 403, 408, 409
Nilaweera, Dixon, 72, 198–99
Nonlinear thinking, 14, 19, 294, 374, 390–92, 394. See also Chaos theory
Normative orientations, 189, 330–32, 336–45, 347, 351, 354, 371, 373, 385, 402, 409

Objectivity, 13–14, 22, 278–80, 309–10, 315–18, 319, 321, 323, 403–4
Officials, 8, 57, 62, 68, 86, 96–97, 103, 164, 198, 218, 220, 372, 411. See also Farmer-official relations
Olson, Mancur, 327–30, 353
Open systems, 14, 291, 294–96, 314, 358
Opposites, 21, 318–25, 348, 398–99

Paddy Marketing Board (PMB), 148, 211, 214, 223–25, 253–54, 271
Pareto optimality, 17, 342, 345
Participatory development, 25, 102, 273, 301, 410
P. B. Ratnayake, 156, 169–71, 259
Perera, Jayantha, 201, 256
Perera, K. D. P., 49, 170, 236, 262, 265, 365
Personal responsibility, 24, 167, 397, 400, 408
Phenomenological thinking, 403–7
Plato, 20, 311, 318, 404
Politics, 31, 38–39, 41, 55–56, 80, 100–01, 107, 108–9, 126–27, 166, 171–72, 177, 196, 207, 219, 248, 264, 270–71, 362
Polonnaruwa IO program, 238, 260, 263, 267
Positive-sum dynamics, 4, 17, 82, 86, 134, 285–88, 318, 324, 327–28, 332, 341–45, 357, 355, 363, 364, 365, 375, 377, 379–80, 386, 387, 393, 409, 410
Positive-sum thinking, 15, 24–25, 282–83, 284–88, 291, 294–96, 303, 310, 327, 341–45, 351, 363, 375, 386, 409, 410. See also Both-and thinking
Possibilities, 4, 24, 297–301, 399–402
Post-Newtonian thinking, 25, 272, 373, 383, 386, 396, 399, 402, 403, 405, 407
Power, 290–91, 382
Prisoner's dilemma, 328, 355
Priyantha, 216, 230, 259
Probabilistic thinking, 319, 324, 337, 357, 360
Probabilities, 297–301, 319, 399–402

Probability distributions, 338–40, 359, 365, 384, 385, 397, 400–402
Project (level) committee, 195, 201–3, 208–9, 210–14, 218, 234, 244, 248, 252, 254, 264, 270
Project steering committee, 72, 104, 105, 107
Punchibanda (short), 175, 177, 178, 197, 205, 220, 223, 245
Punchibanda (tall), 118, 154, 165–66, 175, 178, 197, 202, 221, 223, 248, 263, 264, 292, 340, 386

Quadrants, reflecting normative orientations, 341, 343–45, 347, 356, 366, 370, 376, 378, 380, 409
Quantum mechanics (quantum theory), 13–14, 19, 277, 279, 280, 291, 297–98, 304, 309, 324, 399, 400, 403, 404
Quantum wave functions (qwiffs), 338, 400, 407–8

Rajapakse, Menike, 163, 176, 178, 220
Raj de Silva, Ariyabandhu, 135, 141, 201, 202
Ranasinghe Perera, 157, 168–69, 173, 195, 202
Rationality, 240, 326–28, 339, 351, 353, 355, 381, 405, 410
Ratnasiri, 130–31, 155, 157, 169–70, 194, 196–97
Ratnayake (Gongolla), 97–98, 101–2, 166, 170, 176, 178, 197, 221–22, 241, 243, 244, 247, 270–71, 386
Ratnayake (Weeragoda), 146, 147, 183–85, 190–91, 216–19, 249, 252, 263, 271
Razaak, 208, 227, 234, 236, 237, 255
R. B. Senekaarachchi, 34, 36, 39, 41
Reductionism, 18, 296, 302–3, 313, 325, 327, 352, 365, 373, 388, 394, 396, 397, 409
Rehabilitation work, 31, 46, 48–51, 59–61, 87, 124, 127, 129–30, 147, 218–19, 220–21, 251; by farmers, 82, 92–93, 153, 177, 180, 195, 204, 209, 220–21, 241, 243, 255, 268
Relativity theory, 14, 24, 277, 291, 304–10, 389, 404
Right choices for wrong reasons, 40, 44, 51, 297, 369
Rigor vs. relevance, 284, 294–96, 410
River analogies, 20, 277–78, 305, 312–13, 395, 408
Rotation: of farmer-representatives, 184, 191; of water deliveries, 56–57, 62–63, 67–68, 72, 75, 79, 82, 83, 107, 116, 138, 188–89, 247
Rural Development Committee, x, 32–33. See also Cornell University

Sanda, 58, 85, 86, 103, 340
Sarath Silva, 158, 249, 257, 259
Sarath Wijesiri, 123, 183–87, 191, 194, 202, 206, 210–11, 231, 226, 245
Saving: of money, 150, 185; of water, 64–66, 68, 72, 75, 77, 107, 108, 150, 162, 182, 188, 192–93
Secretary of Lands (Abeywickrama), 133, 140, 171, 198, 207, 238, 246, 257
Self-help (self-reliance), 83, 88, 89, 92–93, 130–31, 193, 219, 347, 372, 373, 410
Self-interest, 17, 288, 326–27, 339, 350–51, 354–56, 365, 377–80, 385
Selfishness, 78–79, 86, 103, 199, 338–39, 343–45, 348–50, 364, 378, 379, 408
Self-organizing systems, 14, 314, 374, 392–93
Senanayake Samudra reservoir, 5, 26–27, 91, 142, 152, 174
Shramadana, 56–58, 63, 64, 67, 72, 83, 88, 89, 94, 95, 98, 100, 108, 113, 120–21, 124–25, 128, 131, 150, 158, 182, 204, 205, 227–29, 248, 254, 268
Sinhalese-Tamil relations, 6, 8, 50, 59, 65, 71, 75, 76, 82, 98–99, 101–2, 109, 116, 119–21, 138, 154, 158, 166, 177, 239, 246, 249, 308, 360, 405–6
Sita, 54, 90, 123–24
Small Farmer Development Program, 37, 205, 207, 224, 226, 369
Smith, Adam, 19, 280, 351, 377
Social energy, 16, 292, 303, 367–74, 381, 386, 395, 408, 410
Social science, 11–13, 16–18, 19–20, 303, 311–12, 324, 326, 330–32, 366, 375, 388–89, 396–411; essentialist, 311–15; existentialist, 311–12, 314
Social Science Research Council, ix–x, 280–82, 318
Solidarity, 303, 368, 378–79, 384, 410
Strange attractors, 374, 393
Strangers, 343
Structural analysis, 330–36, 347, 354
Subesinghe, T. B., 34, 41, 71, 206, 256
Subjectivity, 315–18, 319, 321, 323
Synthesis, 283–84, 295–96, 316, 318, 319

Tamil program, 115, 135, 141, 143–45, 154–55, 167–69, 172, 173, 194
Technical Assistants, 36, 50–51, 64, 68, 76, 79, 81, 98, 99, 101, 130, 162, 166, 180, 181, 195, 209, 215–16, 219–20

Tennessee Valley Authority (TVA), 28, 161
Terrorism, 11, 168, 172, 174, 175, 201, 202, 210, 214–15, 217, 234, 236, 239–40, 263
Thermodynamics, 279, 291, 322, 358, 393
"Tigers" (LTPE), 11, 166, 207, 249, 259, 260, 262, 265, 266, 405
Tilakaratne (Project Manager), 211–13, 226, 235, 246, 253, 256, 263, 264, 270
"Tragedy of the commons," 327–30, 353
Trip reports, 23, 51, 52–53, 140–41, 236–37,
Turnover of responsibility to farmers, 266–67, 269–70

Uhana, 54, 62, 80, 149, 179–82, 230, 249
Uncertainty, 24, 324, 395, 399, 403
U.N. Research Institute for Social Development, 273
U.S. Agency for International Development, ix–x, 5–6, 10, 30–34, 42–43, 45–46, 48, 71–72, 74, 114, 137, 165, 195, 200–201, 205, 233, 236, 238, 248, 265, 359
Uphoff, Norman, role of, 52, 197, 366
Utility functions, 287, 336. See also Interdependence of utility functions

Values. See Ideals; Normative orientations
Vander Velde, Ed, 39, 41, 43, 46, 48, 49, 73, 299–300
Vidanapathirana, 112, 113, 171, 270
Violence, 11, 39, 65, 76, 94, 119, 140–41, 215, 234, 236, 239, 240, 249, 260, 266, 405–6

Waratenne, 55–57, 80, 81–82, 86, 159–63, 179, 200, 203, 271, 340
Water: distribution of, 29–30, 182, 213, 221, 264, 271, 332, 361–62; efficiency of use of, 57, 64, 146, 215–16, 235, 264, 268; tax (O&M fee), 126, 146, 149, 176, 178, 180, 203, 242, 256, 262. See also Rotation; Saving
Weeragoda, 112, 113, 119, 128, 138, 146–47, 149–51, 183–85, 216–19, 249
Wickramasinghe, Lakshman, 34, 36, 39, 41, 43, 45, 46, 112, 113, 117, 128, 133, 140, 198, 300, 364
Wijayaratna, C. M., 34, 36, 39, 41, 43, 45, 46, 47, 54, 59, 66, 73, 107, 112, 134, 198, 202, 236, 238, 239, 240, 243, 248, 250, 255, 257, 258, 300
Women's participation, 44, 54–55, 58, 89–90, 114, 163, 178, 220, 269
World Bank, 136, 137, 171, 195, 237, 238
Wrong choices for right reasons, 45, 47

Yala (dry) season, 9, 44, 21, 56, 72, 76, 105–6, 116, 138, 190
Yaya palaka, 55, 57, 64, 86, 245
Yields, of rice, 132, 137–38, 217, 222, 235, 291, 294–95

Zero-sum outcomes, 134, 282–83, 284–88, 303, 318, 324, 328, 332, 363, 385, 387, 393
Zero-sum thinking, 4, 15, 25, 279, 282–83, 291, 294–95, 298, 301, 303, 310, 317, 323, 327, 342–43, 347, 355, 363, 364, 410. See also Either-or thinking